Lasers and Electro-Optics Research and Technology

High Power P-P Lasers

LASERS AND ELECTRO-OPTICS RESEARCH AND TECHNOLOGY

Additional books in this series can be found on Nova's website
under the Series tab.

Additional e-books in this series can be found on Nova's website
under the e-book tab.

LASERS AND ELECTRO-OPTICS RESEARCH AND TECHNOLOGY

HIGH POWER P-P LASERS

VICTOR V. APOLLONOV
AND
SERGEY M. SILNOV

nova publishers
New York

For permission to use material from this book please contact us:
Telephone 631-231-7269; Fax 631-231-8175
Web Site: http://www.novapublishers.com

Library of Congress Cataloging-in-Publication Data

ISBN: 978-1-63117-767-5

Published by Nova Science Publishers, Inc. † New York

CONTENTS

INTRODUCTION

Since the creation of the first lasers in the early 1960s the search has been under way for new principles of fabrication of coherent radiation sources, expanding their range of applications in scientific and applied research, as well as in military equipment. In this case, the goal is, as a rule, to increase the energy of a single pulse, while reducing the duration and increasing the average optical power due to the periodicity involved in the radiation generation.

Creation of high-power lasers has opened a new era in the development of fundamental research and in the emergence of new technologies in various fields of practical use. This is due to the unique qualities of laser sources of high-power coherent radiation. These qualities include:

a) high monochromaticity (i.e., a narrow linewidth), which opens new ways to high-resolution spectroscopy;
b) high spatial and temporal coherence (i.e., oscillations of light waves behave in such a way that a distinct interference pattern is observed), which has given a strong impetus to the development of holography and optical methods of information processing;
c) relatively high specific energy, which can excite the laser medium;
d) possibility of varying the time during which the energy stored in the laser medium can be released within the shortest possible time (from milliseconds to femtoseconds);
e) possibility of varying the time structure of radiation from tens of Hz to tens of GHz;
f) small divergence, which enables tight focusing.

Due to these properties, the power density of laser radiation, which can affect the matter, reaches a giant value of 10^{20} W/cm^2. This fantastic opportunity has been thoroughly investigated by scientists around the world in the past decades.

Fabrication of high-power lasers with a large integral beam cross section by using amplifying channels is a complex scientific and engineering problem. To date, the research teams in the U.S. and Japan have designed and actively tested multichannel solid-state lasers with the output level of a few tens of kilowatts. These lasers are extremely complex devices that require ongoing technical support. On the other hand, diode-pumped solid-state lasers, built on traditional optical systems and having an output power of 1-3 kW, are already quite commonplace devices in the West. The existing solutions make it possible to manufacture lasers, which require virtually no maintenance over the entire service life (thousands of hours).

Obviously, one of the most effective solutions in fabrication of high-power solid-state laser is the multi-channel principle of their construction. The laser in this case is assembled in the form of an array of identical channels, each of which ensures laser generation. Small cross section of the channels allows for efficient and easy-to-implement pumping of the energy into the active medium and heat release from the medium. According this scheme, a 105-kW laser (Northrop Grumman Corporation) was fabricated. Further improvement of the output parameters of multichannel laser systems can result from a significant reduction in the optical thickness of the amplifier, and therefore from the employment of the disk geometry of amplifying channels.

Already at the beginning of the 20th century, K.E. Tsiolkovsky, founding father of rocketry, foretold in his memoirs that the space vehicles of the future will, of course, be launched with the help of electromagnetic waves guided by an external energy source.

It was the epoch of great discoveries: Radio was invented, and the scientific community was thinking about the power of radio waves and their new applications. The laser was not yet known to the world, but the potential long-distance transmission of energy was already the subject of scientific research and foresight.

At present, we consider a new approach to the problem of creation of a laser jet engine (LJE), which is based on the resonance merging of shock waves (SWs) generated by an optical pulsating discharge (OPD), produced by the laser. To obtain an OPD, we suggested using high-power pulse-periodic laser radiation.

The jet thrust of the proposed project device is implemented under a high-power high-frequency pulse-periodic laser radiation. For this purpose use can be made of carbon dioxide lasers, chemical lasers and solid-state laser systems.

In the future, high-power high-frequency pulse periodic lasers will find applications in the field of ecology. Besides, they will be used to transmit power over long distances, to obtain high-power plasma radiation in the range of tens to hundreds of angstroms, to realize arbitrary geometry of the breakdown in a three-dimensional space, etc.

A new class of rocket engines – a LJE – objectively belongs to the most promising rocket engines of the foreseeable future. The LJE is substantially more cost-effective than conventional chemical fuel engines. At the initial stage of the flight, the LJE employs atmospheric air as a working medium, and in outer space use is made of a small space-borne gas supply or an easily sublimated substance. In this case, the specific costs of the cargo launching to outer space may be reduced down to 160–330 USD/kg.

The LJE-based solution of a series of very interesting and important problems is predicted, in particular:

(i) Designing of a space interceptor that would destroy debris and other dangerous space objects, such as asteroids, meteors, etc.;
(ii) Launching of macroobjects with a very high acceleration that is orders of magnitude greater than the mass;
(iii) Realization of super-long orbital-scale conducting channels for addressing the global challenges of ecology and energy.

The idea of laser fusion, associated with the possibility of providing conditions for thermonuclear reactions in the region of focusing of laser radiation on a solid target, gave birth to a new area of research, i.e., physics of the interaction of intense laser pulses with matter, and plasma, obtained by this method, was called "laser plasma." This direction in the last 50 years has progressed rapidly through the rapid development of laser physics and technology. Long-term efforts on experimental and theoretical study of the properties and characteristics of the laser plasma extended the domain of effective applications, which initially involved only laser fusion.

Instead of laser radiation, the solution of the controlled thermonuclear fusion (CTF) problem can rely on the beams of high-energy particles, which provides a much more efficient transfer of energy to the deuterium-tritium drop. The most promising is a beam of heavy multiply charged ions. Because of the high charge of the particles, it is easier to accelerate them to high energies. Secondly, it is easier to obtain higher currents, because the greater the mass, the smaller the influence of the repulsion of the ions in the beam due to space charge. Third, the beams of heavy multiply charged ions will transfer with maximum efficiency the energy to the target. In this case, it is an essential issue of how to obtain primary beams of heavy multiply charged ions for subsequent acceleration. A simpler way to implement the inertial controlled fusion reaction is to produce first heavy multiply charged ions with high-power laser beams and then to use them for irradiation of a deuterium-tritium target to achieve CTF.

A laser-plasma generator (LPG) of multiply charged ions produces a large number of heavy ions in the regime of short periodic pulses, which is of interest for ion accelerators operating in the pulse periodic regime. The source of this type is also promising for research in the field of heavy-ion fusion. The fundamental idea of a LPG of multiply charged ions was put forward 50 years ago by a group of authors from the Moscow Engineering Physics Institute (MEPhI). The LPG is based on the physical phenomenon of generation of highly excited states of atoms by a high-power laser pulse focused on the surface of a solid target. With expanding a high-temperature laser plasma into vacuum a high-power flux of charged particles is produced. The advantages of LPGs as compared to other types of generators of ion pulses include:

- generation of beams of multiply charged ions of virtually all elements of the periodic table,
- generation of intense, short (1 – 100 μs) ion pulses with extremely high brightness.

The main difference of the laser plasma from other high-temperature plasma objects is the high energy density, caused by the ability of lasers to concentrate energy during a short time in a small volume (10^{-6} cm^{-3}) of matter.

To date, as a result of a large series of experimental and theoretical studies scientists managed to identify the role and nature of the absorption of laser radiation in a plasma, the mechanisms of energy transfer between different components, the main properties of kinetics of the ionization state and dynamics of expansion of the laser-plasma torch. A significant contribution to the experimental research has been made, in particular, due to the development and application of laser mass spectrometry, allowing one to determine the charge composition of the plasma, the properties of the processes of ionization, acceleration, and recombination in laser plasma. Numerous studies have established that the laser plasma is an intense emitter of charged and neutral particles: the electrons, ions, and atoms. The mechanisms of the energy spectra of ions and neutral particles have been studied in detail, and laser sources of particles, which have found application in accelerators, cyclotrons, and mass spectrometers, have been developed and technically implemented. According to the results of the studies of laser plasma, jointly with the Joint Institute for Nuclear Research (JINR) (Dubna), laser sources of atoms and ions for a collective heavy ion accelerator (KUTI) have developed and successfully tested: The use of such an accelerator makes it possible to load and accumulate ions upon their injection into electron rings. Joint work of the MEPhI with the Laboratory of Nuclear Reaction at the JINR has made it possible develop a low-background laser-plasma ion source for a highly sensitive mass spectrometer in order to search for super-heavy elements in nature and sources for cyclotrons. The efforts of the Alikhanov Institute for Theoretical and Experimental Physics (ITEPh) are aimed at developing a laser source of multiply charged ions of heavy elements for inertial fusion. Jointly with the MEPhI and JSC "SELMI" (Sumy, Ukraine) producing instruments for micromorphological and element analysis, there was developed a series of industrial EMAL-1 and EMAL-2 laser mass spectrometers.

In conclusion, it is important to note that the range of possible applications of multiply charged ions produced in the interaction of high-power pulsed and pulse periodic radiation with matter is expanding every day. The next step is its significant expansion through the introduction of high-power pulse periodic laser systems with a high repetition rate in wide practice. This will provide the solution to the problem of a plasma screen and allow the delivery of considerable fluxes of charged ions of average power into the matter, thereby creating prerequisites for production of more powerful fluxes of charged ions for their effective use.

Prof. Sergey M. Silnov and Prof. Victor V. Apollonov.

Part 1. High-Power Pulse-Periodic Lasers for Generation of Plasma Torches

Chapter 1

1.1. PRINCIPLES OF CONSTRUCTION OF COHERENT HIGH-POWER LASER SOURCES

Since the creation of the first lasers in the early 1960s the search has been under for new principles of development of coherent radiation sources, expanding the field of their applications in scientific research and technology. An example of this are femtosecond (1 fs = 10^{-15}s) lasers which at a pulse energy of hundreds of millijoules make it possible to obtain radiation with a power up to 10^{12}W. In the beam focus the intensity is 10^{20}W/cm^2, and higher, which is close to the relativistic intensity at which the oscillation energy of an electron in the light field is comparable to the rest energy of an electron. In CW lasers the power has already reached the value of 10^5-10^6 W. The cross section (aperture) of the beams of such lasers is about tens of centimeters. These lasers are very large facilities with complex systems of energy pumping into the active medium, with sophisticated systems of heat release and generation of output radiation. Even in carbon dioxide lasers, which have a relatively high efficiency (more than 10%), it is needed to ensure efficient removal of large heat fluxes from the entire volume of the active medium. Otherwise, overheating of the gas mixture will lead to the inhomogeneity of the mixture and, as a consequence, to the deterioration of the output quality. Continuous overheating of the medium can cause degradation of the gas mixture, the thermal destruction of the cavity mirrors and other optical elements. In semiconductor injection lasers an increase in power by increasing size of the *p-n* junction leads to the disintegration of the output beam into fragments, at which the intensity distribution in the beam cross section accidentally changes during emission.

Creation of high-power lasers with a large cross section of the output beam (wide aperture lasers) is a challenging scientific and technical problem of modern laser physics. One of its possible solutions involves the so-called multichannel laser principle. In accordance with this principle, the laser is assembled in the form of an array of identical channels, each of which produces lasing (Figure 1.1). Such an array of lasers forms an optical source with a composite aperture ensuring laser beam combining. The small cross section of the channels allows efficient and simple pumping of energy into the active medium as well as efficient heat removal from its bulk.

Figure 1.2 shows the photograph of a multichannel technological laser, one in a series of lasers of this type. With the protective housing being removed, one can see glowing blue channels of waveguide CO_2 lasers arranged in a circle. The active medium is pumped by a discharge in an alternating electric field produced by the electrodes, i.e., the plates on the outer surface of the quartz tubes of the channels. Air cooling of tubes is used in 500-W lasers; oil cooling is used in 1-, 2.5-, 4-kW lasers.

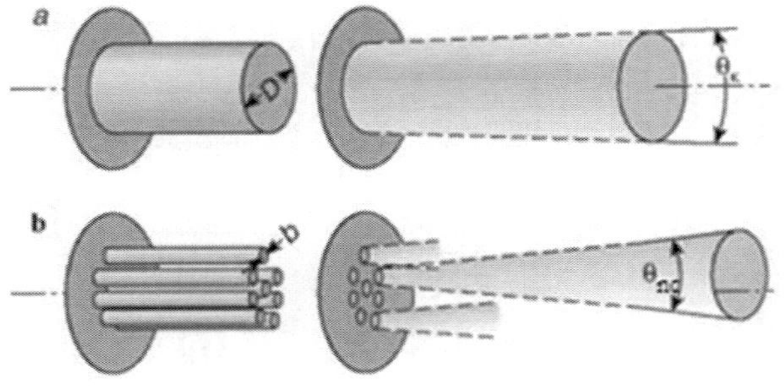

Figure 1.1. Multichannel laser principle: (a) wide-aperture laser with output beam diameter D; (b) multi-channel laser with the same composite aperture size D, the diameter of the channel is b.

Figure 1.2. 500-W laser system: number of channels, 20; inner diameter of the channels, 7mm; channel length, 1.6m. In 2.5- and 4-kW lasers the number of channels is *N*=61.

In this case, the wavelength of the emitted high-power laser radiation is 10.6 μm.

Despite many structural advantages of multibeam lasers, high radiation directivity is a complex physical problem. Indeed, the divergence angle of coherent laser radiation Θ_K is close to or greater than the diffraction angle Θ_d, which is inversely proportional to the beam diameter. For a wide aperture laser the means (Figure 1.1):

$$\Theta_k \geq \Theta_d\,, \Theta_d = \frac{\lambda}{D} \tag{1}$$

where λ - is the wavelength, and D - is the diameter of the aperture.

In a multichannel laser, each channel generates a wave, whose phase is independent of the phase of the waves of other channels. The fields of the channels are incoherent, and their intensities are added. In the incoherent regime the resulting intensity I_{nc} of a multichannel laser has the form

$$I_{nc} = N^* I_0 \tag{2}$$

where I_0 - is the intensity in one single channel, and N - is the number of channels.

Radiation directivity of a multichannel laser in this case is determined by the beam divergence of each channel of diameter b. Consequently, the angle divergence of its radiation I_{nc} in the case of incoherent channels is

$$I_{nc} = \lambda/b \quad (3)$$

In a multichannel laser, whose aperture is made up of independent channels of diameter b<D, the divergence Θ_{nd} is much greater than that the divergence of a wide aperture laser with the same aperture D:

$$\Theta_{nc} \gg \Theta_k \quad (4)$$

Quite a different situation is observed, when it comes to a multichannel laser beam in the case of phase-locked collective generation, when phases of radiation from the channels coincide. The output beam is formed due to the interference of the waves whose phases on its axis are equal. The resulting intensity Iк on the axis of such a beam is N^2 greater than the intensity of one channel:

$$I_k = N^{2*}I_0 \quad (5)$$

The angular divergence in the axial region of the beam in the case of phase-locked generation in the channels is close to the divergence of a wide aperture laser with aperture diameter D. Therefore, a key challenge for multichannel lasers is the attainment of the phase-locked regime of collective oscillation.

Let us consider the physical principles of phase locking of individual lasers. These principles form the basis for the development of modern multichannel lasers with high radiation directivity.

1.1.1. Interference of Radiations from Many Phase-Locked Sources

The main properties of the multichannel laser beam formation follow from consideration of the classical problem of superposition of waves from many sources. Let the wave frequencies coincide and the electric field amplitude be equal to E_0. Then, for the *n*th source we can write the expression

$$En(z, t)=E_0\cos(at-kz+\beta_n) \quad (6)$$

where β_n – is the phase of the wave of the *n*th source. The resulting field E(z, t) is expressed as a sum

$$E(z,t) = \sum_{n=1}^{N} E_n(z,t) \quad (7)$$

The intensity of radiation as the time-averaged power density is determined by the integral:

$$I \sim \int_0^T E^2 dT, T = \frac{2\pi}{\omega} \tag{8}$$

For a single source up to a factor the intensity has the form $I_0 = \frac{E^2 0}{2}$.

After substitution of (6) and (7) into (8), the intensity in the case of superposition of the waves from N sources reduces to the form:

$$I = NI_0 + I_0 \sum_{n=1}^{N} \sum_{m=1}^{N} \cos(\beta_n - \beta_m), n \neq m \tag{9}$$

In (9), the second term is called interference.

Let the wave phases (β_n be a random variable in the interval [0, 2π], that is, the sources are incoherent. Then, summation over many *N* sources will make the interference term vanish. The resulting radiation intensity I_{nc} of incoherent sources is equal to the sum of the intensities I_0 of the waves from different sources [see (2)].

Let the sources be coherent, that is, the phases of their radiation are not random. In this case, the sources are phase-locked, that is, the phase of their radiation coincide: $\beta_n = \beta_m$. Under these conditions, the interference term in (9) is different from zero. After simple transformations, expression (9) reduces to (5). Coherent summation of the fields at which the resulting intensity $I_к$ is N^2 greater than the intensity of one channel.

Consider the intensity distribution, Iк, in space upon superposition of waves from many phase-locked coherent sources located at the same distance d from each other (Figure1.3). Let the wave phases have the parameters (β_n=0, n=1,2,3, N). Far away from the source (L>$(\lambda Nd)^{1/2}$), the wave propagating from the sources can be considered plane and the resultant wave at an angle is equal to the sum:

Iк(α) - is the angular dependence of the resulting field intensity in the far-field region, L>$(\lambda Nd)^{1/2}$.

I_0(α) - is the angular dependence of the intensity for one source, and

Δα - is the width of the interference maximum.

$$E(z,t) = E_0 \sum_{n=1}^{N} \cos\left(\omega t - kz - \varphi(n-1)\right) \tag{10}$$

The phase shift of φ the waves from adjacent sources is proportional to the path difference of their propagation. When observed in the far-field region, where L>>N d, the phase shift has the form

$$\varphi = kd\sin\alpha, \tag{11}$$

where $k = 2\pi/\lambda$ – is the wave number.

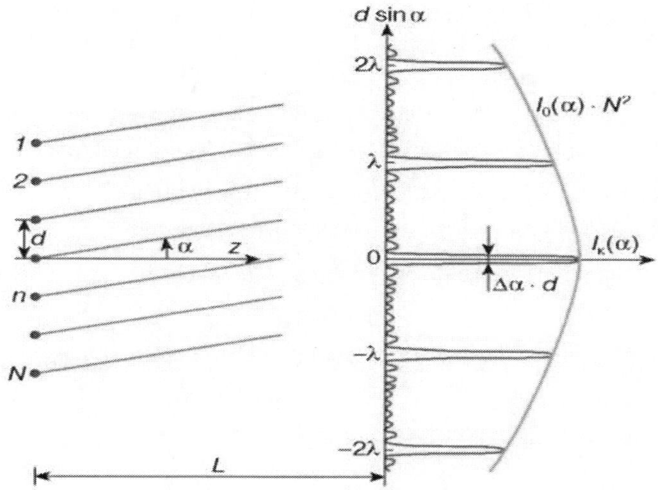

Figure 1.3. Interference of waves from the array of coherent sources.

The use of Euler's formula for the transition from the trigonometric to exponential function with an imaginary argument reduces expression (10) для E(z, t) to the sum of terms of geometric progressions with the denominator $e^{tn\varphi}$. After transformations we can obtain

$$E(z,t) = E_0 \frac{sinN\frac{\varphi}{2}}{sin\frac{\varphi}{2}} \cos(\omega t - kz - \frac{\varphi(N-1)}{2})$$

The intensity I_K of the resulting field is a function of the angle α:

$$I_k(\alpha) = I_0 \frac{sin^2 N\frac{\varphi(\alpha)}{2}}{sin^2\frac{\varphi(\alpha)}{2}} \tag{12}$$

This function reaches a maximum equal to $N^2 I_0$, when $\varphi_m = 2m\pi$, i.e., during the propagation at an angle αmax, which in view of (11) is determined from the condition

$$d\sin\alpha_{max} = m\lambda, \tag{13}$$

where m=0, ±1, ±2, … – is the order of the interference maximum.

If the phase difference φ is such that $N\varphi/2=p\pi$, where p is an integer and not a multiple of N, then the intensity $I_\kappa(\alpha)=0$. Hence, the condition for the interference minimum is given by

$$d\sin\alpha_p = \pm\, p\lambda/N, p=0,N,2N,\ldots \quad (14)$$

The angular distribution of the intensity $I_\kappa(\alpha)$ upon interference of waves from the N phase-locked coherent sources is presented in Figure 1.3. The interference maximum value at some angle α is proportional to the radiation intensity from a single channel, $I_0(\alpha)$, at this angle, which decreases with its growth due to the diffraction divergence. For sources with an aperture size b, the characteristic scale of a decrease in $I_0(\alpha)$ is Θ_{nc}, which is expressed by formula (3).

The angular width $\Delta\alpha$ of the interference maximum is defined as the half of the interval between the two minima closest to this maximum. At small angles α the width $\Delta\alpha$, according to (14), has the form

$$\Delta\alpha = \frac{\lambda}{Nd} \quad (15)$$

When N>>1, the denominator of the derived expression is close to D-the size of the composite aperture formed by an array of periodically located coherent sources.

The function $I_\kappa(\alpha)$ is called the radiation pattern, where the interference maxima are its lobes. Thus, a periodic array of phase-locked sources generates a radiation pattern of several narrow lobes, where the radiation energy is focused. The angular width of these lopes is close to the diffraction angle of the beam divergence Θ_d, whose diameter is equal to the size D of the entire array as a whole [cf. (1) and (15)].

Narrowing of the radiation pattern lobes of the emitter array with increasing size of the entire aperture is a characteristic feature of the interference of waves from many coherent sources. A coherent phase shift between the emitters of the array can change the position of the pattern in space. Because of the reversibility of wave propagation, the angular distribution of sensitivity for the array of receivers coincides with the radiation pattern of the same array of emitters. These remarkable properties of composite antennas are widely used in radars, astrophysics and space communications. With increasing size of the receive and transmit antenna array, and consequently, with decreasing angular width of the lobes of the radiation pattern can increase accuracy of the target location. After reaching the orbit an umbrella deployable antenna on the satellite is used to establish communication with the Earth. Astrophysical radio telescopes receiving electromagnetic radiation from space represent a set of many spaced receivers. Their signal is first summed in accordance with (7), and then the power is measured. In the RATAN-600 radio telescope in the North Caucasus, these receivers are placed along a 600-m-diameter circle. The radio telescope of the Siberian Branch of the

Russian Academy of Sciences located in the foothills of the Sayan Mountains, the crossed arrangement of 256 receivers forms an integral antenna of size 622m. To control the position of the radiation pattern of these antennas the corresponding phase shifts are introduced into electrical signals from the receiver to the adder. Such astrophysical antennas with an aperture of a few hundred meters offer high accuracy in determining the new space radio sources, which carry information about the evolution of the universe.

1.1.2. OPTICAL COUPLING

In the radio range, the phase control of the array elements, in particular, the phase-locked regime is implemented by means of phase shifters in their supply circuits. In the optical range, where the electromagnetic radiation frequency is on the order of 10^{15}Hz, the electric phase control of waves in the channels is impossible. To this end, to obtain phase-locked lasing in a multichannel laser, use is made of optical coupling between the light fields of the channels.

For example, in the arrays of injection lasers, generation channels represent a dielectric waveguide. If the refractive index in the channels is larger than in the gaps between them, then the field is confined in them due to the effect of total internal reflection at the side walls of the channels. The electromagnetic field penetrates through the walls to a distance of the order of the wavelength (Figure 1.4a). Because the spacing between the waveguides is a few μm, and the wavelength is λ=0.8μm, the fields penetrate into the neighboring waveguides. As a result, optical coupling appears, in which adjacent channels impose on each other the same phase of waves propagating in them. Because the waves are phase-locked in all the channels of the array, multichannel lasers are characterized by the phase-locked generation regime. Such coupling between the neighboring channels is rather weak. Slight deviations of the parameters of the laser channels, their length, width, spacing, etc. lead to the disintegration of the array on the specific areas where the radiation phases are independent of each other.

Figure 1.4 shows the change in the refractive index n and the light field E in the cross section of the array. When the distance between the plane of the exit apertures of the channels and the coupling mirror is equal to $Z_T/2$, the scheme forms a Talbot cavity.

'Strong' optical coupling is typical of the neighboring channels in the array of Y-couplers (Figure 1.4b).

In such systems, the radiation from each channel is decoupled. Because of the efficient exchange of radiation between the channels, the injection lasers with Y-couplers are highly stable in the case of phase-locked generation. However, significant losses of in the region of channel branching hold back the development of such schemes.

High stability of phased-locked generation is achieved in a multichannel optically globally coupled laser, when each channel is coupled with each channel. In arrays of injection lasers such coupling appears when in a periodic structure made of waveguides and spacings between them a standing wave of the electric field is formed in the direction which is transverse to the axes of the waveguides (Figure 1.4c). This wave rigidly couples the phases of the fields in the entire array of channels and under certain conditions provides phase-locked multichannel generation.

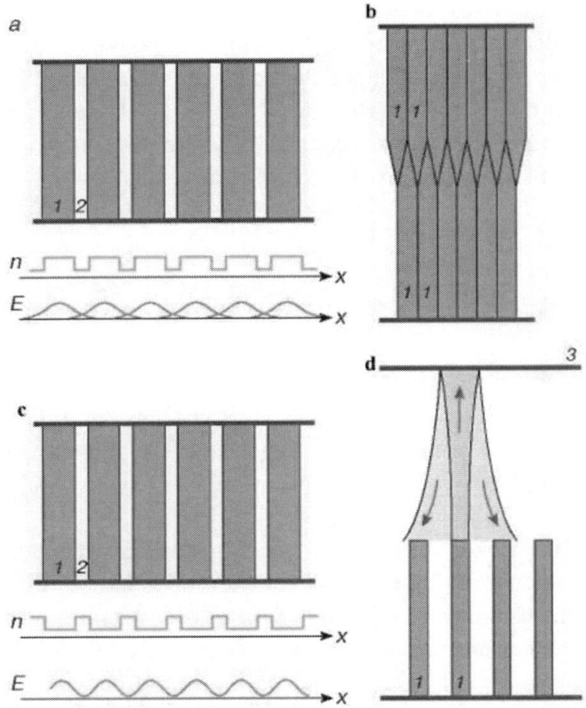

Figure 1.4. Schemes of optical coupling between the channels of the laser array: *1*- channels, *2* - spacings between the channels, *3* - coupling mirror. (a) Coupling of neighboring channels through the field penetration between the dielectric waveguide channels; (b) strong coupling of neighboring channels in the array of semiconductor lasers with a Y-coupler; (c) global optical coupling in the array of semiconductor lasers, the coupling is formed by a standing wave in the direction perpendicular to the axes of the channels; and (d) diffractive optical coupling between the channels of the laser array in a common resonator.

Another method of global coupling is based on diffraction exchange of radiation between the channels. In this case, the array of channels is placed between the mirrors of the common cavity (Figure 1.4d). The light beam emitted by the channel is broadened due to diffraction when it propagates to the coupling mirror and back. As a result, part of the beam penetrates neighboring channels. Thus, there appears optical coupling covering a significant part of the adjacent channels. Diffraction coupling applies to both miniature injection lasers and to waveguide carbon dioxide laser with an aperture of a few centimeters. To increase the diffraction coupling power, the coupling mirror should be placed far enough from the plane where the windows of output channels are located. This results in a significant increase in the diffraction losses in the laser. Using the diffraction effect of the light field self-imaging in a periodic array of emitters can significantly reduce these losses.

1.1.3. Talbot Effect

Discovery of the Talbot effect is very significant, especially for the modern scientific community. Self-imaging of a periodic diffraction grating was first observed in 1836 by

Henry Fox Talbot, a British inventor and photography pioneer. In his article published in the 'Philosophical Magazine' in 1836, he described experiments in which he found a periodic change of colors in the image of the diffraction grating when he moved away from it the focusing lens used for observations. In his work there were neither quantitative measurements, nor attempts to explain the phenomenon observed.

The Talbot effect went unnoticed for nearly fifty years before being rediscovered by Lord Rayleigh who was interested in self-imaging in diffraction gratings. The great physicist immediately realized that the effect is spectrally sensitive. He repeated Talbot's experiments, by carefully measuring the distance of the grating self-imaging in green and red light.

These measurements helped Rayleigh to establish the dependence of the self-imaging distance Z_T on the wavelength λ and the period of the structure d:

$$Z_T = \frac{2d^2}{\lambda} \tag{16}$$

He also explained theoretically the effect. In his paper devoted to this issue, Rayleigh, recognizing the priority of H.F. Talbot in the discovery of the effect, literally reproduced the first experiments of Talbot. The near-field diffraction effect was called the Talbot effect, and the regular distance Z_T for the image of the grating to be repeated was called the Talbot length.

Physically, the Talbot effect can be explained as follows. Let a plane light wave be incident on a grating with a periodically varying transmission coefficient. After the grating the wave amplitude E_0 changes periodically along the direction perpendicular to the grating grooves: $E_0(x)=E_0(x\pm nd)$. The periodic function $E_0(x)$ can be written as a sum of trigonometric functions with periods d, d/2, d/3, and so on:

$$E_0(x) = \sum_{s=0}^{\infty} A_s \cos(k_s x), k_s = \frac{2\pi}{d}s \tag{17}$$

The light wave after the grating, according to (6), can be written as

$$E_0(x) = \sum_{s=0}^{\infty} A_s \cos(k_s x)\cos(\omega t - kz), k_s = \frac{2\pi}{d}s \tag{18}$$

or

$$E_0(x) = \frac{1}{2}\sum_{s=0}^{\infty} A_s[\cos(\omega t - kz - k_s x) + \cos(\omega t - kz - k_s x)] \tag{18'}$$

This expression is a superposition of harmonics - plane waves propagating at angles $\pm\gamma_s$ to the OZ axis (Figure 1.5). The angle is γ_s<<1, since the grating spacing d is much larger than the wavelength λ and $k_s<k$.

Let us consider in (18) two fundamental (s=0) and first (s=1) harmonics. For some time t the front of the fundamental harmonic shifts to the distance z. The front of the first harmonic will pass for the same time the distance z, but at an angle γ_1 to the axis. The path difference d, which is acquired by the harmonics along the Oz axis, has the form $\delta = \frac{z}{cos\gamma_1} - z$

Substituting $\cos\gamma_1 = \frac{k}{\sqrt{k^2+k_1^2}}$, we obtain $\delta = z\left(\frac{\sqrt{k^2-k_1^2}}{k} - 1\right)$.

Since k_1<<k, then approximately the path difference is $\delta = \frac{k_1^2}{2k}z$. Hence, the phase shift $\Delta\varphi_1$ of the first harmonic relative to the fundamental harmonic will be

$$\Delta\varphi_1 = \frac{k_1^2}{2k}z \tag{19}$$

One can see from (17) that the phase shift $\Delta\varphi_s$ for *s*th harmonic is s^2 times higher – $\Delta\varphi_s = \Delta\varphi_1 s^2$.

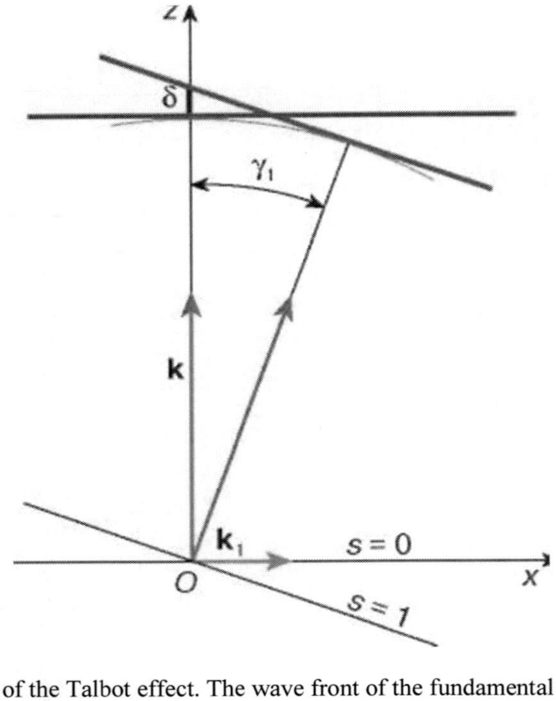

Figure 1.5. Explanation of the Talbot effect. The wave front of the fundamental (*s*=0) and the first (*s*=1) harmonics.

If at the distance Z_T the phase shift $\Delta\varphi_1$ of the first harmonic is 2π, then it will be multiple of 2π for all other harmonics. Because of the periodicity of the trigonometric functions, the expression for the light wave E(x, Z_T, t) at a distance z=Z_T coincides with (18) at *z*=0. As a result, at the distance Z_T the initial distribution of the field amplitude $E_0(x)$ is repeated. Thus, the Talbot length Z_T is a minimum propagation distance of a wave with a periodic modulation of the amplitude at which the phase shift of all harmonics is multiple of 2π. The expression for Z_T is found from the condition: $\Delta\varphi_1(Z_T)=2\pi$. Hence, using (19), we derive expression (16) for the Talbot length Z_T. If the coupling mirror is mounted at a distance $Z_T/2$ in a common cavity (Figure 1.4d), then radiation from channels, which is first reflected from this mirror and then returned to the array, will again coincide with the distribution of field emitted by it. In this geometry, the common cavity with a periodic array of channels is called the Talbot cavity. In the Talbot cavity the reflected radiation returns completely into the amplification channels, which reduces significantly losses in a multichannel laser with strong optical coupling.

1.1.4. Collective Modes

In an array of optically coupled lasers collective generation from the entire array is established. In this case, the field amplitude and phase in the laser channels take certain values. A change in the amplitude and phase in the aperture of the entire array is called collective modes or supermodes. If the field distribution in the cross section of each channel under conditions of optical coupling is the same as in its absence, the collective modes are uniquely determined from a set of the values of the field amplitude En and phase βn on the axes of channels - $2N$ values for an array of N channels. In the case of the in-phase mode, the amplitude En and phase β_n remain virtually unchanged from channel to channel in the entire array, except for the edges where the diffraction effect comes to the fore.

If amplification is absent in the channels, the collective modes are damped by losses caused by reflection of radiation from the mirrors, by diffraction and absorption of radiation. The smaller the losses for a collective mode, the lower the lasing threshold, i.e., the minimum gain for the emergence of lasing on this mode. For a multichannel laser it is important that the in-phase collective mode, i.e., radiation from all the channels is phase-locked, had the lowest lasing threshold. In this case, if the threshold value is significantly less than the lasing threshold of all other modes, the multichannel laser is characterized by high selectivity. In this laser the in-phase mode is reliably generated even under conditions of small natural variations in the array parameters: the period of the location of the channels, their length, and the gain.

The angular distribution of the intensity Iκ(α) in the far-field region $(L>(\lambda Nd)^{1/2})$ yields a radiation pattern of the laser in the case of a collective mode. Figure 6 presents the radiation pattern for a two-dimensional array of 36 channels located at the corners of the squares, and of 37 channels at the vertices of regular triangles. Shown is the radiation pattern, corresponding to the positive range of α_y. angles. One can see a narrow global intensity maximum in the central lobe at $\alpha_x=\alpha_y=0$. Its width with respect to $\Delta\alpha$ is much less than the angular divergence Θ_{nc} of the radiation from one channel and is close to $\Delta\alpha$, defined by the formula (15) for N=6. For an array with a square arrangement of the channels the global maximum is surrounded by eight smaller value lobes; in the case of triangular arrangement – by six lobes, at the vertices of a regular polygon with the center at $\alpha_x=\alpha_y=0$.

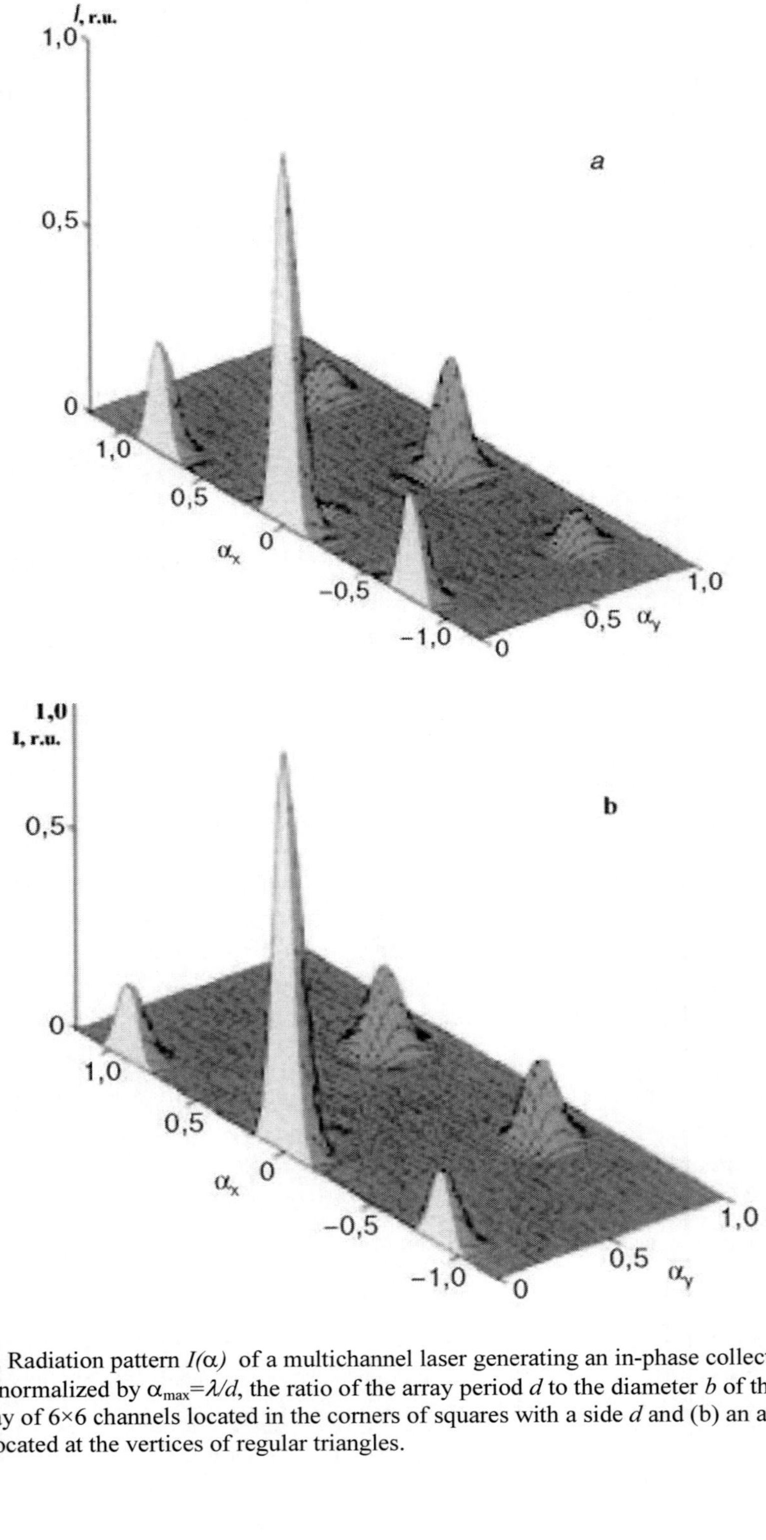

Figure 1.6. Radiation pattern $I(\alpha)$ of a multichannel laser generating an in-phase collective mode. The angle α is normalized by $\alpha_{max}=\lambda/d$, the ratio of the array period d to the diameter b of the channel is 4. (a) An array of 6×6 channels located in the corners of squares with a side d and (b) an array of 37 channels located at the vertices of regular triangles.

1.1.5. CONCLUSION

Several variants are known for the implementation of a high-power solid-state laser with a synthetic aperture ensuring diffraction-limited quality of radiation and intensity on the beam axis, corresponding to expression (5).

If one aims to reach an output power of 100 kW or higher, the use of an elementary source, i.e., the previously described laser with the output power of 1 kW, can lead to significant technical difficulties. This is due to the need to align simultaneously a very large number of channels. Probably, it is more reasonable in this case to fabricate, on the basis of the previously described laser, an emitter with an output power of 5-10 kW and use it as an elementary source in the laser array. This laser can be realized in two ways: as a high-power laser and an oscillator-amplifier system.

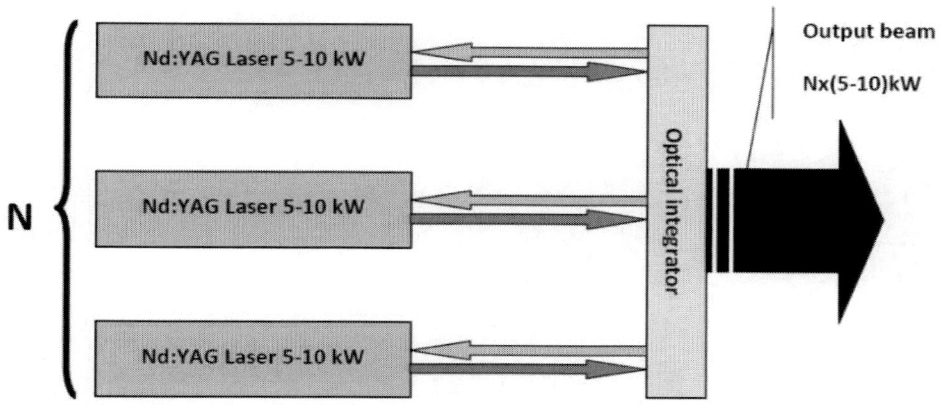

Figure 1.7. Laser with optical coupling between channels.

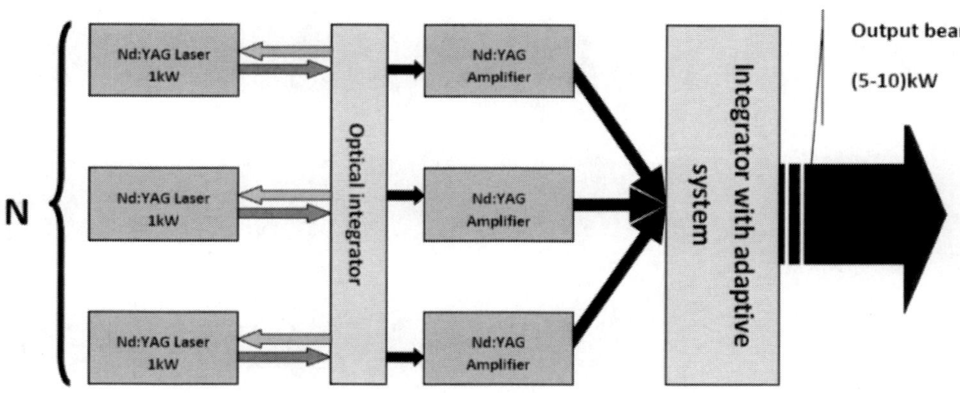

Figure 1.8. A combined scheme of a high-power laser.

Depending on the variant use can be made of two approaches to the implementation of a synthetic aperture laser:

1) In the case of a high-power laser, the synthetic aperture can be based on the scheme shown in Figure 1.4b, i.e., advantage can be taken of optical coupling between the channels (Figure 1.7).
2) In case of the oscillator-amplifier scheme it is necessary to use a combined scheme, i.e., optically coupled oscillators (Figure 1.1b) and adaptive systems in the amplifying channels to compensate for the difference of the phase shifts that occur in the various amplification channels (Figure 1.8).

Chapter 2

1.2. High-Frequency Pulse-Periodic Lasers

Currently, in technological processes with high average power (over 1kW) lasers, advantage is taken only of sources operating in two regimes–continuous and pulsed-periodic with a repetition rate from a few to hundreds of Hz at pulse durations within tens to thousands of microseconds and even milliseconds. In this case, the discussed processes usually rely on purely thermal mechanism, because use is made of the possibility of a laser source to deliver a sufficient amount of energy to a small area of the surface of a material being processed. High-frequency pulse-periodic *Q*-switched laser systems with a high average power, which provide the required pulse train durations from a few to hundreds of nanoseconds, allows implementation of a fundamentally different mechanism of interaction of radiation with matter–sublimation (ablation) ensuring a local energy release not only in space but also in time. This results in an explosive local evaporation of a substance without an intermediate liquid phase. This mechanism can significantly extend the range of technological applications of laser sources. However, it has found no real use in practice because of the great complexity of its implementation in the case of the mean powers required in a significant number of technologies. In general, such mean-power lasers are used in micro surface treatment, product marking, scribing, term cleavage, micro drilling, and laser-induced chemical etching.

It is known that for CW radiation, the threshold power density of evaporation is defined as:

$$q_{cw} = \frac{kT_{vap}}{Ar_o} \quad (1)$$

where A is the absorption coefficient ($A=1-R$), k is the thermal conductivity, T_{vap} is the evaporation temperature, and r_0— is the radius of the beam cross section in the focal plane. At $T_{vap}=3000^0$K, A= 0,5 ($\lambda = 1.06$μm), k=0,32W/cm×K, and r_0=10μm, we obtain for steel $q_{cw} = 1.9 \cdot 10^6$W/cm^2. The corresponding laser power is:

$$P_{p\text{-}p} = qS \quad (2)$$

where S is the area of the laser spot, and thus $P_{p\text{-}p} \approx 6.0$W.

For a pulse-periodic laser source with the same parameters and the corresponding thermal regime, the threshold power density can be expressed as:

$$q_p = \frac{kT_{vap}}{2A}\sqrt{\frac{\pi}{\alpha\tau}} \tag{3}$$

(a is the thermal conductivity 3.45×10^{-2}cm^2/s for steel), which gives the value of $q_p \approx 0.9 \times 10^6$W/cm^2.

In addition, we can calculate the pulse power, P_p, and the average power Pp-p:

(f is the pulse repetition rate), $P_{p\text{-}p} \approx 2.8$W and $t = 10^{-4}$ s, f=50Hz.

Thus, on the basis of the above-said, we can conclude, in general, the following:

$$\frac{\bar{P}_{p-p}}{P_{cw}} = f\tau\frac{r_0}{\sqrt{\alpha\tau}} \tag{4}$$

at above stated parameters $\frac{r_0}{\sqrt{\alpha\tau}} \approx 1$ and $\frac{\bar{P}_{p-p}}{P_{cw}} \sim f\tau$. Then for f=250Hz and $t=10^{-4}$s

$\frac{\bar{P}_{p-p}}{P_{cw}} = 10^{-2}$ for f=1kHz and nanosecond pulses we have $\frac{\bar{P}_{p-p}}{P_{cw}} = 10^{-6}$.

Thus, for a repetition rate of 30kHz and pulse duration of 100ns, i.e., the regime characteristic for the solution to most engineering problems, the ratio is 10^{-3}. Note that the physical processes in the case of pulse-periodic cutting are very similar to those of drilling, as the cut is a set of individual holes. Based on the above-said, we can conclude that the pulse-periodic regime of laser operation is more advantageous for most processing operation than the CW regime. Then, let us answer the question: what regime of laser operation, continuous or pulse-periodic, is preferable in terms of quality of materials processing?

For high-quality processing of materials it is necessary to ensure the following conditions:

(i) The minimum amount of the liquid phase formed during drilling or cutting. The amount of the liquid phase is proportional to $\sqrt{\alpha\tau}$.

(ii) The maximum amount of liquid phase removed from the drilling or cutting region by vapor pressure. The recoil vapor pressure is proportional to.

Both of these factors indicate that the pulse duration for the pulse-periodic regime should be as short as possible based on the mean power requirements.

Now we perform the same comparison in terms of requirements on the accuracy of scanning systems used for materials processing. It is evident that for processing conditions (stable product quality) to be stable, it is needed to ensure constant power flux density q. The required q for scanning systems is equal to:

$$q_s = \frac{hV_{sk}}{d^0}L_p \tag{5}$$

or with respect to the laser power P this condition is expressed as:

$$\frac{P}{V_{sk}} = hd_0L_p = const = B \tag{6}$$

On this basis, we can once again confirm our earlier conclusion: from the energy point of view, the pulse-periodic regime of laser operation is more preferable for laser cutting or drilling than the CW regime. This means that the stable conditions imply the constant laser power P and beam scanning speed V_{sc}. This is especially important for high-speed processing of materials of complex shape. In this case, it is impossible to maintain the same speed on straight lines and sharp turns. If V_{sc} is not a constant and varies according to some law $V_{sc}(t)$, the condition $P=BV_{sc}(t)$ cannot be met fin the case of the CW laser power.

Nowadays, laser technologies are widely used in the manufacture of products made of carbon steel and non-metal materials, such as composites, plastics, glass, rocks, wood, etc. Among the variety of different lasers, solid-state lasers and gas laser systems based on the CO_2 molecule have found the most frequent use in industrial applications.

1.2.1. Solid-State Lasers

Solid-state lasers are most widely used in processing of the materials made of carbon steel and non-metal materials. The use of solid-state lasers for machining nonferrous metals and their alloys is less common. This is mainly due to significant reflectivity of nonferrous metals, which reduces the laser radiation-to-heat conversion efficiency and performance, thereby making the materials processing impossible. It is known that for nonferrous metals and alloys to be processed, more appropriate are high-peak-power Q-switched lasers ensuring ablative material removal mechanism. Experiments on cutting of nonferrous metals in the high-frequency pulse-periodic regime show their high efficiency and much higher quality of machining. However, the power of the available solid state lasers operating in this regime is low – typical values of the output power are no more than 100W. Low average power does not provide acceptable performance. Increasing average power will open the way for industrial application of these lasers. Comparisons performed for the same average power showed that the speed of cutting of nonferrous metals in the high-frequency pulse-periodic regime exceeds the speed of cutting in the CW regime by more than an order of magnitude in some cases. The absence of melting of the material ensures the quality of the edges comparable with the quality of mechanical machining without a cold-worked layer, thereby making additional processing unnecessary. Fabrication of industrial solid-state lasers operating in the high-frequency pulse-periodic regime with an average output power of 500-1000W and higher will significantly expand the scope of application of these lasers.

1.2.1.1. High-Frequency Intracavity Loss Modulation

This approach is realized in flashlamp-pumped solid-state lasers with an average output power up to 1kW. Such lasers can be fabricated by using the existing experience in the field of calculation and design of laser resonators, which provide their operation at the stability boundary with simultaneous optical decoupling of intracavity elements. In industrial applications, stability can be reached with the help of an active rear mirror, which provides stabilization of the fundamental mode in the cavity. The facility relies on a standard lamp-pumped Nd:YAG laser (Figure 2.1), which has two 6.3×130mm active elements. The active elements are mounted in the cavity. In standard configuration, the laser provides an output power of 500W with radiation beam quality M^2=40. The experimental setup uses the active cavity, which includes two active mirrors – a rear mirror with a variable radius of curvature (Figure2.2) and an output mirror – with variable reflectivities varying along the radial coordinate (Figure 2.3). Using the system of active mirrors has made it possible to significantly improve laser radiation quality. The parameter M^2=15 was obtained with decreasing power down to 350W (Figure 2.4). Therefore, the increase in the power density in the focal plane of the lens exceeded one order of magnitude. Besides, improving the divergence of the output radiation creates prerequisites for the normal operation of standard acousto-optic modulators used in single-mode lasers. Thus, the Q-switched regime can be realized in the frequency range from 3 to 30kHz at an output power of 250W. When using a similar approach, the Q-switched regime at an output power up to 1kW, with a slight deterioration in quality of the output radiation, has been implemented.

1.2.2. Carbon Dioxide Lasers

If low-average-power Q-switched solid-state lasers exist in principle, high-frequency pulse-periodic CO_2 lasers with an output power that is suitable for technological purposes are absent. However, the use of high-frequency pulse-periodic CO_2 lasers will fundamentally expand the scope of their applications. One of the most promising applications is the use of CO_2 lasers for glass cutting. Continuous-wave CO_2 lasers provide high-quality and high-speed cutting of flat glass of virtually any thickness (the method of splitting non-metallic materials was proposed and patented by V.S. Kondratenko). In recent years, the technique of laser splitting of different materials is increasingly used in mass production. However, this method ensures only a straight cut and is completely unsuitable for complex laminated glass. High-frequency pulse-periodic devices emitting pulses of short duration deliver a cut along an arbitrary trajectory, without softening a material. The possibility of high-frequency pulse-periodic operation of lasers of virtually any power has been demonstrated both theoretically and experimentally. Experimental testing under laboratory conditions suggests the principle possibility of transferring the available technology to industry. The basic requirement for industrial lasers is the high reliability and long-term trouble-free operation. In this connection, it is desirable to ensure the low serviceability of equipment. In this regard, the development of industrial laser sources requires additional applied research to obtain maximum stability of the output characteristics, reliability and service life of the laser. The main direction of the research should comprise optimization of optical schemes of laser resonators operating in the high-frequency pulse-periodic regime and search for solutions that provide high radiation resistance of optical elements of lasers.

In addition, it is also advisable to conduct additional research aimed at the commercial development of constructive and technological implementation of the technique of phase-locking of individual lasers (in order to increase the output power without deteriorating the spatial and temporal characteristics of the laser output) in metal and glass processing. The high-frequency pulse-periodic regime fundamentally differs from CW and low-frequency operation modes by the mechanisms of interaction of laser radiation with matter. In the future this will require applied research aimed at the refinement of the technologies of cutting, scribing, marking, laser-induced chemical etching, attainment of high-power secondary radiation from plasma, etc. for different boundary conditions of interaction with a wide range of materials. Experiments show that in the case of high-frequency pulse-periodic laser sources, it is needed to significantly increase the ranges of variation of many interaction parameters.

1.2.3. High-Frequency Pulse-Periodic Regime

The above objectives can be achieved through physical foundations, elaborated by our team in this area since 1993. The emphasis is on the two main approaches to implementation of the high-frequency pulse-periodic regime.

1.2.3.1. High-Frequency Intracavity Loss Modulation

In lasers with a high average output power (GDL,HF/DF,COIL) use is commonly made of unstable resonator configurations owing to the large cross section of the active medium[8]. In resonators of this type, externally injected low-power beams may exert a significant effect on the characteristics of output radiation [9].

One way to realise the control regime is the self-injection of radiation extraction from the resonator and return of a part of radiation after changing its spatial-temporal characteristics. The transition to the transient lasing mode is effected through the modulation of the self-injecting beam. Earlier, a study was made of laser versions with radiation self-injection into the paraxial resonator region. However, analysis showed that the power of the beam injected into the paraxial beam region should be comparable with the output laser power to efficiently control the resonator of a continuously pumped laser, unlike pulsed systems with regenerative amplification [3].

The self-injection of a part of output radiation through the resonator periphery is more efficient: on return to the paraxial resonator region, the injection power significantly rises due to the large number of passages to play the dominant part in the formation of output radiation.

In the case of a traditional resonator, the role of waves converging to the resonator axis was found to be insignificant, because their source is a narrow region with a small relative area at edge of the output mirror; accordingly, the power of the control wave injected into the resonator is low. This wave has a large divergence, and only its small part (of the order of $1/N_f$, where $N_f>>1$ is the Fresnel number) participates in lasing.

The effect of injection wave on the resonator characteristics can be enhanced by matching the beam phase with the resonator configuration and increasing the radiation power returned. In this case, the propagation direction and the wavefront curvature of the injection beam should be so matched with the resonator configuration that the injection beam concentrates, after a relatively large number of passages through the resonator, near the optical resonator axis and transforms to a divergent wave that forms the output radiation. The injection beam energy should be high enough to exceed, after its arrival to the resonator axis, the saturation energy of the active medium.

The configuration [2] was realized in a gas-dynamic CO_2 laser with the following parameters: the length of the active medium L_a=1.2m, the unsaturated gain coefficient g_0=0.6m^{-1}, the time it takes the active medium to transit the resonator τ=0.92x10^{-4}s, the relaxation time τ_p =2.76x10^{-4}s, the total go-round resonator time τ_f =4.2x10^{-9}s, the luminescence lifetime τ_l=5s, the resonator magnification factor M=1.45, the diameter of output laser aperture a=0.08m (Figure 2.1).

The laser resonator is made up of two spherical mirrors with rectangular apertures, which provided a geometrical amplification factor of 1.45. The active medium travels across the optical resonator axis. In what follows all theoretical and experimental data are given for a laser with the above parameters.

A part of the output laser radiation was diverted by an inclined metallic mirror to the injection beam formation system consisting of two spherical mirrors with conjugate focal planes. In the vicinity of the focal plane there formed the waist of the branched part of the laser beam, and a modulator was placed near the waist. The modulator location was so selected that the laser beam completely filled the aperture of the modulator. The maximal modulation frequency in our experiments was equal to 33kHz.

A mirror 4 focused the radiation onto the calorimeter. The duration of an individual pulse was about 150-200ns. We emphasise that the recorded pulse duration was limited by the measuring path bandwidth equal, as noted above, to 50MHz. The amplitudes of individual pulses exceeded the average value of output power by factor –10. The average output power was measured with a calorimeter cooled by running water. It is noteworthy that the average output power in the pulse-periodic mode was equal to the output power in the CW laser-operating mode. Good agreement between the experimental and theoretical data for frequencies ranging up to 25-33kHz testifies to the adequacy of the proposed model and the possibility of employing this method at higher frequencies to convert a CW laser to the operating mode similar to the Q-switching mode.

1.2.3.2. Pulse-periodic mode for HF/DF laser

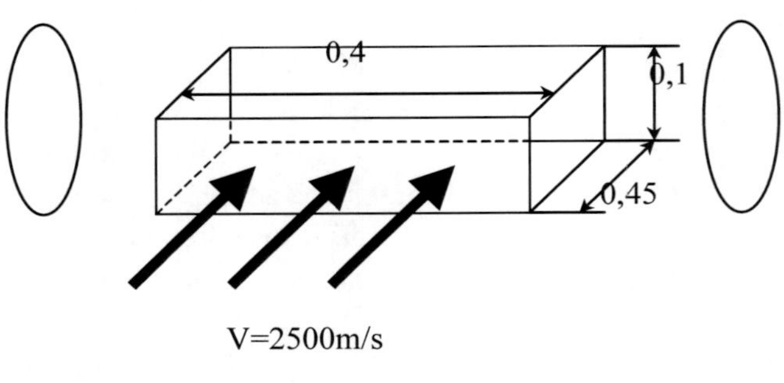

Parameter	Value	Parameter	Value
La	0,4m	τ	0,18*10^{-4}s
Lr	1,0m	τ_p	(2-3)*10^{-5}s
g_0	5-10 m^{-1}	M	1,4
C	3.10^{8}m/s	τ_f	3,3.10^{-9}s
A	0,08m	τ_l	20/330s

Where:
La – active media length;
g_0 – unsaturated gain coefficient.;
g – saturated gain coefficient.;
c – light velocity;
τ - time of active media travel across of the resonator;
τ_p – relaxation time;
τ_f – round trip time;
τ_l – luminescence time;
M – magnification of resonator;
a – diameter of aperture.

For the configuration above [2] one can conclude:

Pulse-periodic mode is valid for the modulation frequency 60-600KHz.The depth of modulation will be chaotic and structure –bundles of pulses for the frequency gap: 100-300KHz. The duration of bundles will be controlled by filling factor. For the frequencies more than 300KHz the depth of modulation will reach 100%.For frequencies more than 450KHz one can get a classical regime of single pulses train. Duration of single pulses will be in the gap: 100-250ns.The ratio Ppeak/Paver goes to the value 1000-10000.

1.2.3.3. Pulse-Periodic Mode for COIL

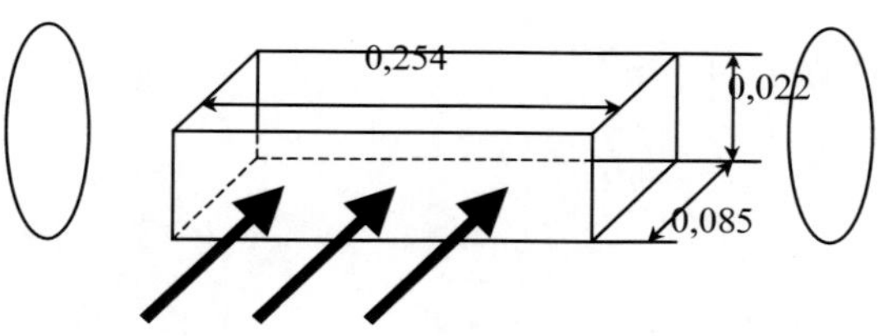

Parameter	Value	Parameter	Value
La	0,254m	τ	$0{,}71.10^{-4}$s
Lr	1,5m	τ_p	$(10\text{-}12).10^{-5}$s
g_0	$56{,}9\text{m}^{-1}$	M	1,2
C	3.10^8m/s	τ_f	5.10^{-9}s
a	0,04m	τ_l	2s

All parameters of the system have the same definitions as for table mentioned above All the data were taken from the papers [2,4] one can conclude:

Pulse-periodic mode will start for the modulation frequency more than 20KHz.The depth of modulation will be chaotic and structure–bundles of pulses. The duration of bundles will be controlled by filling factor. For frequency more than 100KHz the depth of modulation will reach 100%.For frequencies more then 150KHz –classical regime of single pulses train. Duration of single pulses will be less then 150ns. The ratio Ppeak/Paver goes to the value 100-1000.

1.2.3.4. *Q*-Switched Regime for Industrial Applications

Regime of Light Supported Detonation Wave for "Impulsar"/"Lightcraft" applications
Typical velocities of LSDW V~5-10km/s.
Typical thresholds for free space and for near surface conditions: 10–100MW/cm^2 (λ=10.6 μm).
Expression for velocity [5]:

$$V = [2 \cdot (\gamma^2 - 1)]^{\frac{1}{3}} \cdot \left(\frac{I}{\rho_0}\right)^{\frac{1}{3}} 1830 \left(\frac{I}{P_0}\right)^{\frac{1}{3}} \text{ [m/s]} \tag{7}$$

I –laser radiation intensity, ρ_0- gas density, I' [MW/cm^2], P_0 [atm]. In our case V~5–10km/s.

The length of laser spark for LSDW for unlimited duration and for constant intensity- Z_P
Expression for Z_P

$$Z_p \frac{d_b}{Ff} = 0.013 \left(\frac{W}{P_0}\right)^{1/2} \left\{1.93 + InP^0 + \text{In}\left[\left(\frac{W}{P^0}\right)^{1/2} (6.3 + InP_0)^{3/4}\right]\right\}^{3/4} \tag{8}$$

d $_b$ [cm] diameter of the beam, F_f – focal length, W [MW] –laser pulse power, P_0 [atm.] air pressure. For P_0=1 и 0.1atm. From (1) we can get:

$$Z_p \frac{d_b}{F_f} = 0.013 \left(\frac{W}{P_0}\right)^{\frac{1}{2}} \left\{3.32 + \frac{1}{2}\text{In}\left(\frac{W}{P_0}\right)\right\}^{3/4} \tag{8a}$$

$$Z_p \frac{d_b}{F_f} = 0.013 \left(\frac{W}{P_0}\right)^{1/2} \left\{0.67 + \frac{1}{2}\text{In}\left(\frac{W}{P_0}\right)\right\}^{3/4} \tag{8b}$$

The time of laser radiation is on- t=t_r and up to the moment when the regime of LSDW is over, Z_P for this case is:

$$Z_p = 1.32 \left(\frac{F_f}{d_b}\right)^{\frac{2}{5}} W^{\frac{1}{5}} \cdot t_r^{\frac{3}{5}} \frac{1}{(AP_0)^{\frac{1}{5}}} = 0.67 \left(\frac{F_f}{d_b}\right)^{\frac{2}{5}} \left(\frac{W}{P_0}\right)^{\frac{1}{5}} \cdot t_r^{\frac{3}{5}} \tag{9}$$

Here t_r [μs].
From (8) and (9) one can get the time of LSDW decay due to the laser beam expansion:

$$t_r \frac{d_b}{F_f} = 0.0014 \left(\frac{W}{P_0}\right)^{1/2} \left\{1.93 + InP_0 + \text{In}\left[\left(\frac{W}{P_0}\right)^{1/2} (6.3 + InP_0)^{3/4}\right]\right\}^{5/4} \tag{10}$$

For P_0=1atm. and 0.1atm expression (10) can be written as:

$$t_r \frac{d_b}{F_f} = 0.0014 \left(\frac{W}{P_0}\right)^{\frac{1}{2}} \left\{3.32 + \frac{1}{2}\text{In}\left(\frac{W}{P_0}\right)\right\}^{5/4} \tag{10a}$$

$$t_r \frac{d_b}{F_f} = 0.0014 \left(\frac{W}{P_0}\right)^{\frac{1}{2}} \left\{0.67 + \frac{1}{2} \ln\left(\frac{W}{P_0}\right)\right\}^{5/4} \quad (10b)$$

Laser energy absorbed during t_r is:

$$\frac{q}{P_0} \cdot \frac{d_b}{F_f} = 0.0014 \left(\frac{W}{P_0}\right)^{3/2} \left\{1.93 + \ln P_0 + \ln\left[\left(\frac{W}{P_0}\right)^{1/2} (6.3 + \ln P_0)^{3/4}\right]\right\}^{5/4} \quad (11)$$

For P_0=0.1 и 1atm. expression (7) can be written as :

$$q \frac{d_b}{F_f} = 0.0014 \left(\frac{W}{P_0}\right)^{\frac{3}{2}} \left\{3.32 + \frac{1}{2} \ln\left(\frac{W}{P_0}\right)\right\}^{5/4} \quad (12a)$$

$$q \frac{d_b}{F_f} = 0.0014 \left(\frac{W}{P_0}\right)^{\frac{3}{2}} \left\{0.67 + \frac{1}{2} \ln\left(\frac{W}{P_0}\right)\right\}^{5/4} \quad (12b)$$

Length of spark- $Z_P(d_b/F_f)$, time duration and energy of pulse $t_P(d_b/F_f)$, $(q/P_0)(d_b/F_f)$ for gas pressure P_0=1, 0.1atm can be presented by:

For P_0=1atm	P_0=0.1atm.	
$Z_P=0.0336W^{0.57}$ F_f/d_f ,	$Z_P=0.0667$ $W^{0.62}$ F_f/d_f	
$t_P=0.0068$ $W^{0.618}$ F_f/d_f,	$t_P=0.01$ $W^{0.7}$ F_f/d_f	
$q=0.0068$ $W^{1.62}$ F_f/d_f,	$q=0.01$ $W^{1.7}$ F_f/d_f.	(13)

Estimations for the case of German`s experiments [6]with the following parameters:

q=280J, t_r~30mks, W(peak)~10MW.

One can get:

1. For P_0=0.1atm. - Z_P=0.28cm, t_r=0.05mks, q=0.5J;
2. For P_0=1atm. - Z_P=0.12cm, t_r=0.028mks, q=0.283J. (14)

One can get from here that for W~10MW regime of LSDW was not effective. In that case the absorption process had the next stages: optical breakdown – LSDW-SbRW-standard absorption in expanding plasma.

In our case we need to find: W и t_r, optimal for the energy absorption.

For P_0=1atm.	For P_0=0.1atm.	
$W=21.7$ $(q\ d_f/F_f)^{0.617}$	$W=15$ $(q\ d_f/F_f)^{0.588}$	
$t_r=0.0455$ $(q\ F_f/d_f)^{0.381}$	$t_r=0,0665$ $q^{0.411}$ $(F_f/d_f)^{0.588}$	
$Z_P=0.194$ $q^{0.352}$ $(F_f/d_f)^{0.648}$	$Z_P=0,357$ $q^{0.365}$ $(F_f/d_f)^{0.635}$	(15)

All that is valid , if OPD conditions are corresponding to:

$$2,5 \cdot M_0 > f > 5,88 \cdot (1 - M_0)^{1.5}. \quad (16)$$

Here: $M_0=V_0/C_0$ – gas velocity in comparison with OPD ($C_0 \approx 340$m/s – velocity of sound in air) and pulse frequency $f = f \cdot t_d$, $t_d = R_d / C_0$ – dynamic time , $R_d = (Q_r / P_0)^{1/3}$ [m] – dynamic radius, Q_r [J] – energy of spark, P_0 [pa] – gas pressure. Reasonable values of M_0 ~0.65–0.95.

q, J		10	100	1000	280 German`s experiments[6]
$P_0 = 0.1$ atm.	t_r, mks	0,17	0,439	1,13	0,665
	W, MW	58,1	225	873	411
	Z_P, cm	0,83	1,97	4,52	2,86
$P_0 = 1$ atm.	t_r, mks	0,11	0,263	0,63	0,39
	W, MW	90	372	1540	702
	Z_P, cm	0,44	0,98	2,21	1,41

1.2.3.5. Optimal Conditions for QCW Mode

For W_C –fixed one can find the area of $\{Q_r, f\}$, where the optimal conditions can be satisfied:

$$f_1\ [\text{kHz}] < 40 \cdot M_0 / (Q_r/P_0)^{1/3},\ P_0\ [\text{atm}] \quad (17)$$

$$f_2\ [\text{kHz}] > 93 \cdot (1 - M_0)^{3/2} / (Q_r/P_0)^{1/3} \quad (18)$$

$$Q_r > Q_{r\,min}, \quad (19)$$

$$f_3\ [\text{kHz}] = W_C/Q_r. \quad (20)$$

For $M_0 = 0.8$ and $P_0=1$atм. (18) – (19) can be presented by:

$$f_1\ [\text{kHz}] < 32 / (Q_r)^{1/3}, \quad (21)$$

$$f_2\ [\text{kHz}] > 8.3/(Q_r)^{1/3}, \quad (22)$$

Laser radiation parameters for launching: vehicle mass-M_a ; launching is along of straight line S_t~200km ; height h~50km; final velocity- V_a=5km/s.

The conservation law for our case:

$$0.5 M_a V_a{}^2 + M_a G h + F_r S_r = F_a S_t \quad (23)$$

$$0.5 M_a V_a / M_a G h \approx 25$$

$$F_t(t) = \int_0^{R_S} \left(\frac{P_{r(t,r)}}{P_0} - 1 \right) P_0 r dr$$

During the time Δt for vehicle we have:

$$\Delta P_a = M_a \Delta V_a = \int_o^{\Delta_t} F_t dt = tf \int_{t_i}^{t_{i+1}} F_f dt = tf\delta P_a = tF_a \quad (24)$$

$\Delta P_a = \delta P_a\, f\, t = F_a t = J_\kappa \cdot q \cdot f \cdot t = J_\kappa \cdot W_c \cdot t.$

Here W_c – average power of pulse-periodic laser,

$F_a = J_r \cdot W_c$

By using $Fa = J_r \cdot W_c$ from (17) one can get main parameters of the task untill the vehicle would get the velocity-V_a.

$$\frac{M_a}{w_c} = \frac{J_{rc} S_t}{0.5V_a^2 + Gh + G_r S_t} \approx \frac{2J_{rc} S_t}{V_a^2} \quad (25)$$

$$\frac{M_a}{W_c} \approx \frac{J_{rc} \cdot t_a}{V_a} \text{ [kg/MW]} \quad (26)$$

$$t_a = \left(\frac{2M_a S_t}{J_{cr} W_c}\right)^{\frac{1}{2}} \quad (27)$$

Here: t_a – time of acceleration,S_t –distance of acceleration V_a –final velocity, $J_{rc} \approx (J_{r0}+J_{r1})/2$ (J_{r0} – for the beginning and J_{r1} for the height-h).Average value: $J_{rc} \approx 0.75J_{r0}$ (J_{r0}=2000H/MW), in case of acceleration of vehicle by QSW. For the reference: J_{r0}=500 and J_{r1}=100 H/MW.

Some estimations for the following parameters:

S_t ~200km, V_a=5km/s, h=50km from (18) and (20) one can get:

In the case of QSW mechanism: M_a/W_c=25kg/MW, t_a=80s

In the case of experimental conditions [6]: M_a/W_c=5kg/MW, t_a=180s

The comparison shows that the improvements factor of the launching conditions in our case goes to the value 5, which is just pessimistic estimation only not taking in to account any possibility of launching conditions further optimization.

Laser-radiation self-injection (modified regenerative regime) with the transformation of the temporal and spatial structure was tested on wide-aperture CO_2 lasers with an output power up to 10kW. Currently, the temporal structure of the radiation is modulated either using a disk modulator (mechanical modulation), or a deformable-mirror modulator. For industrial applications, we plan (within the framework of the proposed project) to use a bimorph deformable mirror made of SiC as a modulator. This bimorph mirror was developed by our team. We have already demonstrated theoretically and experimentally the principle possibility of its use for modulation. One of the schemes of implementation of a high-power, high-frequency pulse-periodic CO_2 laser is shown in Figure 2.5. In this figure: M1 and M2 are the mirrors of an unstable resonator; the radiation is coupled out from the cavity outside the edges of the mirror M2; the output-coupler M3 directs a part of the output beam in the injected-beam-forming system (IBFS); using the same mirror the beam formed by the IBFS returns to the resonator; and in the IBFS the laser beam is modulated with respect to power and acquires the required phase distribution. This scheme has been implemented in a gas-dynamic CO_2 laser. The laser resonator consists of two spherical mirrors of rectangular shape, providing a geometric magnification of 1.45. The active medium is moved perpendicular to the optical axis of the resonator. The experimental time dependences of the output radiation of the above-described self-injection laser are shown in Figure 2.6. The average output power was measured with a calorimeter cooled by running water. The beam was focused on the calorimeter with the help of the mirror M4. In CW regime, the signal was recorded whose noise component did not exceed 5% of the constant level of the signal. At a modulation frequency of about 27 kHz and a modulation depth of 2%−3%, quasi-stationary modulation (Figure 2.6) is realized; in this case, intensity fluctuations, which are consistent with the modulating signal (the deviation of the output power is about three times of the average value), are observed in laser radiation. With increasing modulation depth up to a level of 7%−8%, pulse-periodic regime of laser operation was observed (Figure 2.6). In this case, a train of 5−10 pulses was generated within one cycle of the opening of the modulator. The duration of a single pulse was about 200 ns. It should be noted that the duration of the recorded pulse was limited by the bandwidth of the measurement path, which was 50 MHz. The excess of the amplitude of the individual pulses over the average value reached the value of 15−25. The average output power in the pulse-periodic regime was equal to the average output power in the CW regime, which is extremely important for the motivation of the transition to the new regime [1, 2].

1.2.4. Conclusion

Our work demonstrated, both theoretically and experimentally, the feasibility of converting high-power wide-aperture laser radiation to the pulse-periodic lasing mode by way of self-injection of the modulated fraction of output radiation without a sacrifice in output radiation power in comparison with the stationary lasing mode. In this case, the peak output power can exceed the average output power by more than an order of magnitude. For pulse-periodic modulation with a pulse length of 150–200ns the peak power was greater than 100kW. Applicability of the method and variations of temporal structure of radiation for other type of high power lasers (HF/DF, COIL) have demonstrated the possibility of Q-switched mode realization with very much similar to GDL shape of radiation [7].

The regime of light supported detonation wave for theoretically investigated temporal structure of laser radiation with much higher output power has demonstrated a very high efficiency for “Impulsar”/”Lightcraft” applications. QSW mode based on effect of shock waves joining had been suggested and analyzed.

Experimental confirmation of the possibility of obtaining the high-frequency pulse-periodic regime in high-power wide-aperture gas lasers, and realization of reflective coatings that are resistant to mechanical deformations allowed us to develop and manufacture a bimorph deformable mirror, capable of operating at frequencies of tens of kilohertz.

The originality and novelty of the above-mentioned approaches are indirectly confirmed by the fact that the implementation of such operation modes is described in detail in a number of theoretical articles; however, experimental works are virtually absent (for example, this regime is realized by reducing the average output power by 100 times for CO_2 lasers).

Real solid-state lasers operating in the pulse-periodic regime with an output power level of hundreds of watts are produced by two companies – 'Lee Laser Inc.' (USA) and 'Powerlase Photonics Ltd.' (UK). The manufactured lasers are pumped by laser diodes, thereby providing 30%–50% efficiency of CW operation. For the pulse-periodic regime, the manufacturers mention the efficiency at a level of only a few percent, i.e. the efficiency decreases by more than 10 times. In our experimental studies we demonstrated high efficiency for solid-state laser, pulse-periodic operation yielding a decrease of 1.5 in efficiency. On the laboratory bench, we carried out preliminary experiments on cutting various materials and showed the possibility of cutting any materials: structural steel, carbide cutting elements, zirconium alloys, diamonds, composites, glass, natural rock, laminated glass, etc.

Thus, from an energy point of view, only pulse-periodic lasers can provide constant flux density in the region being machined and stable quality of laser processing. A well-mastered level of energy performance for pulse-periodic lasers available at the world market is no more than 100W of average power. Our developments make it possible to talk about the feasibility of high-frequency pulse-periodic laser systems with a high average power and about the need for their prompt industrial production to expand the range of technological applications. Currently, it is well understood by laser companies from England, Germany, USA, China and Japan. According to the materials of the 18th International Symposium on Gas Flow and Chemical Lasers & High Power Lasers (GCL/HPL-2010), the methods aimed at obtaining the pulse-periodic regime have resulted in a 800W average output power of a diode-pumped Nd:YAG laser. We realized already in 2000 a high-frequency pulse-periodic CO_2-laser source with a power of 10kW.

1.2.5. DRAWINGS, PHOTOS AND DIAGRAMS

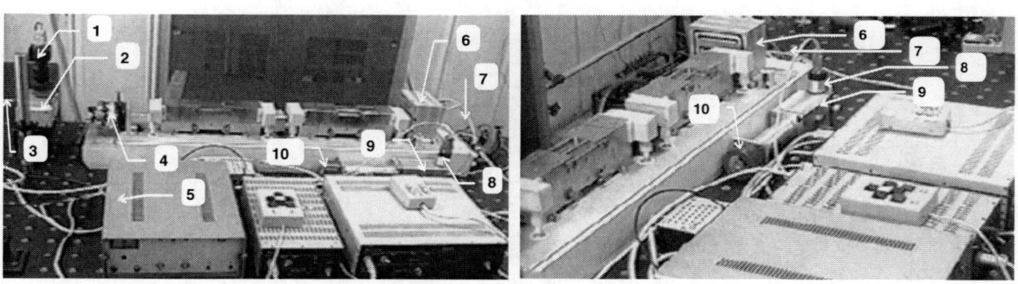

Figure 2.1. Experimental setup: 1 – CCD camera, 2 – optical unit, 3 – brightness sensor, 4 – controllable output resonator mirror, 5 – output mirror control unit, 6 – unit automatically aligning the rear mirror of the resonator, 7 – pressure line, 8 – pressure sensor, 9 – pump, 10 – pump drive, 11 – pressure control unit, 12 – device controlling the unit of automatic alignment of the rear mirror, 13 – control panel.

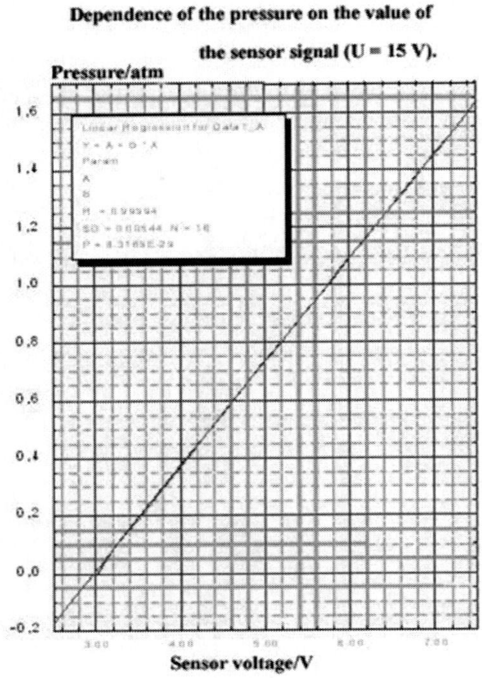

Pump design:

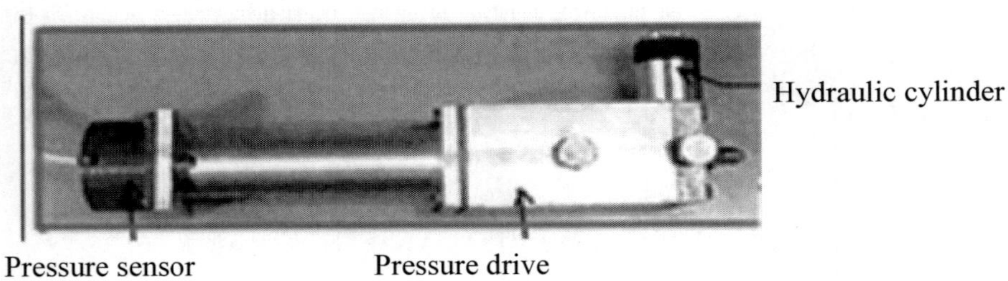

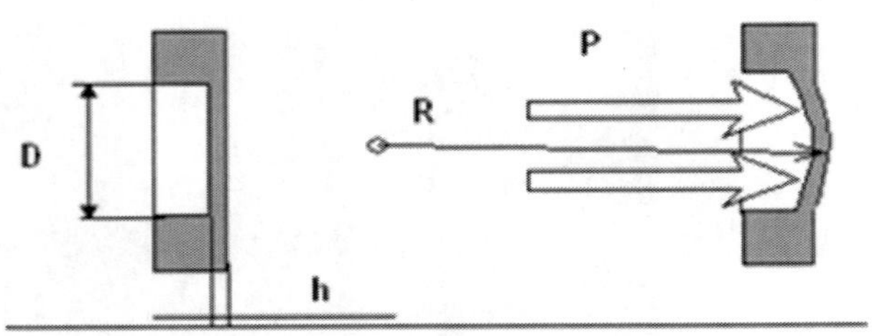

Pressure control device:

Principle of operation of the active rear mirror:

P= -0.5 ... 3 at D=10mm h=0.3mm

Control panel

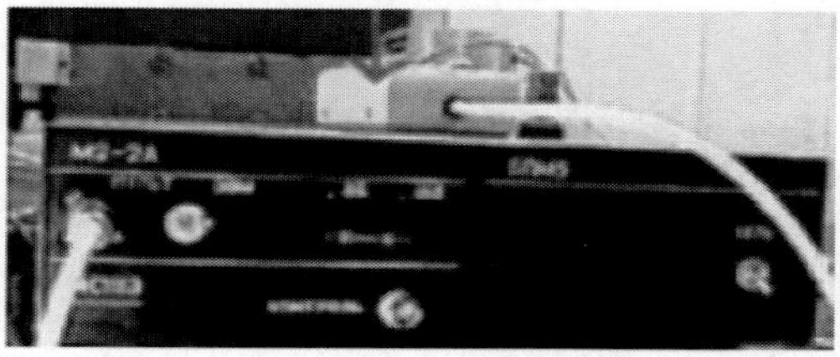

Appearance of the prototype:

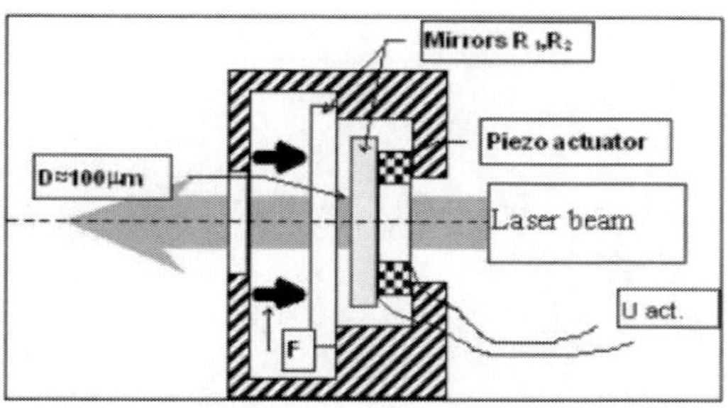

Figure 2.2. Schematic diagram of the active output mirror.

Dependence of the reflection coefficient on the angle and distance between the mirrors.

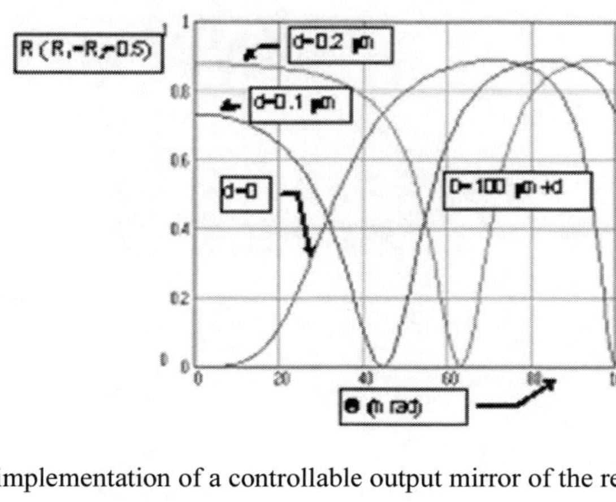

Figure 2.3. Example of implementation of a controllable output mirror of the resonator.

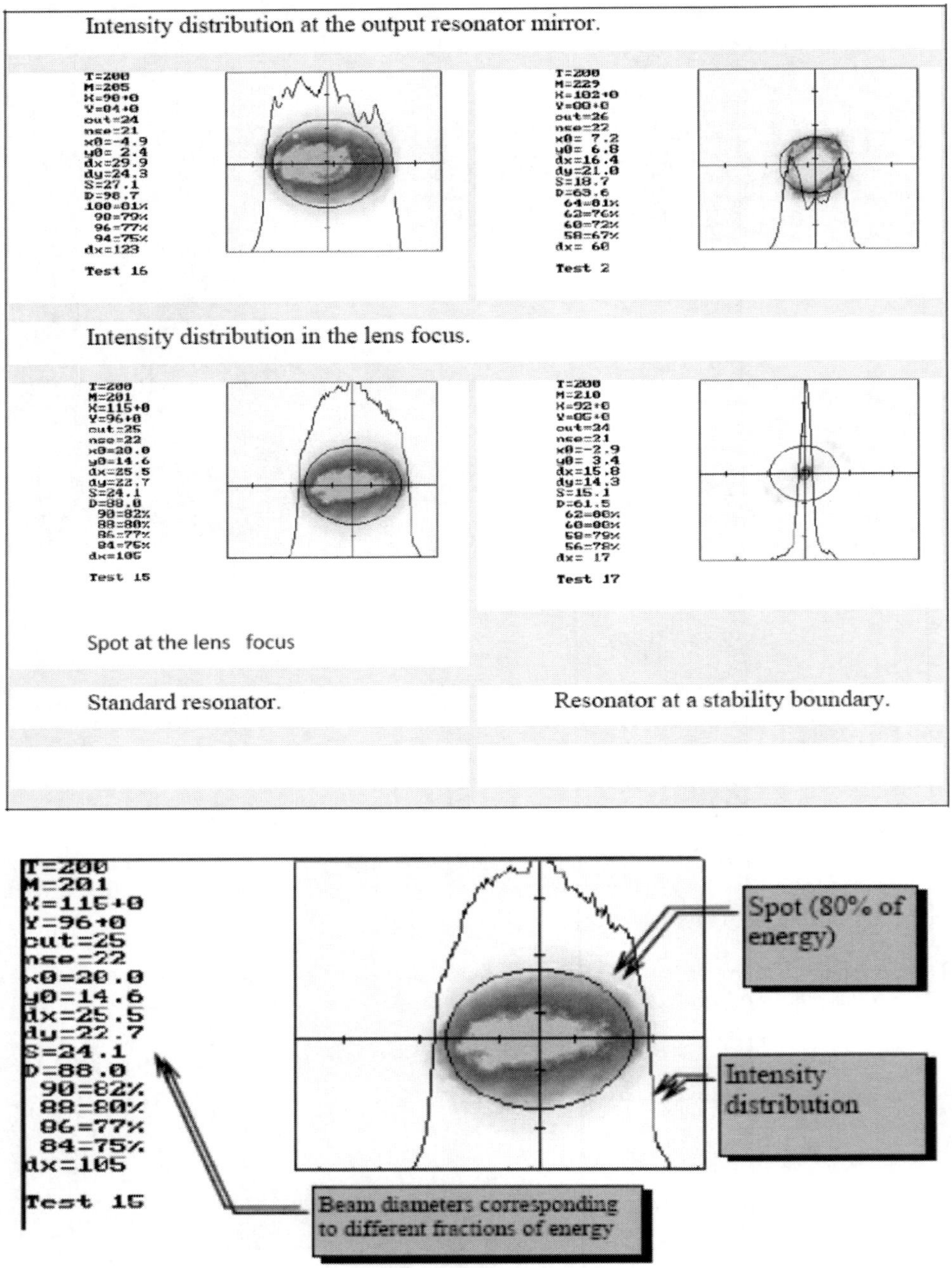

Figure 2.4. Experimental results. The output power is 350W.

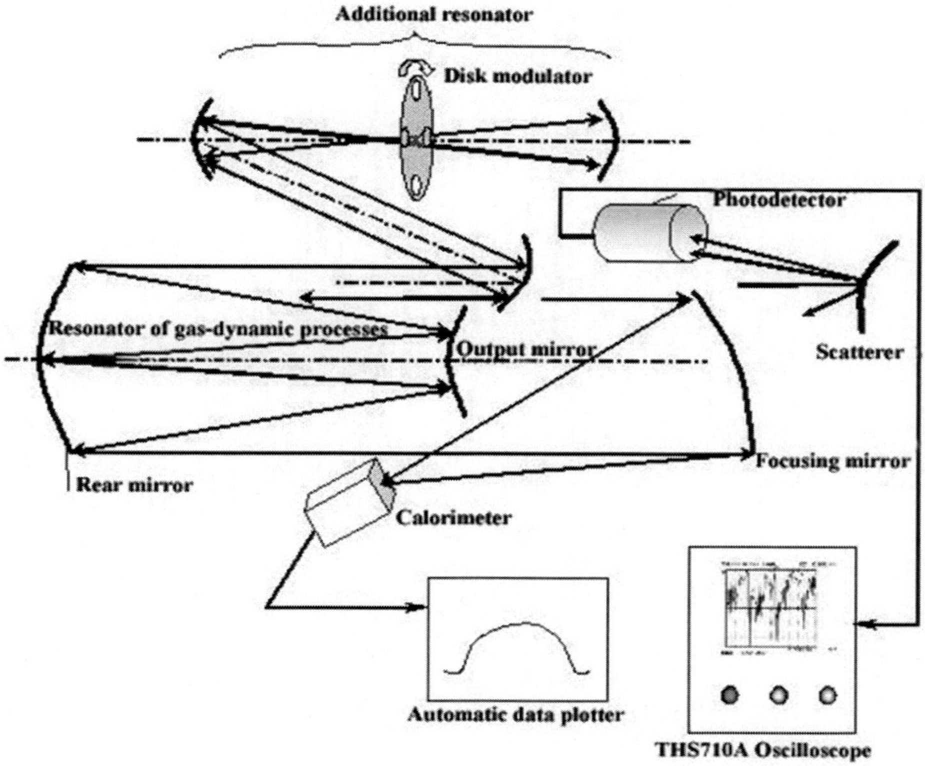

Figure 2.5. Scheme for obtaining the extracavity Q-switched regime in a gas-dynamic CO_2 laser.

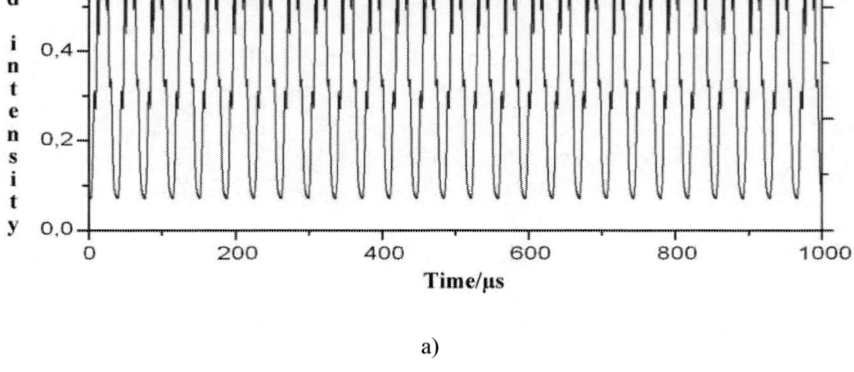

a)

Figure 2.6. (Continued)

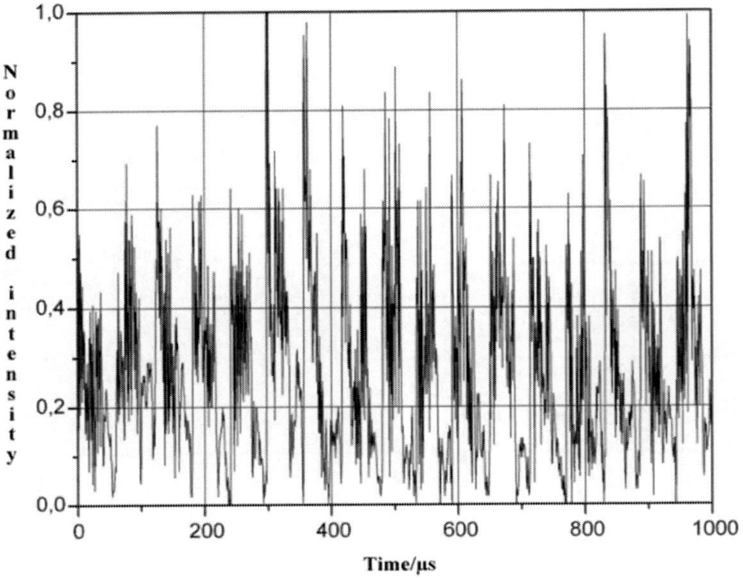

b)

Figure 2.6. Temporal structure of the gas-dynamic laser output (output power of 10 kW): (a) at a modulation depth of ~3%, and a modulation frequency of ~27 kHz (the top graph is the modulating signal, the bottom graph is the output signal of the laser) and (b) At a modulation depth of ~7% and a modulation frequency of about 25 kHz.

1.2.6. References

[1] Apollonov V.V., Prokhorov A.M. "High power laser for industrial and ecological use", *Proc. SPIE,* 3574, 2 (1999).

[2] Apollonov V.V., Kijko V.V., Kislov V.I., Suzdal'tsev A.G., Egorov A.B. "High-frequency repetitively pulsed operating regime in high-power wide-aperture lasers", *Quantum Electron.*, 33 (9), 753 (2003).

[3] Apollonov V.V., Alcock A.J., Baldis H.A., Corcum P.B., Taylor R.S. "Oscillator-amplifier of nanosecond CO2-laser pulse system", *Opt. Lett.*, 5, 333 (1980).

[4] Hager G.D., Kopf D., Plummer D., Salsich T., Crowell P. "Demonstration of a repetitively pulsed magnetically gain-switched chemical oxygen iodine laser", *Chem. Phys. Lett.*, 204, 420 (1993).

[5] TischenkoV.N., Apollonov V.V., Grachov G.N., Gulidov V.I., Zapryagaev V.I., Men'shikov Ya.G., Smirnov A.L., Sobolev A.V. "Optical pulsating discharge interaction with gas: the conditions of stable generation and shock waves joining", *Quantum Electron.*, 34, 941 (2004).

[6] Schall W.O., Bohn W.L., Eckel H.-A., Mayerhofer W., Riede W., Zeyfang E. "Lightcraft experiments in Germany", *Proc. SPIE*, 4065, 472 (2000).

[7] Wallace J. "Gas-dynamic laser enters pulse-periodic mode", *LFW*, p.17, August 2004.

[8] Apollonov V.V., Kijko V.V., Kislov V.I., Suzdal'tsev A.G., Egorov A.B. "High-frequency repetitively pulsed operating regime in high-power wide-aperture lasers", *Quantum Electron., 33, 753 (2003)*

[9] Apollonov V.V., Vagin Yu.S., Egorov A.B., Kijko V.V., Kislov V.I. "Method for obtaining pulse-periodic laser operation mode", Patent No. 2175159 from 28.04.2000.

Chapter 3

1.3. HIGH-FREQUENCY PULSE-PERIODIC DISK LASERS: ADVANTAGES AND PROSPECTS

1.3.1. INTRODUCTION

It has long been customary that as new technologies emerge into the light, potential users and experts start wondering whether these new techs will replace the old, well-established approaches to the solution of known problems. When we look back at the time when the first laser was created, it becomes clear that each new and more advanced technology usually replaces the pre-existing and well-proven technologies. However, a more differentiated and balanced assessment of the many innovations in the field of laser shows that there is no such thing as a perfect design, or an ideal laser. There is always room for improvement and further development. It is possible that in the near future a laser with the disk geometry of the active medium will become the dominant technology. However, despite this, a number of existing laser technologies (with some exceptions, of course) will continue to improve for quite a long time. Based on these considerations, we must continue to develop most types of lasers, each time clearly specifying their undeniable technological niche and knowing well their advantages and disadvantages.

High efficiency and excellent beam quality of disk lasers mean that they can be widely used in modern science and industry for a very large range of applications, including treatment of the surface of dielectric materials in microelectronics, cutting, drilling, welding, polishing and cleaning of the surface, other technological operations with solid and composite materials, etc. Pulse-periodic Q-switched disk lasers and high-average-power, mode-locked laser systems ensure optimal conditions for ablation (sublimation) of a material. A number of advantages of high-frequency pulse-periodic lasers emitting short pulses, as applied to a wide range of industrial technologies, is the basis of many modern concepts of disk lasers. High-intensity light with an insignificant thermal lens effect in the central high-loaded zone of the active medium have led to the lifting of restrictions on the brightness of the pump diode. This has reduced the cost of laser sources, and thus has significantly improved the efficiency of electro-optical conversion, especially in the regime of high average power. The power of the laser source is varied by scaling the cross sectional area of the generated radiation. It should be noted that its ratio to the thickness of the active material in the disk geometry is much larger than the ratio of the cross section of typical core elements of the active medium to their length in conventional solid-state laser systems. This eliminates the problems associated with nonlinear distortions of the geometry of the active medium, and enables realizing high peak values and average powers of laser sources with the disk geometry and the same parameters of the radiation in the far-field region.

1.3.2. Disk and Fiber Lasers

It is clear that each laser source has its advantages and disadvantages. Beam quality of modern solid-state lasers (disk and fiber lasers) is pretty good, and their efficiency is very high. A fiber laser has significant advantages at low, average and peak output powers, while a disk laser, on the contrary, is preferable in those applications which require high and super-high output powers.

Among high-frequency pulse-periodic sources emitting coherent high average and peak output radiation, disk lasers will undoubtedly dominate today and in the future. They provide the best solution for many industrial applications in the multi-kilowatt power range. A disk is a simple and easily excited laser element, which allows cost-effective generation of radiation with improved parameters. It has a lot of potential for the future. The advantage of the disk laser over the fiber laser is already obvious in the kilowatt power range. Having a large radiating surface, the power density of the disk laser is not critical even at high peak powers. On the contrary, in fiber lasers the increase in the peak power negatively affects the reliability of the resonator. The main disadvantage of the fiber laser is also its high sensitivity to the reflected beam, which often appears in the laser due to the interaction of the generated light with matter. If the radiation reflected from the target affects the fiber laser resonator, then it should be immediately shut down to preserve the cavity.

At the same time, the disk laser resonator is insensitive to the penetration of reflected beams. Even in the case of highly reflective materials, technological operations can be performed without the risk of an emergency switching off of the laser. Another advantage of kilowatt disk lasers is their modular design. This design allows one to replace laser modules during after-sales service, thereby ensuring that in the event of failure of the laser, the downtime as well as the cost of repairs will be minimal.

On the contrary, in fiber lasers, the single-unit design of the resonator makes is impossible to replace easily the failed unit without extensive alteration of the system as a whole. At the same time, the use of fiber laser technology offers certain advantages, especially in the power range (up to several hundreds of watts). This laser is suitable for micromachining of dielectric surfaces and composite materials. However, only the disk laser has a high potential in the case of high peak powers, i.e., when it is necessary to provide a train of short pulses with the highest peak power. In a typical disk laser, as opposed to a fiber laser, the power density inside the cavity is far below the critical damage threshold of the laser active medium and surfaces of auxiliary optical elements, even at a high peak power. For this reason, a disk laser allows, among other things, a better use of the laser medium in general, and significantly improvement in laser efficiency.

Thus, in the future solid-state diode-pumped disk lasers will play an increasing role. Laser diode structures are central elements in many new types of lasers. Today, the decisive factor is the acquisition of the necessary experience and skills to design most efficiently new laser systems. Semiconductor-based technologies and applications of laser equipment on new physical principles are key technologies of modern times. For many years, the world has accumulated the experience needed today in this field of knowledge. The challenge is to ensure technological excellence in the application of laser systems, not only today but in the future.

1.3.3. Design and Physical Foundations of Disk Laser Operation

The principle of the disk laser operation is based on the use of an active element in the form of a disk with a cooled surface. High cooling efficiency of the laser medium is provided by a large area of the disc surface, which is important from the point of view of the heat transfer process. Therefore, the average power in the beam can reach quite high values. It is important to note that due to effective heat transfer within the disk area there is no thermal lens effect, characteristic of 'rod' and 'slab' (optical range) geometries of the active element. In this case, cooling is performed through a side surface of the active element; a two-dimensional heat flow, forming a parabolic profile of thermal distortion, propagates through this side surface. The temperature dependence of the refractive index in this case leads to the emergence of a strong thermal lens effect. Such a lens deteriorates the directivity of the output and significantly limits the effectiveness of its action. The thin disk, in contrast, is cooled through a thin contact region on one side and generates a one-dimensional axial heat flow on the other side. As a result, the temperature gradient is distributed parallel to the laser beam, which does not result in the thermal lens effect. In practice, a thin laser element in high-power disk lasers is either connected to a heat sink or is cooled by forced convection.

In both cases, the undesirable thermal lenses are eliminated by the optimum layout of structural elements of the disk laser. Integral heat transfer from the active material to the heat sink depends directly on the area of the pump zone and on the geometry of the heat transfer contact. At a constant temperature of the active laser element, the average output power achieved is directly proportional to the area of the pump zone and, consequently, to the cross-sectional area of the laser beam on the disk. This is a unique feature of disk lasers, which allows one to change the power density at a constant energy of the beam on the disk. This property is particularly important for multi-kilowatt sources, because it makes it possible to adjust the value of the peak output power without changing many other parameters that affect the performance of the overall system.

At first glance, the basic challenge is the disk thickness needed for efficient heat dissipation, because the conventional scheme involves low absorbance of the active disk material. However, the pump scheme allows one to find an elegant solution to this problem. The process of excitation in the active material of the disk can be represented in the form of multi-channel optical configuration consisting of a parabolic mirror and a pump deflection system. The parabolic mirror focuses a collimated beam of the laser cavity containing an active element. After partial absorption of radiation passing through the disk body, some energy of the beam is reflected from a highly reflective mirror located on the backside of the disk. As a result of multiple reflections through prisms and the parabolic mirror in general we deal with 20 iterations of radiation passes through the body of the disk, which ensures high efficiency of absorption of light energy of the pump beam. The optical scheme of such resonators tolerates limitations on the brightness of the pump source. Generally, pumping of a high average power disk laser does not require a beam of high optical quality. That is why both laser diodes and rather complex laser diode structures with a uniform intensity distribution can be used in this case. In terms of the practical use laser diodes with a uniform intensity distribution are more preferable due to much lower energy consumption per one watt of the pump power.

For efficient processing of materials: cutting, drilling, polishing and removal of a thin film from the surface of various materials laser pulses with a high peak power and high repetition rate are needed. Typical pulse duration of a *Q*-switched Yb:YAG disk laser lies within one microsecond. *Q*-switching is provided by means of an acousto-optic modulator. Depending on the resonator geometry it is possible to generate pulses with duration from a few hundred picoseconds to a microsecond. To date, the best results achieved with the help of pulse-periodic disk lasers having high average power are as follows: pulse repetition rate, up to 10 kHz; maximum pulse energy, 100 mJ; peak power, 200 kW; average power, from 100 to 1000 W. However, there are fundamental physical limitations on further scaling of the disk lasers, which we will discuss below.

1.3.4. Laws of Scaling of Disk Lasers

The power of disk lasers is limited not only by the power of the pump and overheating of the medium but also by the losses due to the amplified spontaneous emission (ASE) and the background radiation losses in the resonator. To avoid overheating, the size of the active medium should vary in accordance with the law of power scaling. Then, to avoid large losses caused by their exponential growth during the ASE, amplification of radiation corresponding to transverse round trips should not be large. This requires a reduction in gain *G*. Gain is determined by the reflectivity of the output mirror and disk thickness. Amplification of radiation per round trip, however, should not be substantially greater in magnitude than the radiation loss per round trip along the same optical path. The difference between the gain and losses per round trip of radiation determines the optical energy that is coupled out from the laser cavity. Reduction of the gain at this loss level requires an increase in the disk thickness. In this case, at a certain critical size, the disk becomes optically too thick and cannot be pumped above the threshold without overheating. Some features of scaling can be shown on a simple model. Suppose M is the saturation intensity of the medium. The corresponding optimal thickness of the disk can be estimated as $h \sim T/Mb$. The corresponding optimal lateral dimensions of the disk can be represented in the form: $D \sim T/Mb^2$, where *T* is a parameter of the heat load. Roughly speaking, the loss per round trip of radiation must be scaled inversely proportional to the cube root of the power required: $P \sim n\,T^2 / M\,b^3$.

An additional problem is the efficient delivery of pump energy. In cases with a small-signal gain, single-pass absorption of the pump is also low. It follows that the effective utilization of the pump energy is very necessary for the effective operation of the laser disk. To scale the output power, the medium must be optically thin, which requires a large number of passes of the pump energy through the medium. Besides, the pump energy coupled in through the lateral side of the disk can also be a possible solution for efficient pumping.

To reduce the effects of the ASE, it was proposed to use an optical cover consisting of an undoped material on the surface of the laser disk. This cover allows the spontaneously emitted photons to escape from the active layer and prevents their resonance in the bulk of active material. Rays cannot be reflected from the surface, as in the case of an open disk. This allows the maximum power achievable by the disk laser to be increased by an order of magnitude. Reflection of the ASE from the disk edge should also be suppressed. This can be done through the absorbing layer at the generatrix of the disk cylinder. In the regime when output power is close to maximal, much of the energy is used in the ASE; therefore, the absorbing layers must also have radiators accumulating heat. In the case of the maximum pump density of the disk laser its efficiency is quite low: Most of the pump power is used in the ASE and is absorbed at the edges of the device. In this case, the distribution of the pump energy between several disks can significantly improve the performance of the laser system. Indeed, lasers, consisting of several modules with disk elements in a single cavity, have been repeatedly reported. One of these lasers, fabricated by the TRUMPF Group – a world leader in this class of laser systems – is presented in Figure 3.1.

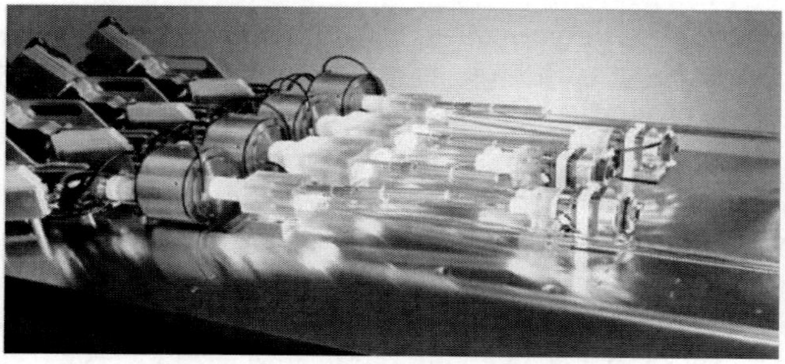

Figure 3.1. Laser consisting of a series of disk modules in a single cavity.

In the quasi-continuous regime, the power can be estimated by scaling the saturation intensity with a duty cycle of the pump and by multiplying the duration of the pump by the pulse repetition rate. At moderate repetition rates (e.g., higher than 1 Hz), the maximum energy of the output pulses is approximately inversely proportional to the cube of background losses b. The undoped cover can increase the average output power by an additional order of magnitude, provided that this cover does not increase the background loss. At low pulse repetition rates (in the single-pulse regime) and sufficient pump power, there is no general limitation on the energy, but the required size of the device increases rapidly with increasing pulse energy, thereby establishing a practical limit of energy. One active element, according to the estimates, can generate an optical pulse with energy of a few thousand joules, depending on the level of internal signal loss in the disk.

Disk lasers, as well as fiber lasers, have a large ratio of the cooling surface area to the gain of the laser. However, these two different concepts of the laser design also differ by the values of the achievable peak power. Beam quality of fiber lasers is determined by properties of waveguides, the refractive index difference between the core and the cladding, and the size of the internal diameter of the fiber that transmits light. On the other hand, beam quality of disk lasers depends on the design of the resonator. With increasing size of the optical pump region (its diameter is usually a few millimeters) at a constant optical power density on the disk surface, scaling of the output power becomes possible. Adjustment of the resonator also has similar features. Self-phase modulation defines the typical limit of nonlinear amplification of ultrashort pulses. It leads to an extension of the spectral line proportional to the ratio of the effective optical path in the material with nonlinear characteristics to the effective cross-sectional area of the beam multiplied by the square of the pulse duration. An electro-optical switch in a regenerative amplifier of the disk laser makes a substantial contribution to the nonlinear characteristics of the gain which is greater than the contribution from the disk. It must be emphasized that the values of the peak power, achieved today in single disk modules, are much smaller than the values obtained in 'rod' and 'slab' solid-state laser systems. However, the level of the average power generated by a single disk module also leaves much to be desired.

1.3.5. Regenerative Amplification of Pulses

At present, in the scientific and technical research and manufacturing processes relying on the use of high average power lasers, only sources operating in two regimes find applications, i.e., CW and pulse-periodic laser sources with a pulse repetition rate up to hundreds of Hz and pulse duration from tens of microseconds to tens of milliseconds. In most processes, purely thermal mechanism of action is mainly implemented, since use is made of the possibility of a laser source to deliver quite a lot of energy to a small area of the surface of the work piece. High-frequency, high-average-power laser systems operating in the *Q*-switched regime, which provides the pulse duration in a periodic train from a few to hundreds of nanoseconds, allows for a fundamentally different mechanism of interaction of radiation with matter, i.e., ablation, ensuring local release of energy not only in space but also in time. This results in an explosive local evaporation of the substance without an intermediate liquid phase. This mechanism can significantly extend the range of technological applications of laser sources. Today, it has found a real use in practice, only at an insignificant level of output power in the range of 1000 W. However, there arises a reasonable question: Why are high-power pulse-periodic lasers not available at the international laser market? The answer is simple: The complexity of their manufacturing consists in the inability of application of the classical modulation methods to laser systems with high average and peak powers. At the same time, it is quite clear that the creation of high-power (>1 kW) and pulse-periodic lasers with frequencies up to 100 kHz or higher (with a peak power exceeding the average power by two to three orders of magnitude) would significantly expand the field of application of laser sources, increase their efficiency, and implement qualitatively new effects.

One of the successful trends in implementation of new methods of high-frequency modulation of the output from high-power laser sources is due to the regenerative amplification of a small signal injected into the cavity of a high-power gas or solid-state laser with the classical rod geometry. Use of intracavity modulators in high-power laser systems is hindered, because high power density inside the cavity leads to the appearance of plasma on the surface elements of the modulator, to the screening of radiation and to the destruction of the optical elements of the modulator. Quite promising is the method of realization of the pulse-periodic regime in high-power lasers, which is based on the use a self-filtering cavity. This cavity is a confocal resonator consisting of two confocal spherical mirrors of different curvatures; in the common focal plane there is an annular output mirror with a hole. Because of the high degree of discrimination of the higher modes with respect to the resonator losses, the lower mode is reliably generated. The spherical resonator mirror is placed outside the active medium and contains a modulation unit comprising spherical mirrors. The laser mode should be formed quickly enough in the leading edge of a giant pulse. The problem of laser micromachining usually involves cutting, drilling, surface cleaning and polishing, as well as removal of thin layers due to the ablation of the material. Micromachining relies on lasers emitting pulses of different duration. The geometry of the beam and the pulse repetition rate are crucial: usually the pulse duration does not exceed a few tens of nanoseconds, and in some cases reaches several hundreds of picoseconds. Disk laser technology provides excellent flexibility in the scaling of the output power by combining it with a high pulse energy and excellent beam quality.

1.3.6. Prospects for Scaling the Power of Disk Lasers

As mentioned above, the design of the disk laser generating a continuous and relatively high power is ideal for cutting and welding metals, where high optical quality of the beam is required. It is important for such industries as automotive, transportation, aerospace and heavy engineering. However, the design of the disk laser is more suitable for a range of new technologies that are currently in demand. Today, a laser with a power up to 16 kW and beam quality of ≥2 mm mrad, fabricated by the TRUMPF Group, is the undisputed leader in this class of laser systems. CW solid-state diode-pumped disk lasers demonstrate highly efficiency. Beam quality of the laser disk is outstanding, which makes it possible to work with a target from a large distance, while providing an extremely high concentration of radiation in the interaction region with the help of the focusing optics. In this design of the laser module the disk size is no more than 4–5 mm, because at larger sizes the energy loss due to the amplified spontaneous emission increases at an absolutely unacceptable rate. The above-presented laser system consists of a set of disk modules in a single cavity. This geometry of the laser system allows implementation of the pulse-periodic regime with a fairly high peak power, but the average power lies within a kilowatt. Parallel operation of disk lasers can increase the peak power of the entire system, but the phase-locking of the disk channels in the pulse-periodic regime requires additional scientific research. Further expansion of the power to the megawatt level of both the average and peak power in both geometries is very problematic.

At the same time, a different approach is known to the implementation of scalable solid-state laser systems, which consists of a set of active elements in the form of 'slabs', followed by phase-locking of the generated radiation. The team of the Northrop Grumman Corporation has created a laser with a power of >100 kW and high laser beam quality equal to the diffraction limit of 1.5 (averaged value) with the exposure time of 300 seconds. The laser efficiency reaches 30%. The authors of the project point to the easy replacement of individual laser channels in case of failure. They also mentions some of the advantages of the parallel structure of the amplifying channels in terms of ease of further increasing output power, if necessary, to 100 kW. Beside, the laser assembled according to this scheme provides the divergence at the level of two diffraction limits from the common (composite) aperture. From general physical considerations, we can assume that for a given power level and a reasonable value of radiation resistance of optical elements, the divergence of the laser radiation at the level of 2×10^{-5} rad can be achieved for the cw generation regime and 0.6×10^{-5} rad for the pulse-periodic regime. A further increase in the number of channels in order to obtain an output power of 1 MW will require the coherent summation of the power from at least 80 channels, which seems an elusive task.

The question arises as to how the average power of a few megawatts can be achieved on the basis of solid-state lasers. And it is this power that is needed to address many problems associated with the removal of debris from near-Earth space, with launching of missiles with the help of lasers, with the creation of long-range conducting channels and others. Fiber lasers are not applicable for these purposes because of the smallness of the area of the exit pupil of the fibers and hence the impossibility of operation of such lasers in the high-frequency pulse-periodic regime with high peak power at an average power of a few megawatts. The laser system based on 'slabs' also seems hard to implement as adjustment of the system and its maintenance in a safe operation mode are comparable to the complexity of working with a multi-element system for the solution to the problem of controlled thermonuclear fusion (CTF) at Livermore (USA). Proceeding from the above, the answer is quite clear: Such a laser can and should be based on the disk laser!

This is due to the indisputable advantages of the disk geometry in terms of the minimal thermal lens in the active media and the high radiation resistance of the disk in the pulse-periodic regime because of the large area of the optical surface to couple out the radiation. But first it is necessary to find a solution to the problem of the suppression of amplified spontaneous emission along the diameter of the disk. In this case, the size of the disk at a megawatt level of the average output will be at least 50 cm, i.e., at least ten times the size of the disk that is used today in existing systems. Radiation from such a laser, obtained during generation in the active medium of a single disk, does not require additional phase-locking. At the same time, such a laser in a single module will be very well combined with a large-diameter telescope for ensuring high peak power density of the laser pulse on space debris. It is known that the disk geometry of a laser was proposed 26 years ago; however, to this day, the solution to the problem of the ASE suppression with increasing transverse dimensions of the active medium in the disk geometry has not been found! Thus, the prospect of creation of a superlaser for new class of cutting edge problems is still open!

1.3.7. Conclusion

Only the future will show what type of laser will be used effectively for the processing of materials and for the solution of other important challenges faced by science and technological advances of the future. However, today we can say that creation of megawatt high-frequency pulse-periodic lasers with a large cross section of the active medium opens up great prospects for their use in solving the problems of launching of small satellites with lasers, formation of super-long conducting channels in space and atmosphere, cleaning of the near-Earth space from the space debris, etc.

Chapter 4

1.4. High-Power Semiconductor Laser Diode Structures and Their New Application

1.4.1. Introduction

At the present, high-power lasers are often treated as laboratory equipment requiring constant maintenance by specially trained personnel. In this regard, the most promising now are semiconductor lasers. To date, they have the record-high efficiency reaching 70%. Modern technologies of fabrication of semiconductor structures have made it possible to significantly increase the lifetime of the laser, which is tens of thousands of hours of continuous operation. Besides the variation of the semiconductor material can change the wavelength range of the radiation from the infrared to the ultraviolet.

Currently, semiconductor lasers are the only lasers used in household appliances, namely, in communication and electronic devices. The cost of semiconductor lasers is still high, but falling from year to year as the production grows up. Semiconductor lasers are very compact and easy to operate.

The main feature of lasers based on semiconductor structures (LD-structures), which allowed them to work at room temperature, is a small amount of the active medium. Therefore, the output power of a single diode is low. Currently, it does not exceed 10 W. To increase the output power, one needs to use at the same time a large number of laser diodes, which form one-dimensional (bars) or two-dimensional (arrays) structures. This direction in the development of high-power semiconductor laser systems is being exploited relatively recently, but very rapidly.

The assembly of laser diodes has almost all the remarkable properties of single semiconductor lasers, such as high intensity, high reliability, long lifetime (up to 10^{12} cycles). These lasers have much smaller weight and size dimensions (up to 10 times) in comparison with other types of lasers. They can be easily fabricated with an independent low-voltage power supply (solar, nuclear) without bulky transformers. Equipment based on laser diode assemblies really becomes a reliable high-performance tool that can be used in industry, medicine, and military hardware.

Almost all developed countries are, in one way or another, involved in the design and development of such laser systems. The leading role is played by the USA, Germany, England, and France. Interest in the creation of these systems is huge. Suffice it to say that just in the last two years, many small companies, specializing exclusively in the production of these lasers, have been established in the U.S. This proves that their commercial benefits come to the fore.

Thus, we can single out now three main areas of use of high-power laser diode assemblies.

1.4.2. Major Trends on the Development of Pulse-Periodic Lasers

1.4.2.1. Pumping of Solid-State Crystal Lasers

The use of LD-structures for pumping solid-state lasers leads to the high output optical quality and generally speaking use to work as a light transformer. The efficiency of such a system is low due to the low efficiency of solid-state lasers; however, pumping by semiconductor lasers makes it maximal. Radiation of semiconductor lasers does not spoil the working medium of crystal lasers, which makes it possible to significantly extend the life of the entire laser system. Weights and dimensions of such systems are small, which allows them to be easily transported. The main purpose is to develop and fabricate diode arrays (bars) with the highest possible power density. Record-high values, i.e., 100 – 115 W from a 1-cm-long single diode array, were obtained by Coherent Inc., USA. The limiting power from one bar, as calculations show, can reach 500 W. At the Department of High-Power Lasers of the A.M. Prokhorov General Physics Institute of the Russian Academy of Sciences a 1-cm-long diode array with an 80-W output was fabricated. Solid-state diode-pumped lasers are being increasingly studied and developed by a large number of firms in the various countries. High-power diode-pumped laser systems are being developed mainly in such major companies as Lawrence Livermore National Laboratory (USA), Rofin-Sinar Technologies Inc. (Germany), Coherent (USA), and Newport Corporation (USA)

1.4.2.2. Coupling Radiation from a Diode Array into a Fiber

An important problem here is the coupling of radiation from a diode array into the fiber for further transport to the work piece to be processed. Radiation losses in such a laser – fiber system reach 20% – 30%. At the same time, 200 ÷ 300-W laser systems are known to be produced industrially in the U.S. and Germany.

1.4.2.3. Phase-Locking of a Diode Array

In our opinion, the most promising application of laser diode assemblies is the creation of a phase-locked diode array. The output radiation of such a laser is a set of narrowly focused interference peaks (in particular one peak) with a small divergence. The estimated output power can reach the power of a conventional diode array producing one set of beams that are *out of phase* with those of another set of beams. This laser is a light source with high efficiency and low divergence that are characteristic of semiconductor lasers and solid-state crystal lasers, respectively. By analogy with the technique of phase-locked antennas in the radio range, the beam divergence in the case of phase-locked diode arrays can easily be increased to 10^{-6} rad without expensive telescopic systems, which provides ultra-long communication and transmission of energy over long distances.

At present, these systems are being actively studied. For example, Closed Joint-Stock Company Energomashtekhnika has been conducting research in this field since 1997 and has produced a prototype of a ~20W phase-locked laser. The possibility has been shown of scaling the process of phase-locking of the arrays and bars of laser diodes. Arrays of laser diodes with an output power up to 500 W have been fabricated. The effect of phase-locking of high-power laser diode structures is being increasingly investigated. According to the press in other countries, the maximum output power in a phase-locked array is reported to be only 1.5 W.

1.4.3. Laser Systems of Propellant Ignition

One of the most important applications of laser diodes is the creation of a system of multiple ignition of rocket propellant in the combustion chamber of a launcher. Apart from the fact that the parameters of the laser system should provide multiple ignition of the mixture with 100% certainty, the laser design must meet the requirements specific to onboard lasers that are currently used in space technology:

1. Compactness
2. Small size and weight
3. Absence of high-voltage in the power supply system
4. Resistance to vibrations and g forces
5. Absence of active cooling (including the closed circuit with a refrigerant and a pump)
6. Long service life.

A new and specific requirement for the launcher-related design of the laser is the need to deliver the laser radiation into the combustion chamber through an optical fiber. Accordingly, the peak and average power of the laser must permit delivering its radiation into the fiber with a minimal chance of the fibre damage.

Since July 2002 works have been on the way for a laser system ensuring multiple ignition of rocket propellant components in the prechamber (or ignitor) of the third-stage rocket motor. In the course of these works the following sources of laser radiation were used:

(1) A solid-state laser operating in a modified spike regime, first proposed and patented by CJSC Energomashtekhnika, with the following radiation parameters:

wavelength– 1.054 μm;
output energy - 4 – 10 J;
temporal shape of radiation - a train of 100 – 150 pulses with an irregular period;
pulse train repetition rate - a single pulse;
duration of the pulse train - 600 – 700 ms;

duration of a single pulse - 200 – 400 ns;
pulse spacing - 4 – 7 ms;
peak output power ~ 100 kW;
divergence - 4×10^{-4} rad;
type of the active element pumping - lamp pumping.

Figure 4.1 shows the oscillograms of a train of pulses (Figure 4.1a) and of single spike (Figure 4.1b) of a laser operating in a modified spike regime. A characteristic feature of the temporal shape is the irregular spacing between neighboring spikes within the pulse train and the high contrast of the spikes with respect to the pedestal (significantly higher than in the pulse of a conventional free-running solid-state laser). It should be noted that the duration of the spikes is by about order of magnitude shorter than in the traditional pulse. These differences in the temporal shape of the modified spike regime made it possible to realize a much higher peak output power compared to a traditional pulse of the same energy.

The laser is a laboratory model with a rather bulky power supply and cooling module. The dimensions of the laser head are as follows: 200 × 200 × 1400 mm. The appearance of the laser head is shown in Figure 4.2.

(2) A solid-state Q-switched pulsed laser:

wavelength, 1.06 μm;
pulse energy, up to 1 J;
peak power, up to 100 MW;
pulse duration, 10 ns;
repetition rate, a single pulse or 10 Hz.
The laser however was a rather cumbersome laboratory bench.

(3) A serial ytterbium fiber laser (YPL1/100/20):

wavelength, 1.062 μm;
average output power, 20 W;
pulse duration, 120 – 150 ns;
pulse repetition rate, 20 – 80 kHz;
maximum pulse energy, 1 mJ;
maximum peak power, 10 kW;
divergence, $10^{-3} \div 3\times10^{-3}$ rad

The laser was significantly smaller in size than the two above-mentioned lasers, the total weight of the power unit and the laser head was 8 kg. Neither water cooling nor a high-voltage power supply was used in the laser.

1.4.3.1. Results of the Experiments

Experiments were performed to study the ignition of different fuel mixtures (oxygen – kerosene, oxygen – alcohol, oxygen – hydrogen) in the prechamber (or ignitor) of the third-stage rocket motor. In the first stage of the research, the radiation was delivered into the prechamber through an optical window, which was installed in the seat of a standard spark plug. The radiation was focused by a lens mounted outside the chamber onto the target placed inside the chamber. In the experiments a metal mesh was installed in the channel of the input laser beam. This target made it possible to successfully ignite the fuel mixture. Due to the large capillary resistance, the target in the form of a mesh effectively protected the sapphire window (through which the laser beam is delivered) from condensation. The disadvantages of this target include the problem of maintaining its thermal stability during the operation of the combustion chamber. A graphite rod, mounted in a brass ring, is inserted into the combustion chamber. This target can be moved along the axis of the combustion chamber, thereby determining the optimal zones to initiate ignition. This target makes it possible to ignite the fuel mixture at very low levels of laser energy. The downside of the target is the possibility of its destruction in the high oxidizing environment; the wall of the subsonic part of the exit nozzle is covered with oxides, special targets in the combustion chamber are absent. The laser beam was focused on the wall with a long-focus lens. This variant is the most reliable one for ignition, as no high requirements are imposed on focusing and there are no changes in the scheme of the combustion chamber.

The experimental results showed a significant advantage of laser(1) over other lasers: the use of this laser made it possible to ignite all the above-mentioned mixtures with a large margin of the radiation energy. In particular, the threshold of ignition for the oxygen – hydrogen mixture was 90 mJ.

Laser (2) ignited only the hydrogen – oxygen mixture (this mixture is known to be the most flammable one), and the ignition threshold was 60 mJ.

Laser (3) ignited all the above-mentioned mixtures. However, reliable ignition was achieved using the full laser power of 20 W, optimal repetition rate of 50 kHz and conical target, which provided cumulation of the material evaporated from the target and, accordingly, its additional heating.

In our opinion, in experiments with two lasers, (1) and (3), the optical breakdown threshold for the gas mixture near the target was not reached; the basic mechanism of ignition was the contact of the gas mixture with heated vapors of the target material formed near it by laser ablation. Because in the case of laser (1) the peak power was about 10 times higher, the higher was the temperature of the vaporized material. Most likely, this parameter played a crucial role in the successful ignition of less flammable mixtures such as oxygen – kerosene and oxygen – alcohol.

Because of a significantly higher peak power, in experiments with laser(2) the optical breakdown threshold of the gas mixture was substantially exceeded. The plasma, produced near the target on the leading edge of the laser pulse, shields it from the rest of the pulse and no target material is significantly evaporated. Therefore, the main ignition mechanism is the ignition of the gas mixture at its contact with the laser plasma. However, based on the experimental results, this type of ignition is less effective because it does not allow one to ignite less flammable mixtures. Based on these experiments, we can draw the following conclusions:

1) The mechanism of laser ablation is more effective to ignite the rocket propellant mixtures than the optical breakdown of the mixture. An additional argument in favor of ablation is the fact that the propellant itself is not involved in the process.

Therefore, hot vapors of the target can be produced either before or during the injection of the propellant components. In the case of the optical breakdown of the mixture, its threshold depends strongly on the gas composition and pressure in the chamber. In this case, the breakdown can occur after filling the chamber with the mixture. Such a regime is ideal in the case of a small combustion prechamber, but can lead to self-ignition of the mixture and to an explosion in a large combustion chamber.

2) The laser pulse energy (and, hence, its peak power) significantly affects the temperature of the evaporated material, and the total energy of the pulse train influence its mass. An increase both in the temperature and the mass of the evaporated material provides more reliable ignition of propellant mixtures. Therefore, in designing an ideal laser for ignition, it is needed to optimize both the peak power of a single pulse and the energy of the pulse train (in the case of a pulse-periodic laser, high power).
3) It is also likely that an important parameter is the laser pulse repetition rate. At the optimal repetition rate additional heating of the target vapors by subsequent pulses is also possible, but in such a way that the target vapors would not fully absorb the incident radiation and allow for ablation in all the pulses of the train. Thus, when working with laser (3), optimal was the frequency of 50 kHz, at which a continuous optical discharge was observed near the target (Figure 4.4).

1.4.3.2. Fiber-Optic Coupling of Radiation and Its Advantages

The use of optical fibers for coupling light into the combustion chamber eliminates the installation of an optical window in the wall of the chamber and delivers radiation through a small hole in the housing, comparable in diameter to those used for the injection of fuel mixtures of components (hundreds of μm). Installation of the optical window is often impossible due to design reasons; besides, an additional risk is due to the possibility of explosion of the window at very high temperatures and pressures in the combustion chamber.

Therefore, in the second stage of the works an original construction has been designed and manufactured to couple optical radiation into the fiber and place the fiber inside the combustion chamber. This design ensures high reliability of the fiber attachment, long service life without re-polishing of the end that is in the combustion chamber, and the optimal delivery of radiation to the target. The optical scheme of the ignition system with fiber-optic input is shown in Figure 4.5. Subsequently, this design was used to deliver the radiation of lasers (1) and (3). The radiation of laser (2) could not be coupled into the fiber due to the high peak power that damages the fiber end.

1.4.4. Conclusion

Thus, a single-pulse *Q*-switched solid-state laser with a pulse duration of ~10 ns is not optimal for igniting the propellant mixtures and causes big problems when coupling the laser beam into an optical fiber. We consider it inappropriate to use such lasers for igniting propellants. Based on the above considerations, it is proposed to test sequentially the lasers developed and designed by CJSC Energomashtekhnika for igniting propellants in the combustion chamber of the rocket booster (the oxygen – hydrogen mixture). The lasers have the following parameters:

I. A solid-state laser with semiconductor pumping:

wavelength, 1.06 μm;
pulse duration, 100 – 200 ns;
pulse energy, 10 – 50 mJ;
peak power, 50 – 500 kW
pulse repetition rate, 5 – 100 μs (the optimal interval is determined for each working mixture);
continuous operating time, 5 s.

Based on the experience in designing solid-state diode-pumped lasers and laser diode arrays at CJSC Energomashtekhnika, we came to the conclusion that such parameters can be demonstrated by a transversely diode-pumped solid-state laser with a passive optical switch (e.g., LiF with color centers) to ensure modulation and by laser diode arrays in phase-locked and out-of-phase regimes. The proposed laser can make use of the original patented diode-pumped laser with the following parameters: pulse energy, 50 mJ; pulse repetition rate, 100 Hz; a special design for fixing the active element and diode laser arrays, which eliminates water cooling. At present, the laser emits pulses with an energy of 50 mJ and a repetition rate of 50 Hz. If long-term continuous operation is required (the laser operates from 1 to 5 s), a substantial increase in the pulse repetition rate (up to 1 kHz) is possible without reducing the pulse energy.

II. A laser diode array in an out-of-phase oscillation mode (intermediate stage):

wavelength, 0.808 μm;
operation mode, pulse-periodic;
pulse duration, 100 μs;
pulse energy, 0.2 J;
peak power, 2 kW;
maximum peak power density during focusing, 2 MW/cm^2;
pulse repetition rate, 1 kHz;
continuous operation time (without cooling), 5 s.

III. An array of laser diodes in a phase-locked oscillation mode:

wavelength, 0.940 μm;
operation mode, continuous;
power, 40 W;
maximum peak power density during focusing, 40 MW/cm^2;
continuous operation time (without cooling), 5 s.

In comparison with diode-pumped solid-state lasers, the laser diode array, operating both in the out-of-phase (cheaper) and phase-locked oscillation modes is even more compact and reliable ignition device. Since, as shown above, the mechanism of laser ablation is more effective to ignite rocket propellants than the optical breakdown, the devices based on the array technology must take a critical position in the solution of the above problems. It should be noted that the first samples of laser diode arrays developed for other purposes, have already been produced at CJSC Energomashtekhnika. Currently, work is underway to ensure phase-locking of laser diode structures.

Estimates show that the radiation of the proposed lasers can be coupled into an optical fiber without the risk of its damaging. A significant increase in the pulse energy and pulse duration at a high repetition rate within 5 seconds or a transition to the continuous regime with sharp focusing of the radiation under conditions of its phase-locking suggests that the proposed ultra-compact laser systems are ideal devices for solving the problem of reliable multiple ignition of various components of the rocket propellant in combustion chambers of the launchers.

Thus, high-power diode arrays are promising sources of radiation, the dominant role of which in the new millennium is ensured by a unique set of basic features and characteristics.

1.4.5. Figures, Photos and Schemes

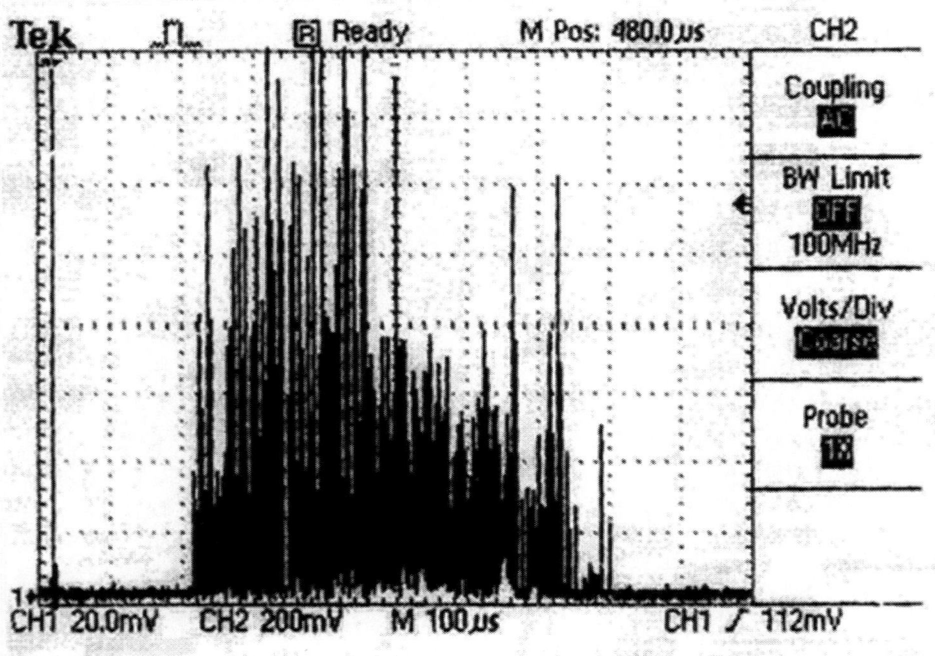

Figure 4.1a. Oscillogram of pulse train (100mks/div).

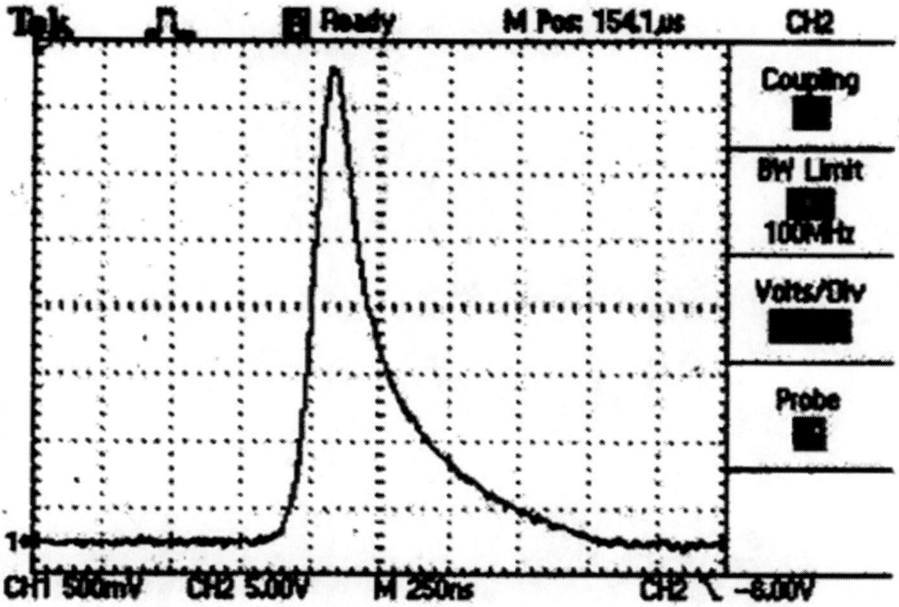

Figure 4.1b. Oscillogram of pulse from train of pulses (250ns/div).

Figure 4.2. Laser head (1).

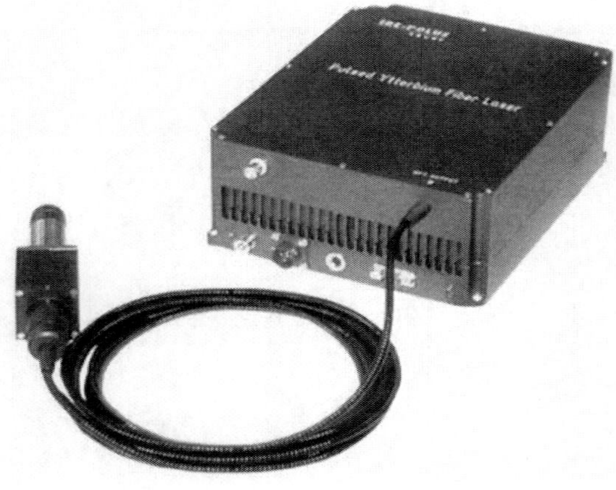

Figure 4.3. Laser head YPL1/100/20.

Figure 4.4. Photo of optical igniter with optical discharge.

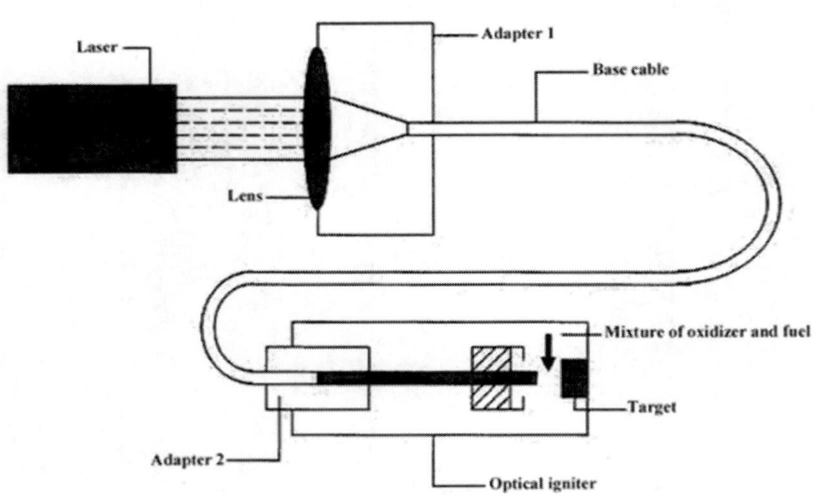

Figure 4.5. Fiber system for delivery of optical radiation to the optical igniter.

Chapter 5

1.5. ELECTRICAL DISCHARGE PULSE-PERIODIC LASERS

1.5.1. SELF-SUSTAINED VOLUME DISCHARGE FORMATION IN CO_2 LASERS

1.5.1.1. Introduction

A SSVD can be established in a gas by creating primary electrons the density of which should exceed a certain minimum value n_{min} throughout the discharge gap [1, 3]. Various methods for preionization of a gas in the discharge gap have been developed for this purpose [4, 7].

Primary electrons are created by these methods directly in the discharge gap, which sometimes causes difficulties in the establishment of conditions necessary for the formation of a SSVD. For example, high voltages are needed for the formation of an initial plasma when the preionization source and the active medium are combined in the same volume [8], where as preionization with ultraviolet radiation may be ineffective because of strong absorption of such radiation in a medium [9], and if soft X rays are used, it is necessary to ensure rigid synchronization of the X-ray and pump sources [10].

In mixtures of gases typical of CO_2 lasers, losses of electrons due to trapping are relatively small at low values of $E = p$, where E is the electric field intensity and p is the gas pressure. The trapping coefficient is considerably less than the absorption coefficient of ultraviolet radiation for the same mixtures [10]. It follows that, in principle, it should be possible to create primary electrons in a density needed for the formation of a self-sustained volume discharge at a considerable distance from an ionization source and then transport the electrons to the gap by drift in an electric field.

A method of formation of a self-sustained volume discharge by filling the discharge gap with a flux of electrons drifting in an electric field, without preliminary ionization of the whole discharge volume, was proposed and implemented [12]. The source of electrons was a plasma formed in an auxiliary discharge initiated under a grid cathode. This method made it possible also to establish a SSVD in a system of electrons with a strongly inhomogeneous electric field in the discharge gap.

1.5.1.2. Physical Model of SSVD Formation

The SSVD formed by filling the discharge gap with primary electrons was studied in the electrode system with a uniform initial distribution of the electric field (the anode with the Chang profile [20] — a flat mesh cathode) with a discharge gap measuring 6 to 10 cm in length. In these experiments use was made of the configuration presented in Figure 5.1. When U_0 was less than the static breakdown voltage of the discharge gap, U_{eT}, use was made of a scheme with pulsed application of the voltage exceeding U_{CT} to the discharge gap. In the latter case, the filling of the discharge volume with electrons from the plasma of the auxiliary discharge was performed either by applying a bias to the discharge gap [21], or by artificially increasing the duration of the leading edge of the voltage pulse applied to the discharge gap [21, 22]. The value of **Tf** was chosen so that the electrons can reach the anode before the onset of the ionization processes in the discharge gap.

The conditions of the SSVD formation in initial highly nonuniform field distribution in the discharge gap were performed using the setup schematically shown in Figure 5.1b. The electrode system (d = 10 cm) was formed by a planar anode 1 (a rectangular plate measuring 18×50 cm in size, curved along a radius of 10 mm) and a cathode 2 with sharp edges (a rectangular brass mesh measuring 10×40 cm in size). The surface area covered by auxiliary discharge 3 under the mesh was slightly larger than the cathode area.

In order to reduce the working voltage [23, 24] and increase the duration of stable combustion of the SSVD [25, 26], easily ionizable substances such as tripropylamine and triethylamine at a vapor pressure of were added into the mixture.

In the experiments we controlled the lower limit for the SSVD ignition with respect to the voltage U_{min} (determined by the voltage of the pump source), and the SSVD stability limit with respect to the $CO_2 - N_2 -$ He gas mixture composition and the energy input. The SSVD uniformity was monitored visually by photos of the glow discharge, as well as by the small-signal gain and its distribution over the aperture of the amplifier.

The electron flux current in the discharge gap was deduced from the voltage drop across a resistor R. The charge carried by the flux was found from the change in the voltage across a capacitor C for R = 0. The contribution to the gap conductance made by the photoionization of the gas in the gap by the auxiliary-discharge radiation was estimated from the charge of the positive ions formed in the gap (these measurements (V.V. Apollonov / Central European Journal of Physics 4, 708 (2003)) were carried out by applying a negative bias voltage U_0).

The following types of an auxiliary discharge were investigated:

(1) A barrier discharge [14] in which a wire insulated with polyethylene was used as the electrode.
(2) A corona discharge initiated by a multipin electrode, stabilized by limiting the current through the pins by a ballast resistor [15].
(3) A volume discharge between a grid (discharge gap cathode) and a metal cathode, initiated as described in Ref. [16].
(4) A volume discharge between a semiconductor cathode (carbon-filled rubber with a bulk resistivity [17] and a grid (discharge gap cathode) initiated without preionization [18].

(5) A multichannel discharge on the surface of a high-permittivity insulator limited by the insulator capacitance [19]
(6) A multichannel discharge on the surface of a semiconductor (carbon-filled rubber [17]) limited by the semiconductor resistance (this auxiliary discharge was realized in the same way as a multichannel discharge on the surface of a ceramic described in Ref. [19]).

In all these auxiliary discharges, an electric field was applied to auxiliary discharge plasma. A special auxiliary electrode was not essential. The auxiliary electrode is the one used to apply the voltage pulse to ignite the auxiliary discharge.

The propagation of primary electrons in the discharge gap (one-dimensional case) in the absence of ionization multiplication is described by the system of equations and boundary conditions for the electric field E(x, t), the density of electrons n(x, t) and negative ions $N_t(x, t)$ [27] :

$$\partial n/\partial t+\partial(n\mu E)/\partial x=-\alpha n\mu|E|; \quad (1)$$

$$\partial N_i/\partial x=\alpha n\mu|E|; \quad (2)$$

$$\partial E/\partial x=-(N_i+n)e/\varepsilon_0; \quad (3)$$

$$\int_0^d E dx=-U_0, \quad (4)$$

$$n(0,\ t)=\text{const}=n(0); \quad (5)$$

$$N_i(x,\ 0)=0; \quad (6)$$

$$n(x,\ 0)=0 \text{ при } x>0, \quad (7)$$

where α , μ, and e are the attachment, mobility and charge of electrons, respectively; ε_0 is the dielectric constant. The X-axis is directed from the cathode (x = 0) to the anode (x = d); the electron density at the cathode surface n(0) is assumed to be constant over time (the escape of the electrons from the cathode to the displacement field is compensated for by the influx from the plasma of the auxiliary discharge). In expressions (1) — (7) the mobility of the negative ions is negligible.

System (1) — (7) was solved using an explicit difference scheme [21]. To simplify the calculations we assumed that α=const.

Figure 5.2a shows the dependences n(x) and E(x) for various time points at U_0 = 400 kV, d = 40 cm, μ = 10^3 cm^2/Vs, n(0)=10^9 см$^{-3}$. It is seen that in the absence of the electron loss through the attachment, stationary distributions n(x) and E(x) are established in the discharge gap for $t>t_0=d/E_0\mu$, where $E_0 = U_0/d$.

The discharge gap accumulates a volume charge of negative ions, which prevents the propagation of primary electrons. Calculations performed for different a and d showed that at the time t = t_0, the electron densities near the anode n_α (d) (at α≠0) and n(d) (at α = 0) are related as by the expression:

$$n_\alpha(d) \approx n(d)e^{-\alpha d} \quad (8)$$

In conventional CO_2 lasers utilizing atmospheric pressure $CO_2 - N_2 - He$ gas mixtures, $E_0 \leq 3$ kV/cm is less than 10^{-2} $см^{-1}$ [11]. Therefore, when d <1 m and E_0 3 kV/cm, the loss of electrons through the attachment does not affect significantly the propagation of primary electrons, and the analysis of the possibilities of the method can rely on the model of a stationary primary electron ($\alpha = 0$ and t> t0).

Approximate steady-state solution to system (1) — (7) at large n(0), when

$$\delta = 3\,\varepsilon_0 E_0/4en(0)d \ll 1 \qquad (9)$$

has the form [28]

$$E(x) = \frac{3}{2} E_0 \sqrt{\frac{x}{d} + \delta^2}; \qquad (10)$$

$$n(x) = \frac{3}{4}\frac{\varepsilon_0 E_0}{ed}\sqrt{\frac{d}{x + \delta^2 d}}; \qquad (11)$$

$$j = \frac{9}{8}\mu\varepsilon_0 \frac{E_0^2}{d}(1 - 3\delta^2), \qquad (12)$$

where j is the current density of primary electrons.

Let us perform estimations. Let E_0= 1 kV/cm, d = 50 cm, then from (11) $n(d) \approx 10^7$ $см^{-3}$ clearly exceeds Nmin, required for the SSVD formation [21]. In this case, condition (9) is fulfilled already at $n(0) \approx 10^8$—10^9 $см^{-3}$. It will be shown below that such low densities of the electrons at the cathode can provide almost any type of the auxiliary discharge.

Thus, our estimates confirm the possibility of obtaining the SSVD by filling the discharge gap with primary electrons, which are necessary for the SSVD formation process by realization of high concentrations of primary electrons at large distances from the cathode. Note some features of this method which follow from the considered model.

It is seen from (11) that at a certain distance from the cathode ($x \gg \delta^2 d \approx 10^{-3}$ d при E_0=1 кВ/см, d=40 см, $n(0)=10^9$ $см^{-3}$) under the condition of (9) n(x) is independent of n(0). Therefore, the discharge gap will be filled by electrons uniformly (in a plane perpendicular to x), even in the case of a significant nonuniformity of the n (0) distribution over the cathode area. This makes it possible to produce a SSVD without small-scale inhomogeneities (typical of systems with UV preionization) caused by shading of the discharge volume by a mesh cathode [29], and to increase the small-signal gain and efficiency of CO_2 lasers.

It follows from (10) that, despite the relatively low concentration of the primary electrons in the discharge volume, the presence of the uncompensated negative charge of primary electrons leads to a significant distortion of the electric field in the discharge gap. As a result, the ionization processes will develop in the first place at the anode, where the electric field can be enhanced by 1.5 times compared to E_0. At the same time, the development of ionization at the cathode is difficult, because the field is close to zero due to the screening of the cathode by the volume charge of primary electrons. Obviously, the effect of the cathode screening when filling the discharge gap with electrons should manifest itself in the real geometry of the discharge gap. Therefore, it is advisable to try to find the conditions for obtaining the SSVD in systems of nonprofiled electrodes, which would greatly simplify the creation of electric discharge lasers.

Distortion of the electric field in the discharge gap by a volume charge of primary electrons should result in a voltage drop (this voltage is used to produce a SSVD), which is an obvious advantage of the proposed method. However, in a system with preliminary displacement U_0, which is used to fill the discharge gap with electrons, and subsequent to apply the voltage pulse with the amplitude U to the discharge gap, degree of enhancement of the electric field at the anode (with respect to the U/d) $K_y=1 + U_0/2U$ may be small, because it is not always acceptable increase U_0 to values close to U_{st}. The degree of distortion of the electric field can be increased by using a scheme with an artificially prolonged voltage front [22]. Figure 5.2b shows the results of numerical solutions of equations (1) — (3), (5) — (7) for n(x) and E(x) at α=0 with the boundary condition

$$\int_0^d E dx = -\frac{U}{2}\left(1 - \cos\frac{\pi t}{\tau_\Phi}\right), \qquad (13)$$

which corresponds to the law slew of the voltage rise in the discharge gap in this scheme. To completely fill the discharge gap with the electrons, it is needed also to satisfy the condition

Tf $2d^2/\mu U$. It is seen from Figure 5.2b that when t= **Tf**$=2d^2/\mu U$, as in the case of constant displacement in the discharge gap, the gain E(d) is reached, which U/d is 1.5 time higher relative to **U/d** at multiple reductions of E(0).

1.5.1.3. Experiments

Figures 5.3a and 3b show oscillograms of the electron charge q(t) extracted from the discharge gap and of the voltage U(t) between the gap cathode and the auxiliary-discharge electrode, recorded for a mixture of N_2 with tri-propylamine obtained for the case when d = 2 cm and the electrode area was S = 200cm^2. The auxiliary discharge was a multichannel discharge on the surface of a ceramic (with the permittivity ε=2000), limited by the capacitance of the ceramic itself [19]. The shape of U(t) was set by the parameters of a special RC circuit. A comparison of the q(t) oscillograms recorded for different polarities of the voltage U(t) indicated that the main rise of q(t) occurred in the region where dU/dt < 0. When a negative bias voltage was applied to the discharge gap, the form of the q(t) oscillograms was independent of the sign of dU/dt. In accounting for these results one should bear in mind that the sign of the electric field in the gap between the ceramic surface and the discharge gap cathode was the same as the sign of dU/dt. Therefore, for dU/dt < 0 some of the electrons were expelled from the auxiliary-discharge plasma into the discharge gap across the grid cathode and drifted toward the anode in the bias field ($U_0 > 0$). For dU/dt > 0 or $U_0 < 0$ it was not possible to drive electrons beyond the grid and in the discharge gap the flux of the current was then entirely due to the photoionization of the gas by the auxiliary-discharge radiation.

Figure 5.3c shows the oscillograms of the voltage U(t) and of the electron current in the discharge gap I(t) for an oscillatory form of the voltage U(t). It is clear from these oscillograms that a considerable current in the discharge gap was recorded only for $dU/dt < 0$, irrespective of the sign of U(t).

It should be stressed that we investigated long-duration auxiliary discharges in order to identify the conditions for the formation of an electron flux from the auxiliary-discharge plasma itself. Under these conditions, in spite of the presence of an easily ionizable substance in the gas, the contribution to the discharge gap conductance of the electrons arriving from the auxiliary-discharge plasma was (for suitable signs of dU/dt and U_0) approximately ten times higher than the contribution made by the photoionization of the gas by the auxiliary-discharge radiation even when the latter was of the surface type. Under these conditions the contribution of the photoionization to the charge of the electrons crossing the gap did not exceed 10% of the contribution of the electron flux for a corona discharge and it was not greater than 1% in the case of barrier volume discharges (these values were obtained for d = 6 cm, U_0 = 30 kV, mixture of N_2 with tripropylamine, P = 1 atm) and the presence of an easily ionizable substance in the gas had practically no influence.

When the auxiliary discharge pulse was shortened (by reducing the inductance or the active resistance in the circuit of this discharge), the contribution of the photoionization to the discharge-gap conductance increased and, in principle, it was possible to form an electron flux from a plasma created by the ultraviolet radiation of the auxiliary discharge in the cathode region (and also from a plasma on the surface of a plasma cathode [30] or a barrier discharge plasma in systems described in Ref. [16]).

As already pointed out, the homogeneity of a SSVD could be improved by selecting the auxiliary discharge characteristics in such a way as to satisfy the condition (9). Naturally, one would then have to ensure a certain minimum density of the electron current in the auxiliary gap.

In this case, the possibility of SSVD igniting in a discharge plasma is determined by the condition $\mathbf{T}_{dd}$ t_0($\mathbf{T}_{dd}$ is the auxiliary discharge duration); otherwise, the near-cathode zone can lose electrons before the ionization processes in the discharge plasma are developed.

Figure 5.4a shows oscillograms of the electron flux current and of the auxiliary discharge current (this was a volume type of discharge) obtained for a mixture of gases of the $CO_2:N_2$ = 1:3 composition at a pressure of p = 0.3 atm when the gap was d = 2 cm and the applied bias voltage was U_0 = 12 kV. The distance between the auxiliary discharge electrodes was 1 cm and the area of the auxiliary electrode was S = 200 cm^2. It is clear from Figure 5.4a that change in the auxiliary discharge current by an order of magnitude altered I(t) by just 10%, indicating that the condition (9) was satisfied.

The measurements showed that the saturation I(t) is reached at

$$I_{dd}/S = i_{dd} \geq (20\text{-}30)i_{lim} = (20\text{-}30)\, \varepsilon_0 \mu E_0^2/d, \quad (14)$$

where i_{lim} is the limiting current density of the primary electron from (12) when Eq. (9) is met. Condition (14) is an empirical criterion, equivalent to (9). Note also that the fulfillment of these conditions becomes simpler when d increases.

Condition (14) can be also easily met when use is made of a pulsed corona discharge as an auxiliary discharge. However, the limiting duration of the stable burning out of both a corona and a volume auxiliary discharge, when condition (14) is met without any special measures taken into account, does not exceed 5−10 μs.

Both volume and corona discharges could be stabilized by utilizing electrodes made of semiconductor materials [18]. Figure 5.4b shows oscillograms of the volume discharge current in a mixture of gases of the CO2:N2:He = 8:32:60 composition with an admixture of triethylamine at p = 1 atm, ignited without a preliminary ionization between a semiconductor material cathode (which was a disk 5 cm in diameter and 2 cm thick) and a grid (discharge gap cathode); it also includes an oscillogram of the charge q(t) of the electrons extracted from the discharge gap in the case when U0 = 5 kV, d = 8 cm, and the distance between the auxiliary discharge electrodes was 1 cm. The density of the auxiliary discharge current obtained in this way was ~50 mA/cm^2 and the duration of the stable burning of the discharge in gas mixtures of various compositions exceeded 30 μs.

In the case of a discharge of the surface of an insulator, which was limited by the capacitance of the insulator itself, our measurements indicated that the electron current in the gap between the insulator surface and the grid was an order of magnitude less than via an auxiliary electrode. Therefore, (14) is equivalent to the condition

$$|dU/dt| \geq (2\text{-}3)\cdot 10^2 \mu E_0^2 \varepsilon_0/dC_d \qquad (15)$$

Condition (15) is easily fulfilled at Tdd=t_0 and Cd=100 pF/cm^2 (Cd is the specific capacity of the dielectric).

Stable burning of long duration discharge, when (14) is met, is realized also in a multichannel discharge grazing along the surface of a semiconductor material limited by the intrinsic resistance of the material due to a choice in the discharge capacitance.

When the necessary condition was satisfied, stable long-term burning of the discharge was achieved also for a multichannel auxiliary discharge on the surface of a semiconductor material, which was limited by the resistance of the material itself, when the charging capacitor was suitably selected It therefore follows that when special measures were taken, all the types of the auxiliary discharge discussed above were capable of forming an electron flux with the maximum parameters specified by Eq. (12).

Figure 5.5a shows the dependence of the electron flux current I(U0), recorded under the same conditions as the oscillograms in Figure 5.4a. Clearly, at low values of U_0 $I \sim U_0^2$ this dependence is in agreement with Eq. (12). When U0 exceed a certain minimum value I, the current rise steeply and this was accompanied by volume radiation in the discharge gap, indicating the onset of the ionization processes and the formation of a SSVD. This minimum value of U0 amounted to (2/3) Uat, as expected on the basis of Eq. (11), according to which the electric field at the anode was enhanced by the space charge of the electron flux by a factor of 1.5 compared with E0. We could assume that the possibility of igniting a SSVD at a U_{st} voltage lower than mentioned in Refs. [31] and [21], was also due to distortion of the electric field by the space charge of the electron flux and in Ref. [21] this flux could be formed both from an auxiliary-discharge plasma under the cathode or from the photoplasma in the cathode region, because the discharge on the surface of a ceramic used in Ref. [21] was of relatively short duration (~1.5 μ).

The influence of the distortion of the electric field on the conditions of formation of a SSVD is manifested also by the results in Figure 5.5b, which gives the dependence of the minimum voltage U_{min} on U_0 for the ignition of such a discharge in the circuit with a preliminary bias of the discharge gap. The sign of this bias was opposite to the sign of U0 in order to avoid formation of an electron flux on application of a voltage pulse to the discharge gap. In this case the auxiliary discharge was of the barrier type. Experiments were carried out on a mixture of the CO2:N2:He = 1:1:3 composition and the other parameters were d = 6 cm and p = 0.5 atm. A voltage pulse was applied to the discharge gap after the delay $T=t_0=d^2/\mu U_0$ relative to the beginning of the auxiliary discharge. It is clear from Figure 5.5b that Umin decreased monotonically on increase in U0 in accordance with the above model.

At low overvoltages across the discharge gap the distortion of the electric field by the space charge of the electron flux had a considerable influence on the conditions of energy deposition in the plasma of the SSVD and on the residual voltage across the gap after the end of the discharge.

Figure 5.6a gives oscillograms of the voltage U(t) and current in a SSVD formed when the density of the electron current was too low for a significant distortion of the field and the condition (9) was not obeyed (d = 6 cm, gas mixture of the CO2:N2:He = 1:1:3 composition, p = 0,5 atm, charging voltage Uch = 45 kV, T_f=0,8 μs). Figure 5.6b shows oscillograms of U(t) and I(t) recorded under the same conditions as in Figure 5.6a, but in this case the condition (9) was satisfied. A comparison of Figs. 5.6a and 5.6b indicated that distortion of the electric field by the space charge of the electron flux resulted in some reduction in the amplitude of the voltage pulse across the discharge gap and also increased the amplitude of the current in the SSVD; it also increased the energy deposited in the discharge by a factor of about 2. This could be important under conditions when it would be difficult to establish a high overvoltage across the discharge gap.

All the forms of the auxiliary discharge investigated by us using a system of profiled electrodes made it possible to form a SSVD when the energy deposited was up to 300 J/liter. The content of CO2 in CO2-N2-He gas mixtures (with the addition of easily ionizable substances) at atmospheric pressure reached 40% when the total content of all the molecular gases was up to 70%, whereas in helium-free mixtures it was up to 30%.

We also established a SSVD in atmospheric air (p = 0,4 atm, d = 10 cm) by the addition of triethylamine, which was evidence of the high efficiency of the method of formation of a SSVD by filling the discharge gap with an electron flux, and this was true also of gases characterized by strong electron trapping.

The best characteristics of a SSVD in a system of nonprofiled electrodes (Figure 5.lb) did not differ from the characteristics of the corresponding discharge observed when the initial distribution of the electric field in the discharge gap was homogeneous. The discharge formed using such electrodes could be ignited employing any of the listed types of the auxiliary discharge, with the exception of a volume discharge utilizing metal electrodes (this type of the auxiliary discharge was difficult to realize using nonprofiled electrodes because of the inhomogeneity of the electric field in the auxiliary gap). The investigated system was characterized by a strongly inhomogeneous initial distribution of the electric field in the discharge gap, which was close to the field distribution between one of the electrodes and a plane of symmetry in a parallel-plate capacitor of limited size (as shown in cross section in Figure 5.1b); the field on the anode decreased monotonically along the normal to the symmetry axis of the discharge gap, whereas the field on the cathode increased reaching its maximum at the edge. Consequently, the static breakdown voltage in the system with nonprofiled electrodes was approximately half the corresponding voltage in the quasi-steady phase of a SSVD. When the discharge gap was subjected to a voltage pulse of amplitude U >

Usi (representing the regime with a deliberately delayed leading edge of the voltage pulse), a spark breakdown at the cathode edge was observed in the absence of an auxiliary discharge. When the auxiliary discharge $U_{st}<U<1,8U_{st}$ was started and the static voltage was in the required range, a spark did not appear during the period T_{dd} even when the formation of such a spark in the absence of an auxiliary discharge was less than T_f. A weak electron flux current was then observed in the discharge gap. When the amplitude of the voltage pulse exceeded required level and an auxiliary discharge was started simultaneously with the leading edge of the pulse a volume discharge appeared in the gap.

It therefore follows that the distortion of the initial distribution of the electric field by the space charge of the electron flux in the case of nonprofiled electrodes resulted, in contrast to the case of profiled electrodes, in an increase (rather than a reduction) of the pulsed breakdown voltage of the discharge gap and the breakdown was of volume nature. A strongly inhomogeneous initial distribution of the electric field in the presence of nonprofiled electrodes was corrected in the process of profiling the discharge gap with electrons, i.e., a dynamic profiling of electrodes by the space charge of the electron flux took place. We shall now consider qualitatively this phenomenon.

Screening of the cathode by the negative charge of electrons reduced several-fold the electric field on the cathode surface, compared with the average field U(t)/d in the discharge gap at a given moment, so that ionization processes could not develop near the sharp edges of the cathode. The field at the anode was then enhanced and the relative magnitude of the enhancement should be highest at the center of the electrode, since the electron flux spread in the course of propagation in a discharge gap as a result of the electrostatic repulsion between the electrons. Bearing in mind also the initial distribution of the field in the discharge gap, we concluded that the ionization processes began to develop simultaneously from the anode to the cathode and from the center to the edges of the discharge gap, i.e., the propagation of an ionization wave, which formed a cathode layer when it closed the discharge gap, was not governed by the initial configuration of the electric field but by that configuration which was formed with the participation of the space charge of the electron flux This made it possible to prevent spark breakdown and to form in a system with nonprofiled electrodes a SSVD with a deposited energy distribution typical of such nonprofiled electrodes. Measurements of the small-signal gain at low pump energies, which represented the distribution of the energy deposited in the discharge in a plane perpendicular to the optic axis, demonstrated that the gain decreased away from the optic axis to the electrode edges. This provided an indirect confirmation of the above description of the formation of a SSVD. It should be stressed that a change in the voltage on the electrodes resulted in a redistribution of the electric field in the discharge gap at a finite rate governed by the electron mobility. Therefore, the condition for screening of the cathode by the space charge of the electron flux (where U/d is the instantaneous value of the field) could be disturbed if the voltage across the discharge gap rose too rapidly. In fact, it was not possible to create a SSVD using nonprofiled electrodes when the voltage was applied abruptly to the electrodes and this was true even when the application of a bias resulted in a preliminary filling of the discharge gap with electrons. Sparks were observed at the corners of the cathode, i.e., in the regions with the maximum concentration of the electric field in the absence of screening. The discharge was stable in the case when the leading front of the voltage pulse was delayed deliberately [21, 22] so that the duration of this front was > 2 μs. for $d = 10$ cm.

A classical example of nonprofiled electrodes with a strongly inhomogeneous initial distribution of the electric field in the discharge gap was the first transversely excited CO2 laser [16]. The present experiments indicated that the electron flux in such a system was formed by a barrier discharge plasma created directly on the cathode surface and this made it possible to satisfy quite readily the saturation (Eq. (9)) and screening conditions. One should stress also that an increase of the stability of a SSVD [22] in systems of the type described in Ref. [16] and considered in the present study, which was observed when easily ionizable substances were added to the gas mixtures, was not due to photoionization of these additives because in the case of regimes used to form a SSVD by filling the gap with an electron flux the contribution of the photoionization to the conductance of the discharge gap during the filling process was unimportant compared with the contribution of the electron flux from the auxiliary discharge plasma and this was true for the majority of the auxiliary discharges. Our investigations made it possible to increase the aperture of SSVD based lasers and amplifiers up to 80 cm and to obtain a radiation energy of CO2 lasers more then 5 kJ for the electric efficiency more then 10%.

1.5.2. High Power Wide Aperture HF /DF Lasers

1.5.2.1. Introduction

The main problem arising in the construction of powerful wide-aperture nonchain HF(DF) lasers in which a chemical reaction is initiated by a SSVD is the implementation of the SSVD itself in mixtures of SF6 with hydrogen carriers (deuterium carriers). The possibility of applying for such purposes the usual methods based on the preliminary ionization of the gas in a homogeneous electric field with the subsequent application of a high-voltage pulse to the discharge gap [32] is limited by the high electro-negativity of the SF6 molecules. A new approach to the solution of the problem of increasing the volume of the active medium and the energy of the radiation emitted by nonchain HF(DF) lasers was found in our earlier studies [33, 34, 35]. In a study [34] of a nonchain HF(DF) laser with preionization by soft x-rays, it was observed that the SSVD characteristics and the laser output characteristics are independent of the presence of preionization if the cathode surface has been subjected to a treatment involving the deposition on it of small-scale inhomogeneities. In other words, if the cathode has a rough surface, preionization is not essential in order to obtain an SSVD in the mixtures of an HF(DF) laser.

The possibility of achieving an SSVD without preionization and with an energy input homogeneous throughout the volume under the conditions of a high edge inhomogeneity of the electric field in the discharge gap of an HF(DF) laser has also been demonstrated [35]. These investigations made it possible to increase the aperture of a nonchain HF(DF) laser based on SF6 - hydrocarbon (deuterocarbon) mixtures up to 27 cm and to obtain a radiation energy of ~ 400 J for an electric efficiency of approximately 4% [39].

An SSVD without preionisation in SF6 and in mixtures containing it, i.e. a selfinitiated volume discharge (SIVD) [36], has a number of properties unusual for volume discharges. The observation of a discharge of this kind opens up unique opportunities for V.V. Apollonov / Central European Journal of Physics 4,708,(2003) the development of simple and effective nonchain HF(DF) lasers operating in both pulsed and repetitively pulsed regimes. Further investigations of the SIVD in order to establish the principal mechanisms determining its development is therefore of undoubted interest.

To review the results of our investigations is in fact the subject of the present paper.

1.5.2.2. Experiments

The SIVD characteristics were investigated in discharge gaps with electrodes having different configurations (Figure 5.7) with the ratio of the partial pressures of the mixture components SF6:C2H6 = 10:1-10:3 and Pd= 0,02 – 0,7 cm bar (P is the pressure of the mixture and d the interelectrode distance). The integral characteristics of the discharge and the dynamics of its development were studied by using a system of plane electrodes illustrated schematically in Figure 5.7a. The SIVD was struck with d = 2 - 6 cm between an aluminium cathode 6 cm in diameter, rounded (1 cm radius) along the perimeter, and an aluminium anode 12 cm in diameter. The surface of the cathode, including the rounded parts, was sandblasted.

In order to investigate the SIVD dynamics, an artificial (transmission) line with a variable number of cells was discharged across this gap [37]. This made it possible to follow, by varying the duration of exposure to the voltage across the gap in the range 40-400 ns at a constant current, the development of the discharge in time and in space by photographing it for various exposures to the voltage. In the remaining SIVD studies in which the above and other electrode systems shown in Figure 5.7 were used, the discharge of a capacitor was used to strike the volume discharge. The discharge gaps (Figs 5.7a and 5.7b) were additionally illuminated from the side by a low-current (I < 3 A) spark enclosed in a quartz envelope. This spark could not ensure the volume photoionisation of the gas, but, as a result of the initiation of electrons at the cathode, it made it possible to stabilize the scatter of the gap-breakdown delay times. The discharge gaps in the configurations shown in Figs 5.7c-5.7f were not additionally illuminated.

The dynamics of the development of the SIVD was also investigated in an electrode system with a sectioned cathode (Figure 5.7b). Here, the main cathode and anode were the same as in Figure 5.7a, but an earthed conductor 1.5 mm in diameter (the initiating electrode) was connected in parallel to the main cathode at a distance 5 mm in a horizontal plane extending from its edge. The vertical distance between the cathode surface and the tip of the conductor was chosen in such a way that, during the initiation of the discharge by a low-current spark, the electrical breakdown of the gap extended initially to the conductor.

A conductor with polyethylene insulation and a core diameter of 1.5 mm served as the cathode in the electrode system in Figs 5.7c and 5.7d (the discharge occurred between the rod and the plane). In these experiments, the SIVD was investigated under conditions such that not more than one cathode spot could be formed on the cathode. In the discharge gap shown in Figure 5.7d, the discharge cross section on the anode was confined by a glass tube 6-8 mm in diameter (confined discharge), placed between the electrodes in such a way that its upper edge touched the surface of the anode.

In the configuration presented in Figure 5.7e, the plane electrodes (the same as in Figure 5.7a) were linked by a dielectric plate 2 mm thick with a transverse dimension of 5 cm. In the electrode system shown in Figure 5.7f, the SIVD was struck for an interelectrode distance (the distance between two aluminium disks 3.6 cm in diameter) of 5.5 cm; the disks were placed in a quartz tube with an internal diameter of 3.8 cm, the electrodes had sharp edges, and the cathode surface was sandblasted.

The SIVD dynamics was studied in discharge gaps with different geometry. The scheme of the setup for investigating the discharge in a gap with a linear cathode geometry is shown in Figure 5.7a. The SIVD with a total duration of ~370 ns was ignited in the SF6: C2H6 = 10:1 mixture at a pressure p = 33 Torr and an interelectrode distance of 4 cm. The electrodes were a 0.5-mm thick copper bar 16 cm in length (a knife-edge cathode) placed edgewise and a disk anode 6 cm in diameter rounded over the perimeter with a radius of curvature of 1 cm. The breakdown was forcedly initiated at the center of the cathode with a weak-current spark limited with an R = 900 Q resistor. This spark could not provide an initial electron density in the gas high enough for the development of a volume discharge, but allowed fixing the position of the primary gap breakdown. The SIVD glow was recorded with a single-frame streak camera tube with a frame exposure time of 20 ns, which was triggered with a variable delay T relative to the instant of gap breakdown.

To study the SIVD dynamics in the plane gap geometry, we used the circuit with a sectioned cathode shown in Figure 5.7b. In this case, the interelectrode distance, the working mixture pressure, and the electrical circuit of the setup were the same as in the previous experiment, except that the cathode was not a copper bar, but a disk 6 cm in diameter rounded over the perimeter with a radius of curvature of 1 cm and subjected to sandblast processing [33, 34, 35, 36]. Insulated conductors 1 mm in diameter were inserted into holes 2 mm in diameter drilled 4 cm apart through the cathode in its plane region. One of the conductors (1) protruded above the cathode surface by 1 mm to ensure the occurrence of the primary gap breakdown at precisely this point. By comparing the oscilloscope traces through the ignitor (7) and monitoring (2) electrodes, we could trace the SIVD spread over the gap.

The dynamics of an individual diffusion channel was investigated using the setup shown in Figure 5.7c. The diffusion channel was simulated by a discharge with a rod (cathode) - plane gap geometry at a pressure p = 16,5 – 49,5 Torr of the SF6: C2H6 = 10:1 mixture for an interelectrode distance of 4 cm. The cathode was the end of a polyethyleneinsulated wire with a core diameter of 1.5 mm, and the anode was a disk 10 cm in diameter. The SIVD development was monitored with the streak cameras, as in the experiments performed using the setup of Figure 5.7a.

In the circuits shown in Figs 5.7a and 5.7b, the capacitor C discharged into the gap via inductance L upon switching spark gap. In the circuit of Figure 5.7c, an artificial line discharged into the gap. The currents in the circuits of Figure 5.7 were monitored with Rogowski coil Rc, the voltage across the gaps was measured with resistive dividers.

A typical photograph of an SIVD in a plane-electrode system is presented in Figure 5.7a. Figure 5.7b shows the corresponding current (1) and voltage (U) oscillograms. It is seen from

Figure 5.8 that the SIVD does not outwardly differ from the usual SVD with preionisation. It consists of a set of diffuse channels, linked to bright cathode spots and expanding towards the anode. By overlapping, the diffuse channels generate an overall diffuse glow. Despite the appreciable strengthening of the electric field on the edge of the gap, this region is scarcely distinguishable and the cathode spots cover the entire surface of the cathode. The SIVD voltage and current oscillograms are also typical for SSVD in electronegative gases. Energy is deposited in the SIVD plasma at a quasi-steady-state voltage The voltage in mixtures of SF6 with hydrocarbons, when the content of the latter is not greater than 17%, depends only slightly on their partial pressure. The dependence of U_{qs} on Pd for Pd = 0,05 to 2,5 cm bar and specific energy inputs in the discharge plasma up to 0.2 J cm^{-3} is described satisfactorily by the expression [38] where $U_{qs} = A + B\ PSF_6\ d$ is the partial pressure of SF6 in the mixture. For pure SF6 and SF6: C2H6 = 10:1, 10:2 mixtures, we have respectively A = 0,72 and 0,79 and 1,1 kV and B = 92,7, 94,8 and 96,4 kV cm bar.

The value of B obtained in the present experiments is close to the known [38] critical reduced electric field strength $(E/p)_{cr}$ = 89 kV/ cm bar (the difference is ~8%). The differences between SIVD and SSVD are manifested in the dynamics of their development. Figure 5.9 presents the results of studies on the dynamics of the development of an SIVD in a plane-electrode system (Figure 5.7a) for the discharge across the gap of an artificial line with a variable number of cells. As can be seen from Figure 5.9, in contrast to SSVD, an SIVD is struck initially in the zone of the maximum amplification of the electric field on the gap edge in the form of one or several diffuse channels linked to the cathode spots. The radiation from the discharge in the remainder of the gap is not detected at this instant. The channels formed first initiate the appearance of the subsequent channels and the SIVD spreads over the gap perpendicular to the direction of the electric field at U_{qs} constant voltage, gradually filling the entire gap. The total number of spots on the cathode increases from the instant of the electrical breakdown of the gap virtually in proportion to time, i.e. in proportion to the energy injected into the discharge when account is taken of the experimental conditions.

Since the SIVD develops at a constant current (with the exception of the leading and rear edges) under the experimental conditions described, the current through channels formed earlier should diminish after the appearance of new ones. The characteristic features of the development of an SIVD listed above suggest the operation of current-limiting mechanisms in SF6 and in mixtures based on it, which hinders the transfer of the entire energy through a single channel. It appears that precisely these mechanisms are in fact responsible to a large extent for the existence of the unusual form of the discharge such as the SIVD, including the possibility of achieving it in gaps with a strong edge inhomogeneity of the field.

The existence of such mechanisms is also indicated by the results of experiments in an electrode system with simulation of the first diffuse channel by the initiating electrode (Figure 5.7b). Figure 5.10a, which presents characteristic oscillograms of the discharge current through the initiating (curve 1) and main (curve 2) cathodes, shows that the current through the initiating cathode does indeed appear appreciably earlier than the current through the main cathode and its maximum is attained at an instant when the main current still continues to grow.

Figure 5.10b illustrates the dependence of the fraction of the energy A, transmitted by the initiating cathode, relative to the total energy W injected into the discharge. The quantity A decreases monotonically with increase in W, reaching saturation, which yet again constitutes evidence in favour of the existence of mechanisms preventing the transfer of the entire SIVD energy through one channel.

However, we may note that that in the experiment under consideration, the simulation of the initial diffuse channel with the aid of a special electrode is not entirely equivalent to the real conditions, because, in order to guarantee that the discharge begins from the initiating cathode, it would be necessary to raise it by approximately 1 mm above the surface of the main cathode. For this reason, the brightness and transverse dimensions of the simulated initial diffuse channel differed appreciably from those of channels in the main gap, whereas under real conditions this was not observed (Figure 5.8).

Figure 5.11 presents a typical dependence of the total number of spots on the cathode N_s on the average specific energy Wsp injected into the plasma. It was recorded in a system of plane electrodes for different values of d and p. Wsp was varied during the experiments by varying the charging voltage and charging capacitance. The duration of the discharge current was also varied. Evidently, for a fixed value of P, N_s depends only slightly on d. Within the limits of error, the points satisfactorily on a linear dependence of N_s on Wsp.

Consequently, for a fixed value of p, the number of spots on the cathode is determined by Wsp and not the density and duration of the current, Wsp actually specifying the total number of spots on the cathode. For a short duration (less than 100 - 150 ns) of the current, the spots may not cover the entire surface of the cathode but only its section near the site of the primary electrical breakdown of the gap. In short discharges, the local density of spots may therefore be greater than in long discharges.

The dependence of N_s on p may be more complex: N_s increases with increase in P and the increase is nonlinear. A diminution of the transverse dimensions of the spots on the cathode and of the diffuse channels on the anode is also observed under these conditions, i.e. the volume occupied by each channel decreases simultaneously with increase in N_s.

One may postulate that such influence of the growth of p on N_s is associated with the increase in the electric field strength (defined as Eqs = Uqs/d) for which an SIVD develops and which is known [1] to determine the probability of the formation of a cathode spot.

The dependence of N_s on the parameter Wsp Eqs plotted for different values of P and d taking into account this hypothesis, is presented in Figure 5.12. It is satisfactorily fitted by the linear function N_s = A + B WspEqs, in which the constant b is in its turn a function of the state of the cathode surface and of the hydrocarbon content in the mixture [39].

A conclusion important for the further interpretation of the mechanisms of the development of an SIVD follows from Figs. 5.11 and 5.12: the volume occupied by the discharge increases with increase in the amount of energy injected into the plasma and almost in proportion to the energy.

Figs 5.13a and 5.13b present photographs of an SIVD in a rod-plane electrode system (Figure 5.7c) for d = 5 cm and P = 34 Torr in the SF_6: C_2H_6 = 10:1 mixture recorded for two energies W deposited in the discharge plasma. In this geometry, the SIVD consists of a diffuse plume, expanding towards the anode, with formation of a bright plasma on the cathode. Under conditions such that new spots cannot be formed on the cathode, the volume V of the glowing part of the plume (an analogue of a diffuse channel) increases with increase in W.

The dependence of V on W obtained under the same conditions as in Figure 5.13 is presented in Figure 5.14. A linear growth of V with increase in W can be seen. For W = 5 J, the volume V is ~ 20 cm^3, whereas the size of the plume on the anode is 3.6 cm. The high stability of the discharge in the gap geometry under consideration, despite the fact that the current density of the cathode is j = 1,2 10^5 A cm^{-2}, is striking. It is noteworthy that the influence of the duration of the discharge current on V is also appreciable in this geometry. This accounts for the considerable scatter of the points in Figure 5.14 because in the experiments the duration of the current varied from 50 to 300 ns.

Figs 5.15a and 5.15b present voltage (U) and current (I) oscillograms for an SIVD in the rod-plane electrode system (Figure 5.7c) for current pulses of 100 and 200 ns duration with identical charging voltage and charging capacitance in the SF_6: C_2H_6 = 10:3 mixture (P = 40 Torr, d = 4,4 cm). Figs 5.15c and 5.15d show similar voltage and current oscillograms for an

SIVD recorded under the same conditions but with confinement of the discharge on the anode by a tube 7 mm in diameter (Figure 5.7d). As can be seen from Figs 5.15a and 5.15b, the voltage and current oscillograms in the rod - plane system do not differ in any way from typical oscillograms obtained in a plane electrode system.

Confinement of the discharge leads to an appreciable change in the type of oscillograms. For a long duration of the current pulse (Figure 5.15d), after the initial fall of the voltage, one observes its increase simultaneously with increase in the current, the current maximum being attained sooner than in the case of an unconfined discharge. The current amplitude also diminishes, whereas the voltage is appreciably higher at its maximum than in the case of an unconfined discharge. Voltage oscillograms of a similar type are obtained when the discharge characteristics are calculated taking into account the attachment of electrons to vibrationally excited SF_6 molecules [40]. However, the calculated oscillograms in Ref. [40] do not agree with the experimental ones. We are aware of other studies in which a similar form of the discharge voltage was observed in SF_6 and in mixtures based on it.

For T = 200 ns, the confined SIVD contracts after the current has reached a maximum. A decrease in the duration of the current makes it possible to obtain an uncontracted confined SIVD (Figure 5.15c). In this case, there is no break in the voltage oscillogram and, following a general increase in the discharge voltage, the 'tail' becomes less tilted. We may note that the modification of the voltage pulse observed in Figs 5.15c and 5.15d following the confinement of the SIVD is associated precisely with the increase in the specific energy input in the plasma because under the conditions of V.V. Apollonov / Central European Journal of Physics 4,708, (2003) the discharge cannot expand during the energy deposition process.

Figure 5.16 presents dependences of Uqs on the energy Wj/2 deposited in the discharge plasma before the attainment of the current maximum, in the rod-plane electrode system for confined (curve 1) and unconfined (curve 2) SIVD; was determined at the current maximum. Evidently, with increase in Wj/2 an appreciable growth U_{qs} is observed in the confined SIVD, significantly more notable than in the unconfined SIVD. The relative increase U_{qs} (compared with U_{qs} for the unconfined SIVD) for the same values of Wj/2 greater the smaller the value of P, i.e. the greater the energy deposited in the plasma per unit volume and per molecule.

The results presented in Figs 5.15 and 5.16 indicate an increase in the electron attachment coefficient in SF_6-based mixtures with increase in the specific energy input in the discharge plasma.

The possibility of obtaining a volume discharge in the electrode systems illustrated in Figs 5.7e and 5.7f is a consequence of the existence of mechanisms restricting the SIVD energy transmitted by an individual diffuse channel and leading to an increase in the volume occupied by the discharge with increase in the input energy. Figure 5.17 presents photographs of a discharge in air and in the SF_6 :C_2H_6 = 10:1 mixture, respectively, at a pressure of 65 Torr obtained in the system comprising plane electrodes linked by a dielectric plate (Figure 5.7e). As was to be expected, the discharge in air develops in the form of the electrical spark breakdown of the gap along the surface of the dielectric plate, the bulk phase being entirely absent. In a SF_6 - C_2H_6 mixture, the discharge is very much of the volume type, its external appearance and oscillograms being completely indistinguishable from an SIVD obtained in the same system in the absence of a plate (with the exception that the photograph shows a black band at the site where the plate is located).

In the system with transverse dimensions of the electrodes close to the internal diameter of the tube (Figure 5.7f), the discharge is also very much of the volume type for the mixture

SF_6: C_2H_6 = 10:1 and P = 15- 65 Torr. When Wsp > 0,05 J cm^{-3}, the surface density of the cathode spots has a maximum in the central zone of the cathode, as in the usual gap with plane electrodes which are not linked by a dielectric surface.

In the subsequent ignitions of an SIVD without change of mixture, an appreciable increase and a decrease in the current are observed in the above electrode system. Since in the case under consideration the SIVD occupies the entire volume of the discharge chamber (Figure 5.7f), this fact indicates that, in the mixture of gases formed as a result of the dissociation of the starting materials, the effective ionization coefficient is smaller than in the initial mixture. We may note that a volume discharge in a tube with an internal diameter close to the diameter of the electrodes had been obtained earlier in a SF_6 - H_2 mixture [41], but the authors [41] did not interpret this result.

1.5.2.3. Discussion

Thus, the following processes are characteristic of SIVD: consecutive formation of diffuse channels (with a simultaneous decrease in the current through the channels which had arisen earlier) during the development of the SIVD in a gap with planar geometry, the expansion of the diffuse-glow zone in the rod - plane gap, an increase in the discharge voltage with increase in the specific energy. As already mentioned, these processes are apparently determined to a large extent by the mechanisms of the confinement of the current in the conducting channel, which depend on the specific energy input. We shall consider two possible mechanisms leading to this state of affairs, namely the dissociation of SF6 by electron impact and the attachment of electrons to vibrationally excited SF_6 molecules.

In the first place we note that, for the ratios E/N close to the critical value, virtually the entire energy deposited in SF6 is consumed in dissociation [45]. According to Ref. [38, 42], the $SF_6 = SF_4 + 2F$ channel is dominant.

The ionization potential of F atoms (17,42 eV) exceeds that for SF_6 (15,7 eV) [43], so that an appreciable contribution by the F atoms produced to the overall ionization should not be expected even at their considerable concentrations. As regards the attachment of electrons, the formation of F^{-1} ions via any of the possible mechanisms is known to be incapable of competing, as shown by estimates [44], with the attachment of electrons to SF_6 molecules. The F excitation threshold is 12.7 eV [43] (the component of the main doublet with a threshold energy of ~0.05 eV may naturally be disregarded in the given instance), i.e. it is in the region of the intense excitation of the SF_6 terms by electron impact [45]. Furthermore, there are grounds for the postulate that inelastic processes with participation of SF_4 molecules also do not influence in any way significantly the energy spectrum of the electrons in the discharge.

Thus, bearing in mind the qualitative nature of the models considered and neglecting therefore the difference between the cross sections for the elastic scattering of electrons by SF_6 molecules and their decomposition products, one may postulate that the influence of the dissociation of SF_6 may only affect the decrease in the reduced field strength E/N with increase in the total concentration of neutral particles N as the specific energy input increases.

It is possible to show that allowance for additional factors will significantly complicate the description without altering fundamentally the result: the dissociation of SF_6 by electron impact and the attachment of electrons to vibrationally excited SF_6 molecules may indeed serve as the mechanisms of the limitation of the current in the conducting channel in the active media of an HF laser. At the same time, the mechanisms of current limitation in the conducting channel are not sufficient for complete understanding of the processes observed in the experiment, in particular the propagation of the discharge into the interior of the gap in the direction perpendicular to the applied field [18]. Figure 5.19 shows the streak-camera images of a SIVD in the gap with a knife-edge cathode obtained at different instants of time. The corresponding oscilloscope traces of discharge current and voltage are given in Figure 5.20. One can see from Figure 5.19 that the gap is broken down at the center (at the region of the auxiliary spark discharge). At this instant of time, the SIVD has the form of one diffusion channel, with a cathode spot already formed. Then, new channels appear near the first channel, which have a significantly lower brightness, and are developing at a voltage close to the static breakdown voltage for SF_6.

The number of new channels increases with time, they 'disperse' from the middle to the periphery of the gap, and their brightness gradually achieves that of the first channel, the channels closest to the first channel being the brightest. All the channels become equally bright with time, while the brightness of the first channel decreases significantly. Upon a further SIVD development, the brightness of the channels at the gap periphery (i.e., the channels removed from the primary breakdown region) gradually increases. However, for $T > 210$ ns the glow becomes uniform again over the cathode length, the glow in the first channel also being restored. This effect will be referred to as a current return (to the channel). Then, we observed the development of discharge instability against the background of the total diffusion glow.

Therefore, the results outlined above confirm directly the fact that the SIVD spreads over the gap at a voltage close to the static breakdown voltage in the form of a sequence of diffusion channels. In this case, the first channel, which was produced upon the gap breakdown, fades away as new channels emerge. Note also that the current return (see Figure 5.19), which is manifested in the equalization of the brightness of all the channels and in the increase in the brightness of the first channel following its initial decrease, is observed only when the energy introduced into the plasma is high enough. This effect vanishes when the energy is lowered or the linear size of the cathode is increased. For high energy inputs (150-200 J/L) typical of HF (DF) lasers, the SIVD spreading over the gap is so fast that the time resolution of our streak camera is insufficient A similar pattern of SIVD development is also observed with the plane discharge gap geometry.

Figure 5.21 shows the oscilloscope traces from the igniter and monitoring conductors (see Figure 5.18b) obtained in the experiment with the sectioned cathode. One can see from Figure 5.21 that the current begins to flow through the monitoring conductor with a significant delay relative to the current through the ignitor.

By the time of onset of the current through the monitoring conductor, the amplitude of the current through the ignitor lowers by more than a factor of two, i.e., in this experiment, too, we observed quenching of the initially produced channel upon the emergence of subsequent channels. One can also see from the oscilloscope trace of the current through the ignitor (Figure 5.21) that a current returns to the first channel after its almost complete quenching in the case of a plane gap as well, provided the energy input is high enough. In the rod-plane gap, the spreading of the frame, which simulated a single diffusion channel, was observed to occur in two stages. During the first 20-30 ns following the gap breakdown, the discharge volume rapidly increased, out of proportion to the energy inputted. Then, the discharge volume increased approximately linearly with the energy.

For a constant total energy W introduced into the plasma, the largest volume V occupied by the SIVD by the instant of energy input termination increased with lowering the mixture pressure p. Figure 5.22 shows the dependence of V on the parameter W/P obtained for different P. One can see that this dependence is satisfactorily, to within the error of experimental determination of V, described by a linear function.

The results presented above confirm assumptions that there exist mechanisms of current density limitation in the diffusion channel of a discharge in SF_6 and SF_6-based mixtures [32, 36, 46, 47, 48, 49] (and also in several other strongly electronegative gases [36, 49]). It would be reasonable to attempt to relate these mechanisms to the specific feature of SF_6, namely, with its strong electronegativity.

Strongly electronegative gases have highly reduced electric field strengths of E/N, where E is the field strength and N is the concentration of neutral molecules, at which an electrical breakdown occurs and the energy is inputted into the plasma. They also have large electronegativity parameters Xa defined as the ratio of concentrations of negative ions and electrons. For this reason, the following features of the charge kinetics inherent in these gases should be taken into account in the study of a volume self-sustained discharge in strongly electronegative gases such as SF_6 and SF_6-based laser mixtures.

First, the electron-impact dissociation of SF_6 molecules and other components of the mixture plays a significant role in this discharge. Indeed, for high E/N close to the critical value $(E/N)_{Cr} = 360$ Td, the average energy of plasma electrons in the quasi-stationary stage of a self-sustained discharge approaches the SF_6 dissociation threshold and more than 80% of the energy inputted into the discharge is spent to the dissociation [44, 52] resulting in the production of an F atom. In the submicrosecond range typical for the discharge under study, decomposition products have no time to escape from the discharge channel, resulting in a local increase in the concentration N of particles, a decrease in E/N, and a decrease in the electric conductivity of the diffusion channel due to the increase in electron losses through attachment.

High negative ion concentrations are attained in the discharge, with the effect that the electron detachment from negative ions can make a significant contribution to the balance of charged particles. Usually the disintegration of negative ions in their collisions with neutral molecules and/or due to associative ionization is taken into account. However, for medium gas pressures (p = 10 to 100 Torr) and a submicrosecond discharge duration, these processes do not make any noticeable contribution to electron multiplication [53]. However, there are firm grounds to believe that the electron-impact detachment of electrons from negative ions can be an efficient channel of the delivery of secondary electrons to the discharge plasma [54]. Third, a significant influence on the parameters of an SF_6-discharge plasma, including its conductivity, can be exerted by dissociative electron-ion recombination because high concentrations of positive ions are achieved in the discharge, and the energy is mainly inputted into the plasma. Ion-ion recombination must also be taken into account, because it may significantly limit the concentration of ions in the plasma.

Therefore, it follows from the above discussion that the main mechanisms of SIVD current density limitation are the dissociation of SF_6 molecules by electron impact resulting in a decrease in the ionization rate and an increase in the electron attachment rate due to the local reduction of the parameter E/N in the diffusion channel, and also electron -ion recombination, which is responsible for the growth of electron losses with increasing current density in the channel. It was shown qualitatively [49] that the current density can be also limited due to the attachment of electrons to vibrationally excited SF_6 molecules. However, the absence of reliable data on the rate constants for these reactions in the literature complicates a quantitative estimate of the role of this process in the balance of charged particles in the plasma.

We performed SIVD simulations to verify whether the above assumptions are consistent with the available experimental data. The channel structure of the discharge was modelled with a set of resistive elements connected in parallel. Similarly to Ref. [55], the concentrations of particles were determined by solving continuity equations for particles of each sort in combination with the Kirchhoff equations for the discharge circuit. The nonuniformity of initial conditions for the development of channels along the cathode length was simulated by prescribing different initial electron concentrations in each of the channels. In addition to the electron-impact ionization of SF_6 and electron attachment to SF_6 molecules, the following processes were included in the calculation.

(1) SF6 dissociation by electron impact. The number of dissociated molecules was defined as Nd = W/qF, where W is the energy inputted into the discharge and qF - 4,5 eV is the energy spent to produce an F atom [41].
(2) Electron detachment from negative ions by electron impact. It was assumed that SF6 - negative ions dominate in the plasma because the charge exchange reactions have no time to occur during the discharge period, while the cross sections for the electron-impact production of other negative ions are too small [52, 53]. The rate constant for the disruption of negative ions by electron impact should not be smaller than the rate constant for elastic electron scattering by molecules [54].
(3) Dissociative electron - ion recombination. The rate constant for this process was estimated assuming that the SF_6 discharge plasma is dominated by SF_5^+ ions (52).
(4) Ion - ion recombination. In this case, the rate constant for the E/N values close to the critical value was taken from Ref. [53].

We simulated an SIVD with nine channels, which corresponds approximately to the conditions of the experiment performed using the setup of Figure 5.18a. Figure 5.23 shows the oscilloscope traces of the voltage, the total discharge current and the current through a single channel for which we prescribed the maximum initial electron concentration, calculated for the energy density inputted into the discharge plasma equal to 80 and 40 J/L. One can see from Figure 5.23a that the current through a single channel has two maxima when the energy input is high. This is in qualitative agreement with the oscilloscope trace of current through the ignitor (see Figure 5.21) in the plane gap geometry and in agreement with the experimentally observed redistribution of the channel glow intensity upon the SIVD development in the knife edge-plane gap (see Figure 5.19). The current return vanishes when the energy input is lowered (Figure 5.23b).

Calculations also show that the electron detachment by electron impact and electronion recombination virtually compensate each other within the accuracy of estimates of the rate constants. The dissociation of SF_6 molecules and other components of the HF(DF)- laser mixture should therefore be considered as the principal mechanism responsible for current density limitation in the SIVD channel. However, it is reasonable to refine the magnitudes of the rate constants. Note that our model (like any zero-dimensional model which neglects the time variation of the SIVD volume) pretends only to a qualitative illustration of the current redistribution in the channels when account is taken of possible mechanisms of current density limitation.

Note also that the electron detachment from negative ions by electron impact should be taken into account in the analysis of the processes determining the instability development of the volume discharge in SF6 and other strongly electronegative gases.

Once again we draw attention to the result that follows from the investigation of the dynamics of a single diffusion channel: an SIVD develops at an approximately constant value of the energy parameter W/(Vp).

1.5.3. Conclusion

We investigated the formation of the SSVD in CO_2-N_2-He mixtures by preliminary filling of the gap with an electron flux from an auxiliary discharge plasma. This method was suitable for large distances between the electrodes. A study was made of the characteristics of some of the discharges which could be used as auxiliary ones. It was established that the formation of a SSVD was affected significantly by the distortion of the electric field due to the presence of a space charge of the electron flux. Dynamic profiling of electrons was found to be possible and a SSVD was attained in systems with a strongly inhomogeneous initial distribution of the electric field in the gap.

The simplicity of the proposed methods is very promising for their applications in CO_2 lasers. Although the main results of the present investigation were obtained for CO_2-N_2- He gas mixtures, the methods can clearly be applied also to other gases like N_2O, for example. Due to our investigations the aperture of SSVD based lasers and amplifiers had been increased up to 80 cm. for the electric efficiency of more then 10%. At the same time the SIVD in SF_6 hydrocarbon mixtures, used as the active media for nonchain HF lasers, has been investigated as well. The following features of the development of the SIVD have been established:

1. After the primary electrical breakdown of the discharge gap, the SIVD spreads in the gap in the direction perpendicular to that of the electric field as a result of the consecutive appearance of overlapping diffuse channels. As the new channels appear, the current owing through the channels formed earlier diminishes.
2. The volume occupied by the SIVD increases almost linearly with increase in the energy deposited in the plasma and, when the discharge volume is confined by a dielectric surface, the discharge voltage increases simultaneously with increase in the current.
3. The hypothesis was put forward that mechanisms exist for the limitation of the current in the conducting channel. They are associated with the specific energy released in the plasma and they prevent the transfer of all the deposited energy through a separate channel. It is shown that such mechanisms may be the dissociation of SF_6 and the attachment of electrons to vibrationally excited SF_6 molecules.

Main result for SIVD based HF(DF) lasers looks very much promising and half of Kj level for energy output is very far from the natural limit for this type of the systems.

1.5.4. FIGURES

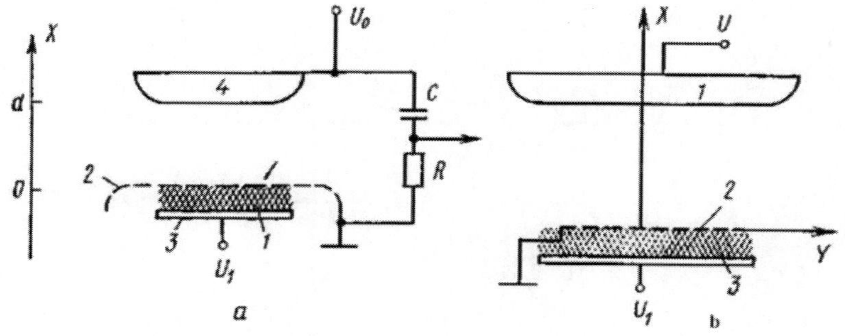

Figure 5.1. Discharge systems with profiled (a) and nonprofiled (b) electrodes.

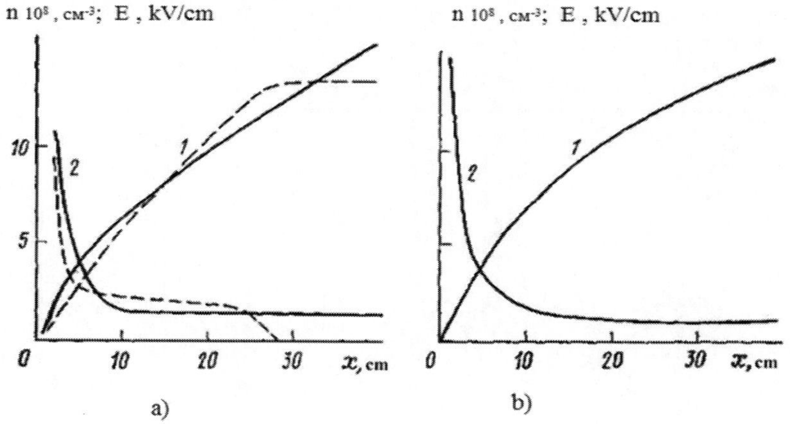

Figure 5.2. Calculated dependences E(x) (curves denoted by 1) and n(x) (curves denoted by 2): (a) t = 2 mks (dashed curves) and 4-6 mks (continuous curves); (b) U = 400 kV, T_f=t= 8 mks, n(0)=10^{10} $cm^{-3.}$

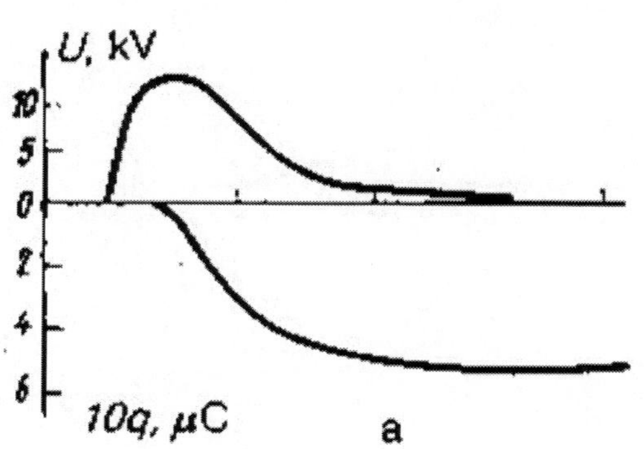

Figure 5.3. (Continued)

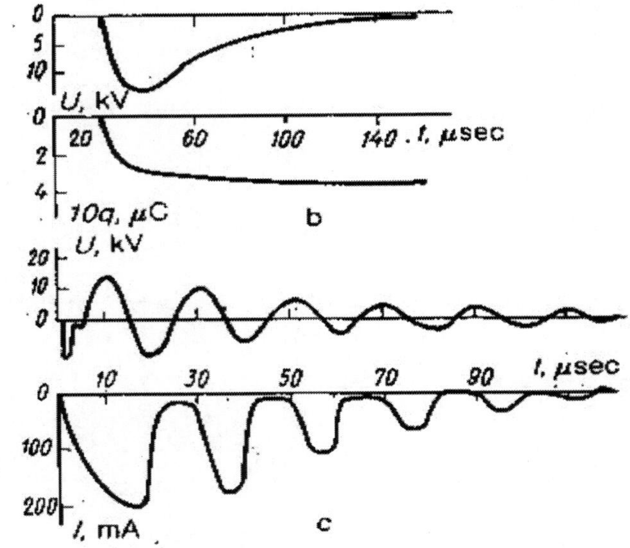

Figure 5.3. Oscillograms of the voltage U(t) on the auxiliary electrode, of the electron charge q(t), and of the electron flux current I(t) in the case of positive (a) and negative (b) polarities, and also in the case of oscillatory nature of the voltage U(t) (c).

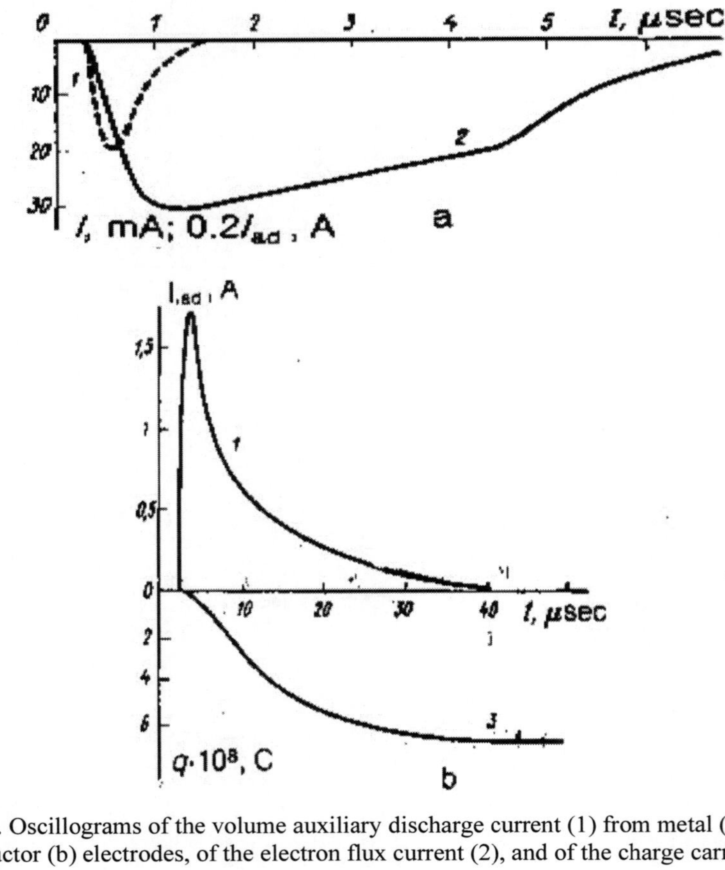

Figure 5.4. Oscillograms of the volume auxiliary discharge current (1) from metal (a) and semiconductor (b) electrodes, of the electron flux current (2), and of the charge carried by this flux (3).

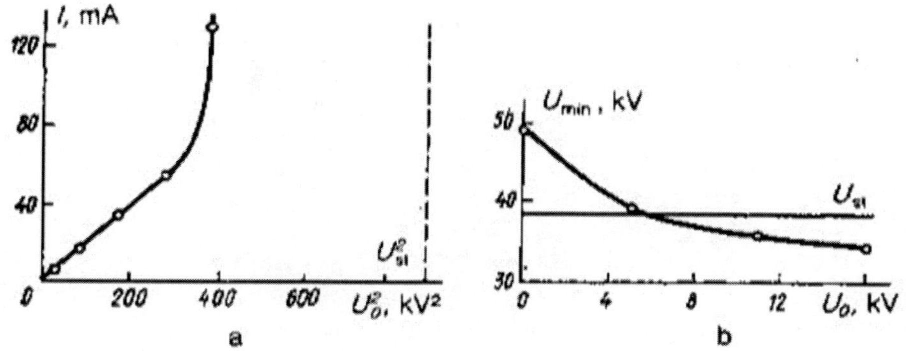

Figure 5.5. Dependences of the amplitude of the electron flux current (a) and of the minimum ignition voltage of a SSVD (b) on U0.

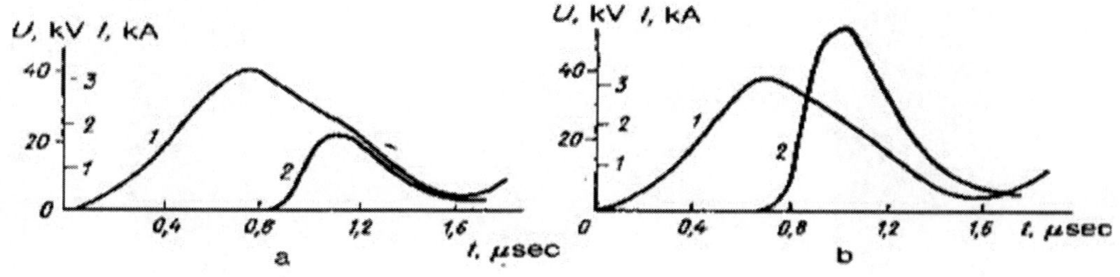

Figure 5.6. Oscillograms of the voltage U (curves denoted by 1) and of the current (curves denoted by 2) for a SSVD when the condition (9) is not satisfied (a) and when it is satisfied (b).

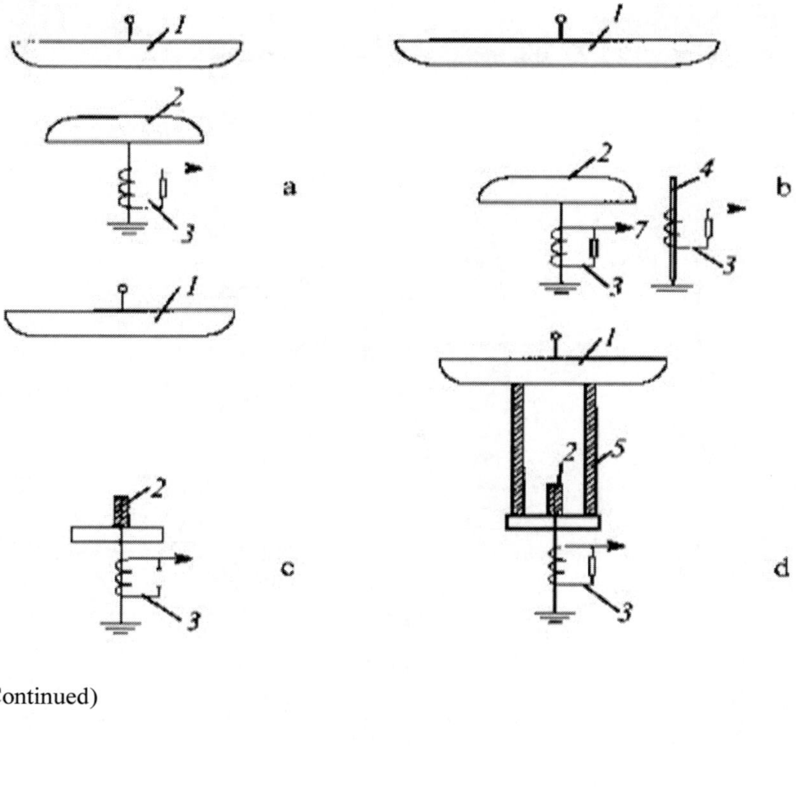

Figure 5.7. (Continued)

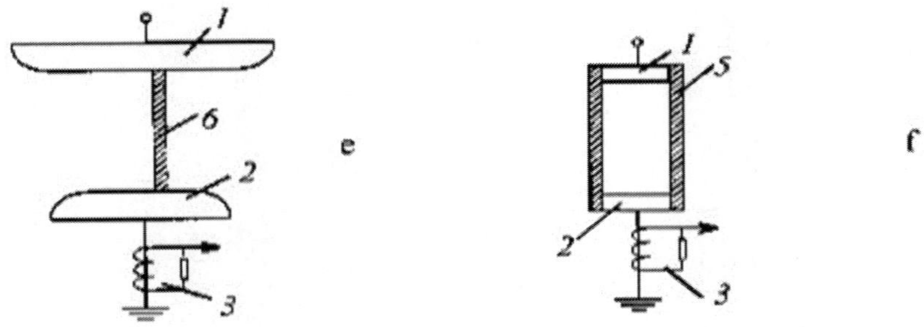

Figure 5.7. Discharge gaps with electrodes having diferent configurations: (1) anode; (2) cathode; (3) Rogowski loop; (4) initiating electrode; (5) glass tube; (6) glass plate.

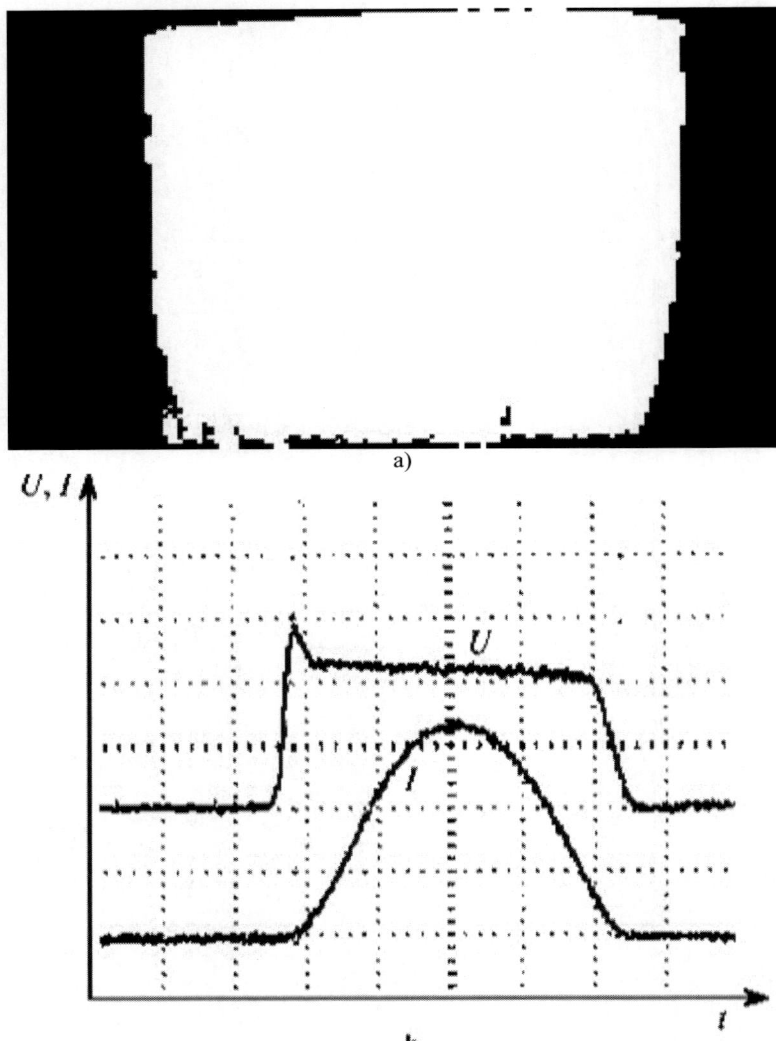

Figure 5.8. Photograph of an SIVD in a system of plane electrodes (a) and typical current (I) and voltage (U) oscillograms for the SIVD; scan|100 ns division-[1] (b).

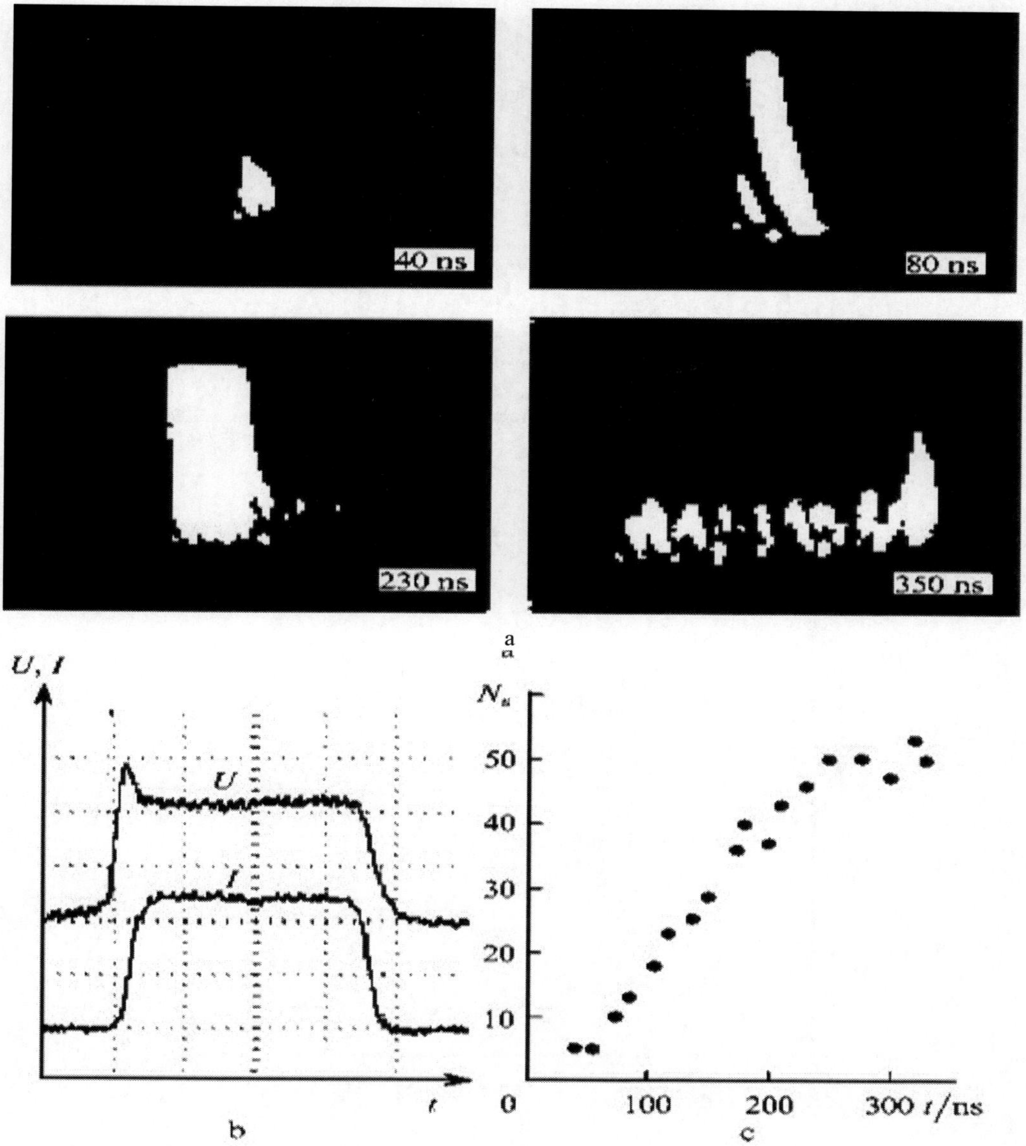

Figure 5.9. Characteristics of an SIVD: photographs of an SIVD in a system of plane electrodes at different times (spark illumination from the left) (a), voltage (U) and current (I) oscillograms for the SIVD during the discharge across the gap of an artificial line consisting of 20 cells (scan - 100 ns division[1] (b) and time dependence of the number of spots on the cathode Ns (c).

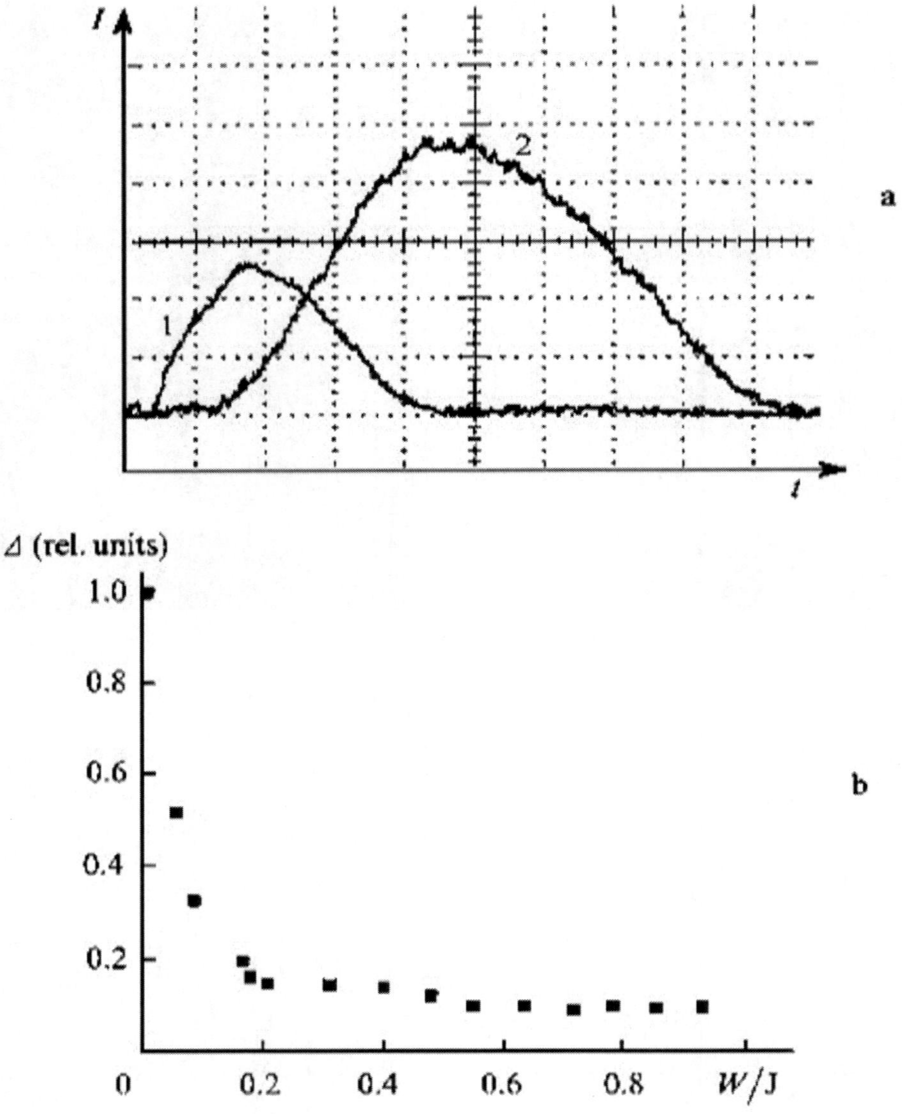

Figure 5.10. Oscillograms of the current through the initiating (1) and main (2) cathodes (39.7 A division, scan 50 ns division)- (a) and dependence of the fraction of energy A transmitted through the initiating electrode relative to the total energy W (b).

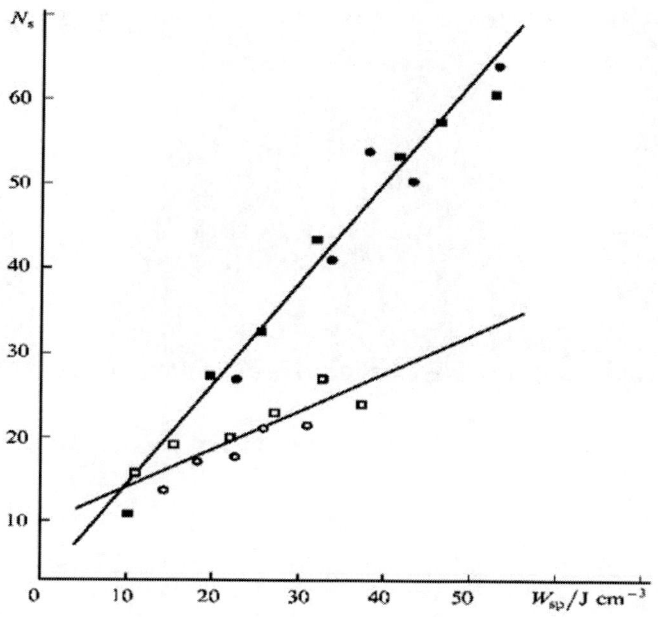

Figure 5.11. Dependence of the number of cathode spots Ns on the specific input energy Wsp for the SF6: C2H6 = 10:1 mixture with: d = 6 cm and P = 33.6 Torr (.); d = 2 cm and P = 33.6 Torr (); d = 6 cm and P = 16.8 Torr(n); d = 2 cm and P = 16.8 Torr (o).

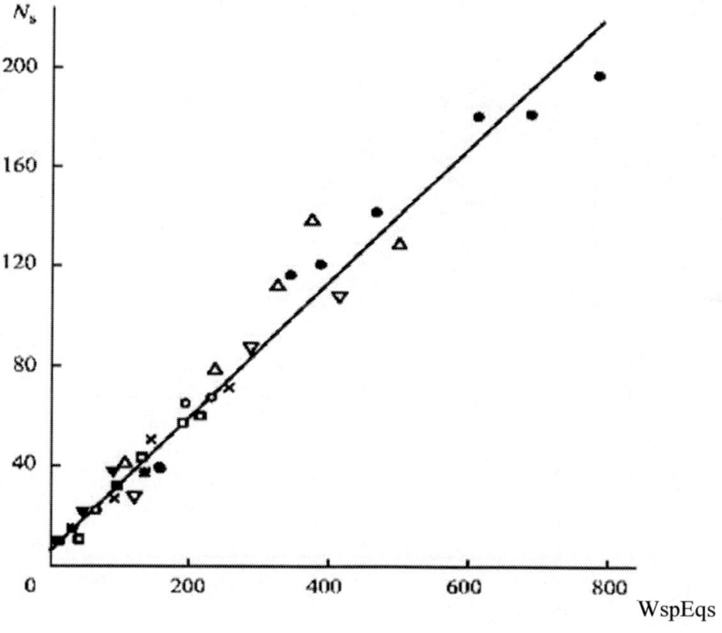

Figure 5.12. Dependence of the number of cathode spots Ns on WspEqs for the SF6 : C2H6 = 10:1 mixture with: d = 6 cm and P = 33.6 Torr (n); d = 6 cm and P= 16.8 Torr (); d = 4 cm and P= 23.3 Torr (*); d = 3 cm and P = 23.3 Torr (); d = 3 cm and P = 33.6 Torr (o); d = 3 cm and P = 50.4 Torr ; d = 3 cm and P = 67.2 Torr (); d = 2 cm and P = 33.6 Torr (x).

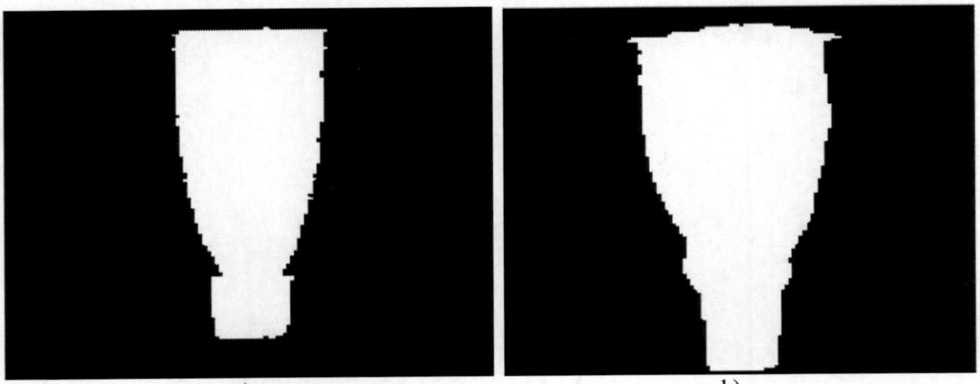

a) b)

Figure 5.13. Photographs of an SIVD in the rod-plane electrode system with W = 0,73 J (a) and 2,33 J (b).

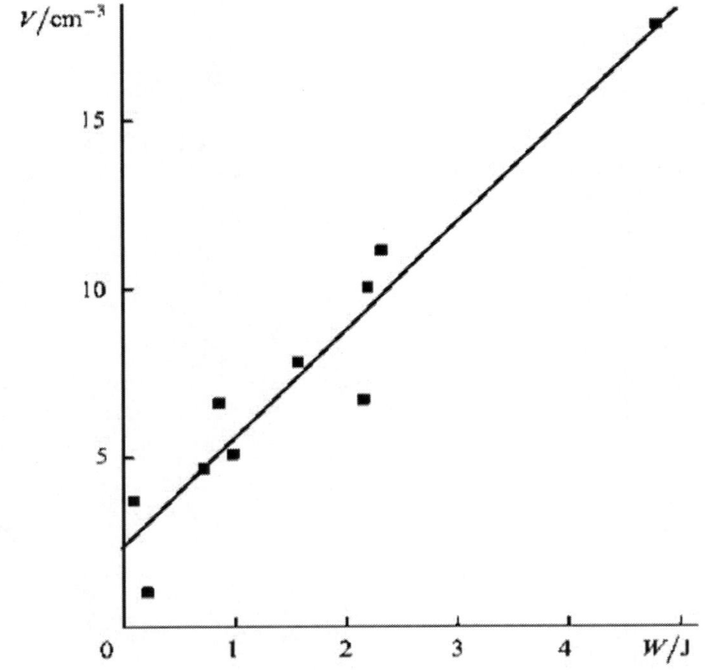

Figure 5.14. Dependence of the volume V of an SIVD on the energy W in the rod - plane electrode system.

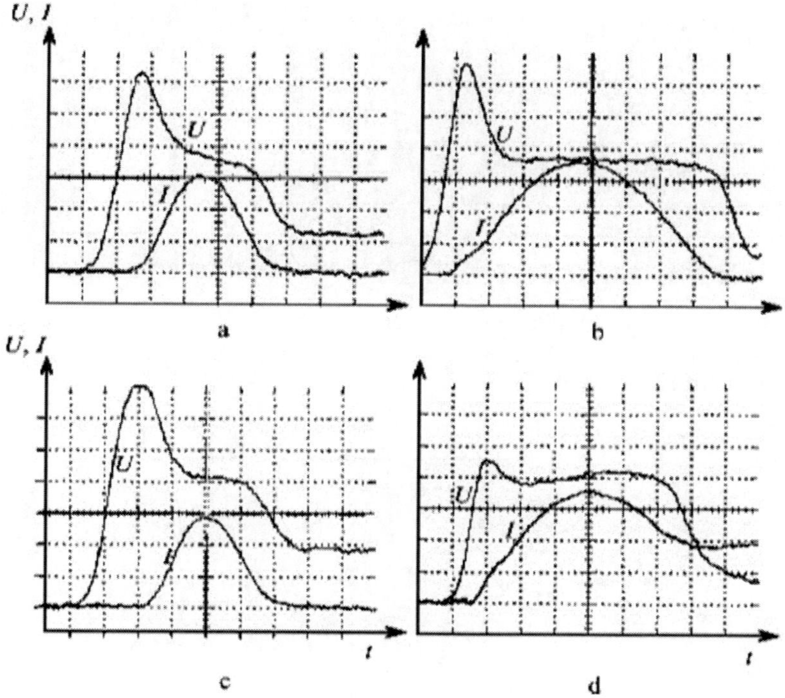

Figure 5.15. Voltage (U) and current (I) oscillograms for an unconfined (a, b) and confined (c, d) SIVD in the rod-plane electrode system with T = 100 ns (a, c) and 200 ns (b, d). Oscillogram scales: 5.1 kV division (a-d), 79.4 A division (b, d), 198.5 A division (a, c); time scan - 25 ns division.

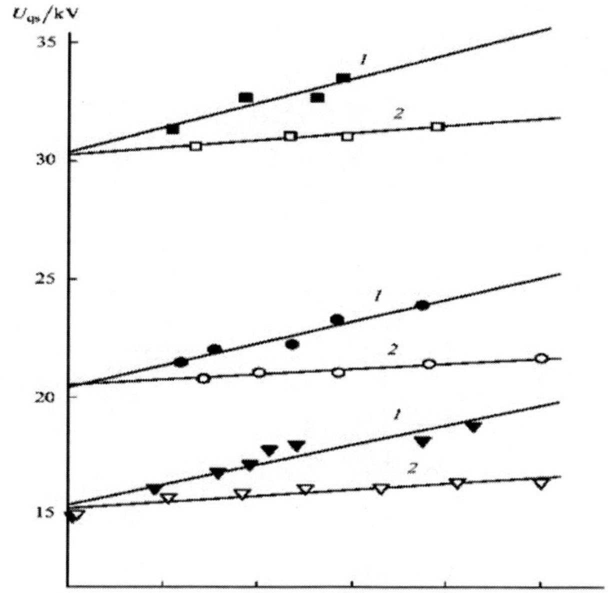

Figure 5.16. Dependence of Uqs on the energy for confined (1) and unconfined (2) SIVD with P = 33.6 Torr (triangles), 45.8 Torr (circles) and 67.2 Torr (squares); $SF_6 : C_2H_6$ = 10:1 mixture, d = 4 cm.

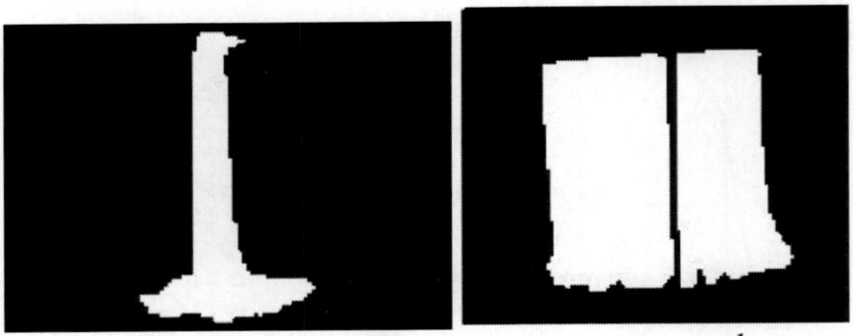

a b

Figure 5.17. Photographs of a discharge in a system of plane electrodes linked by a glass plate in air and in the SF6: C2H6 = 10:1 mixture (b).

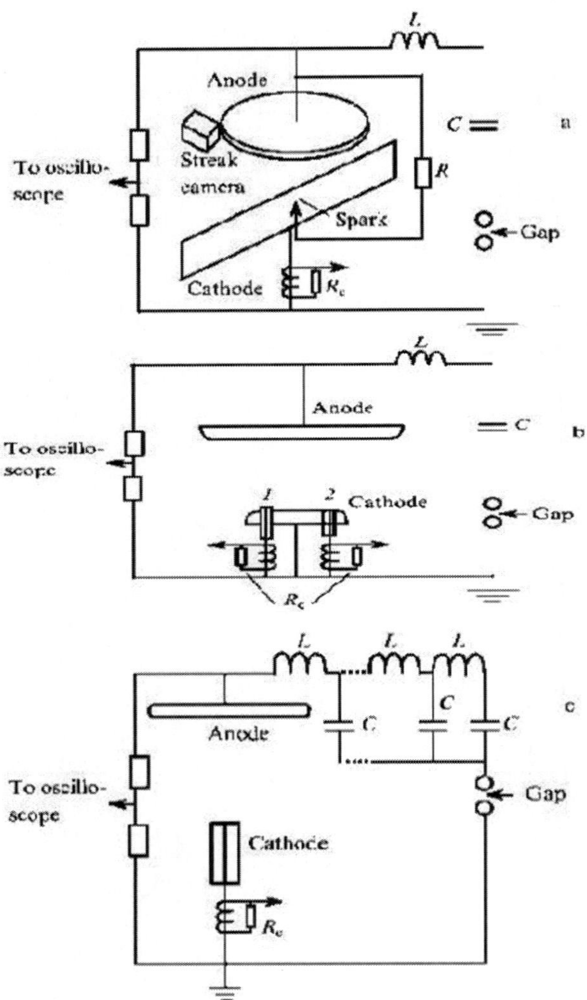

Figure 5.18. Schemes of experimental setups for studying the SIVD dynamics: ignitor (1) and monitoring (2) conductors.

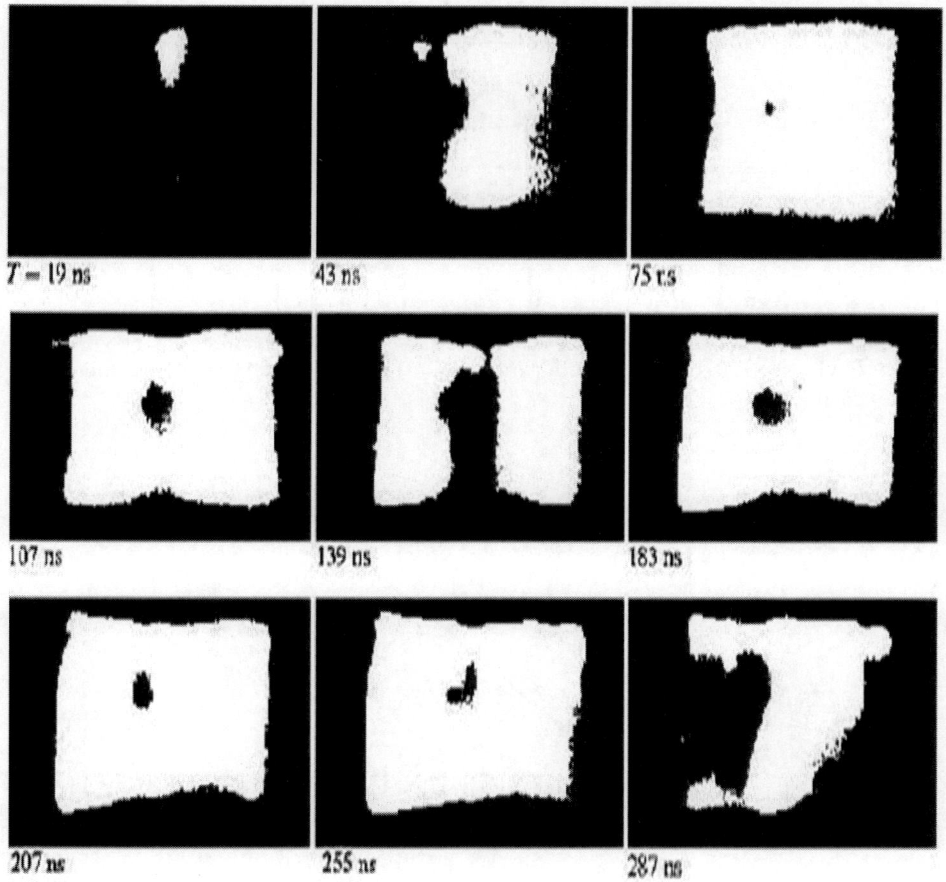

Figure 5.19. Streak camera images obtained at different instants of time.

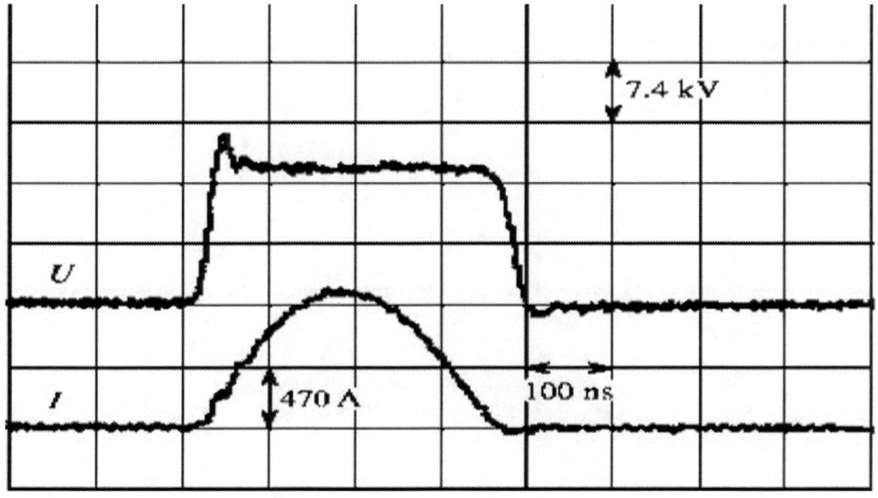

Figure 5.20. Oscilloscope traces of the voltage and current of SIVD.

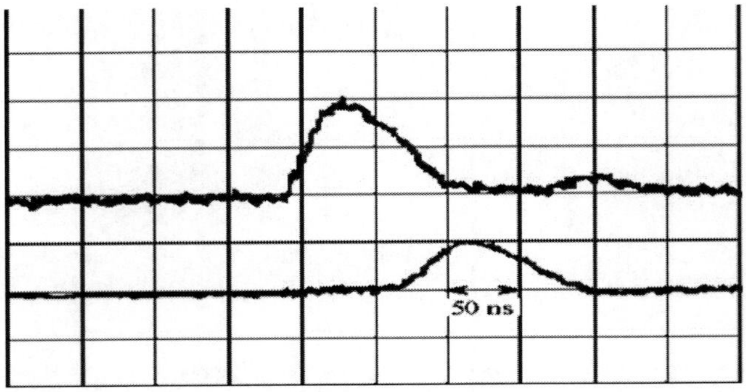

Figure 5.21. Oscilloscope traces of the current through the ignitor (the upper trace) and monitoring (the lower trace) conductors.

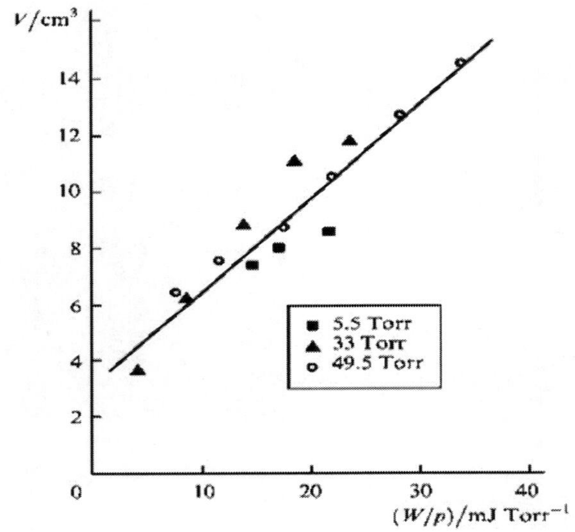

Figure 5.22. Dependences of the volume V occupied by discharge on the parameter W/p for the SF6: C2H6 =10:1 mixture at a pressure P.

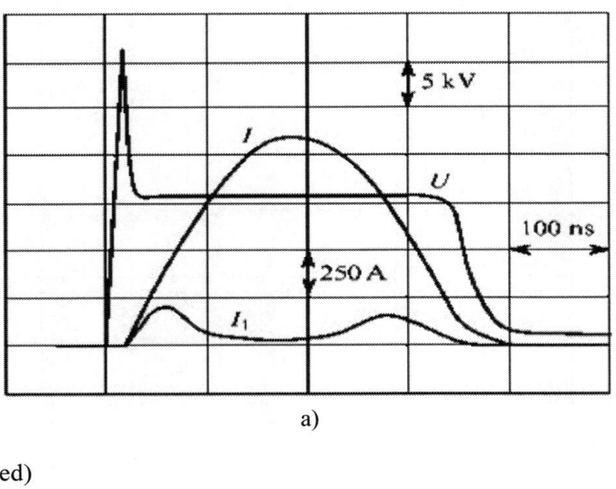

a)

Figure 5.23. (Continued)

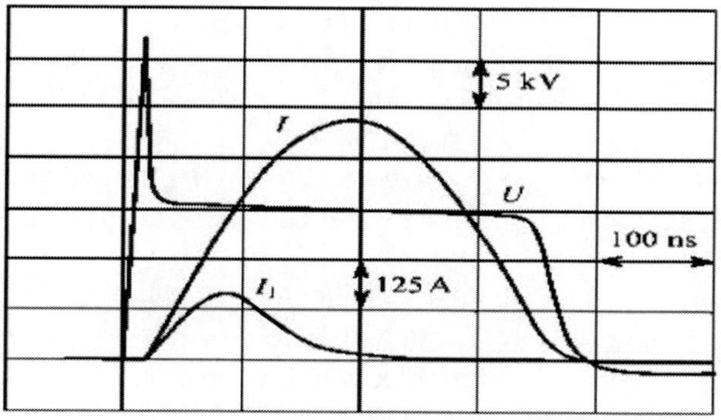

b)

Figure 5.23. Calculated oscilloscope traces of the voltage U, the total discharge current I, and the current In through the first channel for energy density 80 (a) and 40 J/L (b).

1.5.5. REFERENCES

[1] V. Yu. Baranov, V.M. Borisov, A.A. Vedenov, S.V. Drobyazko, V.N. Knizhnikov, A.P. Napartovich, V.G. Niz'ev, and A.P. Strel'tsov: Preprint No. 2248 (in Russian), Institute of Atomic Energy, Moscow, 1972.

[2] A. J. Palmer: "The article you requested is currently in production and will be available online at a later date", *Appl. Phys. Lett.* 25, Vol. 138, (1974).

[3] V. N. Karnyushin, A.N. Malov, and R.I. Soloukhin: Kvantovaya Elektron. (Moscow), Vol. 5, (1978), pp. 555. [*Sov. J. Quantum Electron.* 8, 319 (1978)]

[4] H. Seguin and J. Tulip: "The article you requested is currently in production and will be available online at a later date", *Appl. Phys. Lett.*, Vol. 21, (1972), pp. 414.

[5] A. F. BelyatskiT, D.B. Gurevich, M.A. Kanatenko, and I.V. Podmo-shenski, T. Pis'ma: "Zh. Tekh. Fiz.", *Sov. Tech. Phys. Lett.,* Vol. 6, (1980), pp. 73.

[6] S.C. Lin and J.I. Levatter: "X-ray preionization for electric discharge laser", *Appl. Phys. Lett.,* Vol. 34, (1979), pp. 505.

[7] A. I. PavlovskiT, V.S. Bosamykin, V.I. Karelin, and V.S. *Nikol'skii: Kvantovaya Elektron.* (Moscow), Vol. 3, (1976), pp. 601. [Sov. J. Quantum Electron. 6, 321 (1976)]

[8] A. I. Pavlovskii, V.F. Basmanov, V.S. Bosamykin, V.V. Gorokhov, V.I. Karelin, and P.B. Repin: *Abstracts of Papers presented at Second All-Union Conf. on Physics of Electric Breakdown in Gases*, Tartu (in Russian), (1984), pp. 330.

[9] R. V. Babcock, I. Liberman, and W.P. Partlov: "Volume ultraviolet preionization from bare sparks", *IEEE J. Quantum Electron.*, Vol. 12, (1976), pp. 29-34.

[10] A. V. Kozrev, Yu.D. Korolev, G.A. Mesyats, Yu.N. Novoselov, A.M. Prokhorov, V.S. Skakun, V.F. Tarasenko, and S.A. Genkin: *Kvantovaya Elektron.* (Moscow), Vol. 11, (1984), pp. 524. [Sov. J. Quantum Electron. 14, 356, (1984)]

[11] O. P. Judd: *J. Appl. Phys.*, Vol. 45, (1974), pp. 4572.

[12] R. E. Beverly III: *Appl. Phys.*, Vol. B53, (1991), pp. 187-193.

[13] V. V. Apollonov, G.G. Baitsur, A.M. Prokhorov, and K.N. Firsov: *Pis'ma Zh. Tekh. Fiz.*, Vol. 11, (1985), pp. 1262. [Sov. Tech. Phys. Lett. 11, 521 (1985)]

[14] A. K. La amme: *Rev. Sci. Instrum,* Vol. 41, (1970), pp. 1578.

[15] A. J. Beaulieu: *Appl. Phys. Lett.*, Vol. 16, (1970), pp. 504.

[16] R. Dumanchin and J. Rocca-Serra: C. R. *Acad. Sci. Ser,* Vol. 269, (1969), pp. 916.

[17] V. P. Gorelov, N.N. Minakova, and V.A. Chagin: Elektrotekh. *Promst. Ser. Elektrotekh. Mater.,* Vol. 5, (1972), pp. 20.

[18] V. V. Apollonov / *Central European Journal of Physics* 4 (2003) 708{756 731

[19] P. A. Belanger, R. Tremblay, J. Boivin, and G. Otis: *Can. J. Phys.,* Vol. 50, (1972), pp. 2753.

[20] V. V. Apollonov, N. Akhunov, V.R. Minenkov, S.S. Pel'tsman, A.M. Prokhorov, B.V. Semkin, K.N. Firsov, and B.G. Shubin: *Kvantovaya Elektron.* (Moscow), Vol. 11, (1984), pp. 1241. [Sov. J. Quantum Electron. 14, 835 (1984)]

[21] T. Y. Chang: "The article you requested is currently in production and will be available online at a later date", *Rev. Sci. Instrum.*, Vol. 44, (1973), pp. 405.

[22] V. V. Apollonov, V.R. Minenkov, A.M. Prokhorov, B.V. Semkin, K.N. Firsov, B.G. Shubin, and V.A. Yamshchikov: *Kvantovaya Elektron.* (Moscow), Vol. 11, (1984), pp. 1327. [Sov. J. Quantum Electron. 14, 898 (1984)]

[23] V. V. Apollonov, Yu. M. Vas'kovskii, M.I. Zhavoronkov, A.M. Prokhorov, R.E. Rovinskii, V.E. Rogalin, N.D. Ustinov, K.N. Firsov, I.S. Tsenina, and V.A. Yamshchikov: *Kvantovaya Elektron.* (Moscow), Vol. 12, (1985), pp. 5. [Sov. J. Quantum Electron. 15, 1 (1985)]

[24] V. V. Apollonov, A.I. Barchukov, S.I. Derzhavin, I.G. Kononov, A.M. Prokhorov, K.N. Firsov, Yu. A. Shakir, and V.A. Yamshchikov: *Pis'ma Zh. Tekh. Fiz.*, Vol. 3, (1977), pp. 1073. [Sov. Tech. Phys. Lett. 3, 441 (1977)]

[25] V. V. Apollonov, S.I. Derzhavin, I.G. Kononov, K.N. Firsov, Yu. A. Shakir, and V.A. Yamshchikov: *Pis'ma Zh. Tekh. Fiz.*, Vol. 5, (1979), pp. 1518. [Sov. Tech. Phys. Lett. 5, 643 (1979)]

[26] V. V. Apollonov, N. Akhunov, V.R. Minenkov, S.S. Pel'tsman, A.M. Prokhorov, B.V. Semkin, K.N. Firsov, and B.G. Shubin: *Kvantovaya Elektron.* (Moscow), Vol. 10, (1983), pp. 1458. [Sov. J. Quantum Electron. 13, 947(1983)]

[27] A. A. Kuchinskii (Kuchinsky), B.V. Lyublin (Ljublin), V.F. Shanskii (Shansky), and V. A. Sheverev: *Proc. Sixteenth Intern. Conf. on Phenomena in Ionized Gases*, Dusseldorf,, (1983), (ed. by W. Botticker, H. Wenk, and E. Schulz-Gulde), *Contributed Papers*, Vol. 2, publ. by the Conference Committee, (undated), pp. 200.

[28] A. L. Ward: *J. Appl. Phys.,* Vol. 36, (1965), pp. 2540.

[29] A. H. von Engel: *Ionized Gases,* Clarendon Press, Oxford, (1955).

[30] I. L. Kamardin, A.A. Kuchinskii, V.A. Rodichkin, and V.F. Shanskii: Abstracts of *Papers presented at Third All-Union Conf. on Laser Optics* (in Russian), Leningrad, (1982), pp. 64.

[31] S. I. Andreev, N. M. Belousova, P.N. Dashuk, D. Yu. Zaroslov, E.A. Zobov, N.V. Karlov, G.P. Kuz'min, S.M. Nikiforov, and A.M. Prokhorov: *Kvantovaya Elektron.* (Moscow), Vol. 3, (1976), pp. 1721. [Sov. J. Quantum Electron. 6, 931 (1976)]

[32] Yu. D. Korolev, G.A. Mesyats: "Fizika Impul'snogo Proboya Gazov", *The Physics of the Pulsed Electrical Breakdown of Gases*, (1991).

[33] V. V. Apollonov, S. Yu. Kazantsev, V.F. Oreshkin, N. Firsov: *Pis'ma Zh. Tekh. Fiz.*, Vol. 22, (1996), pp. 60. [Tech. Phys. Lett. 22 1026 (1996)]

[34] V. V. Apollonov, S. Yu. Kazantsev, V.F. Oreshkin, N. Firsov: *Kvantovaya Elektron.* (Moscow), Vol. 24, (1997), pp. 213. [Quantum Electron. 27 207 (1997)]

[35] V. V. Apollonov, S. Yu. Kazantsev, V.F. Oreshkin, N. Firsov: *Kvantovaya Elektron.* (Moscow), Vol. 25, (1998), pp. 123. [Quantum Electron. 28 116 (1998)]
[36] V. V. Apollonov / *Central European Journal of Physics* 4 (2003) 708{756
[37] V. V. Apollonov, N. Firsov, S. Yu. Kazantsev, V.F. Oreshkin: *Proc. SPIE Int. Soc. Opt. Eng.*, Vol. 374, (1998), pp. 3574.
[38] V. V. Apollonov, S. Yu. Kazantsev, V.F. Oreshkin, N. Firsov: II Mezhdunar. Konf. Po Fizike Plazmy i Plazmennym Tekhnolo-giyam, Minsk, (1997), pp. 154. [Proceedings of the Second International Conference on Plasma Physics and Plasma Technologies, Minsk, Belarus, 1997], [Minsk: National Academy of Sciences of Belarus, 1997)
[39] I. M. Bortnik: Fizicheskie Svoistva i Elektricheskaya Prochnost' Elegaza, *Energoatomizdat,* Moscow, 1998. [The Physical Properties and the Electrical Strength of Sulfur Hexa°uoride]
[40] V. V. Apollonov, S. Yu. Kazantsev, V.F. Oreshkin, N. Firsov: IX Konf. po Fizike Gazovogo Razryada, Ryazan, (1998), pp. 58. [Proceedings of the Ninth Conference on the Physics of Gas Discharges, Ryazan, 1998], [Ryazan: Ryazan Radio Engineering Academy, 1998]
[41] L. Richeboeuf, S. Pasquiers, M. Legentil, V.J. Puech: "The in°uence of H2 and C2H6 molecules on discharge equilibrium and F-atom production in a phototriggered HF laser using SF6", *Phys. D,* Vol. 31, (1998), pp. 373.
[42] A. F. Zapol'skii, Yushko: *Kvantovaya Elektron.* (Moscow), Vol. 6, (1979), pp. 408. [Sov. J. Quantum Electron. 9 248 (1979)]
[43] E. B. Gordon, V.I. Matyushenko, P.B. Repin, V.D. Sizov: *Khim. Fiz.,* Vol. 8, (1989), pp. 1212.
[44] A. A. Radtsig, M. Smirnov: Spravochnik po Atomnoii Molekulyarnoi Fizike (*Handbook of Atomic and Molecular Physics),* Atomizdat, Moscow, 1980.
[45] H. S. Massey: *Negative Ions*, Cambridge University Press, Cambridge, 1976.
[46] D. I. Slovetskii, A.A. Deryugin: "in Khimiya Plazmy", *The Chemistry of Plasma,* (1987), pp. 240.
[47] A. A. Belevtsev: Proceedings of the Thirteenth International Conference on Dielectric *Liquids* (ICDL'99), Nara, Japan, 1999, pp. 672.
[48] A. V, Melekhov: *Abstracts of Papers presented at Second Ail-Union Conf. on Physics of Electric Breakdown of Gases* (in Russian), Tartu, 1984, pp. 357.
[49] V.V. Apollonov, S. Yu. Kazantsev, V.F. Oreshkin, A.V. Saifulin, K.N. *Firsov: Izv.* Akad. Nauk, Ser. Fiz., Vol. 64, (2000), pp. 1439.
[50] V. V. Apollonov, K.N. Firsov, S. Yu. Kazantsev, V.F. Oreshkin, A.V. Saifulin: *Proc. SPIE Int. Soc. Opt. Eng.,* Vol. 370 (1999), pp. 3886.
[51] V. V. Apollonov, A.A. Belevtsev, S. Yu. Kazantsev, A.V. Saifulin, F.N. Firsov: *Kvantovaya Elektron.*, Vol. 30, (2000), pp. 207.
[52] V. V. Apollonov, A.A. Belevtsev, K.N. Firsov, S. Yu. Kazantsev, A.V. Saifulin: Proc. XIII Intern. *Conf. on Gas Discharge and their Applications*, Vol. 1, (2000), pp. 409.
[53] V. V. Apollonov, A.A. Belevtsev, K.N. Firsov, S. Yu. Kazantsev, A.V. Saifulin: Proc. XXV Intern. *Conf. on Phenomena in Ionized Gases*, Vol. 1, (2001), pp. 255.
[54] N. Nakano, N. Shimura, Z.L. Petrovic, T. Makabe: "Simulations of rf glow discharges in SF6 by the relaxation continuum model: Physical structure and function of the narrow-gap reactive-ion etcher", *Phys. Rev. E*, Vol. 49, (1994), pp. 4455.
[55] V. V. Apollonov / *Central European Journal of Physics* 4 (2003) 708{756 733

[56] H. Hilmert, W.F. Schmidt: "Electron detachment from negative ions of sulphur hexa°uoride-swarm experiments", *Phys. D: Appl. Phys.*, Vol. 24, (1991), pp. 915.

[57] A.A. Belevtsev, L.M. Biberman: Izv. Akad. Nauk SSSR, *Ser. Energ. Tramp.*, Vol. 3, (1976), pp. 74.

[58] V. V. Apollonov, S. Yu. Kazantsev, A.V. Saifulin, K.N. Firsov: *Kvantovaya Elektron.*, Vol. 30, (2000), pp. 483.

[59] V. V. Apollonov, A.A. Belevtsev, S. Yu. Kazantsev S. Yu., A.V. Saifulin, K.N. Firsov: *Kvantovaya Elektron.*, Vol. 31, (2001), pp. 629.

Chapter 6

1.6. POWER OPTICS AND ITS NEW MANIFESTATIONS

1.6.1. INTRODUCTION

Archaeologists claim that people learned to make mirrors more than five thousand years ago. At home they used polished bronze or silver plates, and only ancient Romans started to manufacture glass mirrors with a tin or lead coating. The technology of mirror production changed many times, but basically all the mirrors were either metal-coated glass or quartz. Never before, the energy of radiation fluxes was so damaging to mirrors because they reflected very little light, for example, from distant stars, from a candle flame or splinter.

Modern problems of interaction of optical mirrors with high-power radiation arose with the advent of first lasers. Virtually in all types of lasers generation occurs in a resonator, which consists of at least a pair of mirrors, and through one of them radiation is emitted. Initially, traditional quartz disks with a mirrored surface suited well the purpose, but over the years passed, the power of laser radiation fluxes has increased thousandfold. Today, the problem of making a mirror that can operate and maintain its high optical performance when exposed to intense radiation has become one of the key challenges in the improvement of high-power lasers and modern high-energy systems. Design, materials testing and technological ideas used in power optics work and are extremely useful for specialists involved in the development of astronomical and space optics, new efficient systems based on laser diode assemblies, and in many other areas of the ever-expanding knowledge.

1.6.2. OPTICS OF HIGH-POWER LASERS

Back in the late 1960s, we drew attention to the phenomenon, which, of course, was to limit the further growth of the power generated by the laser system and, consequently, the intensity of the laser beam on the surface of the irradiated object. More than twenty years of fundamental and applied research devoted to the study of this phenomenon and to the solution of problems associated with it allow a conclusion that its essence consists in the following.

The optical surface of even a very good mirror does not fully reflect radiation falling on it. A small portion of energy (fractions of a percent, depending on the wavelength) is absorbed by the mirror and turns into heat. As the laser beam power increases, even a small amount of it is sufficient to induce thermal stresses in a mirror. They distort the geometry of the reflecting surface, which influence, for example, such an important property of laser light, as the possibility of its propagation over long distances and concentration in a small volume. Now we know that thermal deformation of mirrors not only affect the main characteristics of the beam, but also lead to quenching of generation. And is it possible to manufacture a mirror, which can resist a flux of radiation power of a few megawatts? This question was probably the main one in the world laser community of the early 1970s. The General Physics Institute and enterprises of the defense complexes working under its leadership are carrying out research in the field of so-called high-intensity laser optics or power optics. The main requirement to power optics is a high threshold of optical performance. For each type of a laser system, depending on its applications the permissible value of the elastic distortion of the mirror surface is specified, which, for example, is measured in fractions of a percent of a laser wavelength of a specific laser system. For a CO_2 laser the generated radiation lies in the infrared region of the spectrum, and the wavelength is 10.6 μm; therefore, the elastic distortion shall not exceed fractions of a micron. Even higher requirements are imposed on the optics of the short-wavelength range. The light intensity, at which the distortions of the mirror surface reach some limits for a specified laser source, is an optical performance threshold of the mirror. It means that the better the optical quality of the mirror material, the higher the threshold of optical performance at higher intensities. This threshold level cannot be violated, because elastic deformations of the mirror change the shape of the laser beam, which will ultimately affect the prospects of delivering the light energy to the interaction region. If the light load is further increased, then deformations become inelastic (plastic), which means that irreversible changes are produced on the mirror, i.e., its reflecting surface is damaged and should be replaced. Therefore, it is needed to manufacture mirrors, which could withstand high optical loads for quite a long time, i.e., up to several kilowatts per square centimeter of the mirror surface.

To understand how difficult this task is, it is enough to give two examples. The mirror, which has been in the hands of an experimenter for only a few seconds, can exhibit deformations of the optical surface that is close to the maximum permissible value, because of the uneven hand heating. However, in this case, if the mirror is left to cool, the shape will be restored. In reality mirrors of modern high-power lasers are exposed to power densities of several kilowatts per square centimeter, which is comparable to the energy flux, which is emitted into the environment directly from the unit surface of the Sun. And this means that if we somehow contrived to 'put a laser mirror on the Sun', the shape of its surface should not change significantly even in this case.

To solve this problem and manufacture highly resistant mirrors necessary for high-power industrial lasers, it was necessary to unravel a tangle of issues related to quantum electronics, optics, thermal elasticity and heat transfer, materials science, and many other modern technologies.

It would seem that the first step on this path is obvious: to replace a semitransparent quartz disk on a metal one and couple light out by diffraction through an aperture in the mirror or over its boundary. Metals perfectly reflect laser light, have high thermal conductivity and, therefore, may well remove heat from the region of interaction of the beam with the mirror surface. But pure metals also have disadvantages: (a) high thermal expansion coefficient – they can easily change their size and shape when heated, and (b) relatively low hardness, because of which it is difficult to polish them as well as quartz.

Here it is necessary to say that the modern advances in optical polishing of metals are largely connected with the search for the solution to the problems of laser mirrors. It is funny now to recall an old episode, when physicists first came to opticians with a piece of metal. The request to polish a metal disc in the same room with quartz or calcite was simply wild because metal dust in the workshop, where ultra-precision machining of optical surfaces is performed, was inaccessible. Yet, the way out of this situation was found, i.e., a metallic mirror. And, in fact, after examining all available metals and alloys, scientists managed to increase the optical performance threshold of new mirrors over traditional quartz mirrors by an order of magnitude. But this 'order of magnitude' was not enough. It became clear that the required level of light, and thus of heat loads on the mirror of a high-power laser can be achieved only by using special cooling.

1.6.2.1. Cooling of Mirrors

When cooled with a moving liquid, the heat flow being released is directly proportional to the temperature difference between a hot body and coolant. Thermal power rates measuring kilowatts per square centimeter are not difficult to remove, if the mirror is heated to a temperature of thousands of degrees. But in this case, it is impossible to talk about the high optical quality of the mirror surface. There is a contradiction: convective heat exchange is greater at high temperatures, and for the geometric shape and other characteristics of the optical mirrors to be stable, room temperatures are needed. This contradiction could only be resolved due to more efficient heat removal; however, for room temperatures of optical surfaces of a laser mirror, neither the theory nor the practice was developed at that time.

First, scientists started experimenting with milling deep channels on the back side of a metal disk through which tap water was circulating. They were cut as close to the surface as possible, but this lead to vibration and deformation of the mirror surface due to pressure fluctuations of water. Then, there was an attempt to make the channels shallower and manufacture micro-capillary structures. This approach led to the conclusion that it is better to use specially designed metal capillary-porous structures, similar to a sponge for washing dishes. As a result, so-called felt and powder metal-based structures have been widely adopted. Heat exchange in these structures is very intense, both due to the large surface, and due to the strong mixing of the coolant, which travels in the micro-capillaries. In addition, the matrix with a porous body skeleton functions as a supporting structure: each of its elements is linked with the mirror surface and keeps its original geometry. The coolant in such a structure behaves as a turbulent mountain stream that runs through branched and sinuous gorges. And the matrix of the porous body in this case resembles an openwork design of the bridge with a smooth road surface supported by many piers.

1.6.2.2. Mirror Surface

A thin but very hard layer, which is then polished, must be applied to a highly porous heat exchanger. Only then its surface will serve as a mirror provided that it has a highly reflective coating. The thickness of the coating separating radiation from the coolant should be small, i.e., 100–300μm; otherwise, it will keep the heat absorbed by the reflecting surface of the mirror. Now there are several methods of applying thin layers on a highly porous material, but initially it was quite a challenging task. The problem was solved by using intermetallic compounds, i.e., chemical compounds of metals with each other. Intermetallic coatings are obtained, for example, by deposition from the gas phase. So it is possible not only produce a thin separating layer, but also to restore the mirror if necessary.

Intermetallic coatings have another important property: their structure allows one to manufacture mirror surfaces of high optical quality. If one looks on a polished metal through a microscope, its surface resembles an orange peel: it is covered with tiny hillocks and cavities. To reduce the micro-roughness additional measures are needed, for example, before polishing the metal is doped to make it harder and more fine-grained. But all the same the final size of microroughness is too large: from 0.01 to 0.1µm. In intermetallic coatings the structure is originally very small, about 0.1µm, and after treatment with a diamond cutter or optical polishing very good mirror surfaces with roughness less than a few thousandths of a micron can be manufactured.

Thus, the solution was found. A mirror for high-power CW and pulse-periodic lasers, where the average output power levels are high or very high, is a highly porous heat exchanger and a thin separation layer with a reflective coating. Lasers with such mirrors have been successfully used for welding, cutting and hardening of metals in industry. Besides, they are employed for addressing environmental problems and in military applications.

But it turned out that there is still room for improvement in power optics. By changing the coolant pressure, one can make it boil at room temperature, i.e., in the same conditions under which the final optical machining of the mirror surface is carried out. When the coolant boils, a portion of the absorbed heat is spent on vapor formation; therefore, the heat exchange is tens or hundreds of times higher than in the case of the convective heat exchange. It is only need to remove bubbles from the capillary-porous structure, i.e., the size of the capillaries should be selected accordingly. And then water can be replaced with a liquid metal, for example, the eutectic alloys of sodium, potassium and cesium, which have a low melting point. The heat exchange rate becomes even higher: the heat is now not only removed by the moving fluid, but also is transferred to the coolant itself, which, like all metals, is an excellent heat conductor. Liquid metal coolants enable the removal of thermal powers from a reflecting surface, measuring ten kilowatts per square centimeter of surface.

Using the methods of metallic mirror cooling, one can solve such 'non-mirror' problem as the cooling of large integrated circuits and the development of high-power anodes of X-ray lithography facilities for microelectronics. These devices must withstand the heat load measuring a few tens or hundreds of kilowatts per square centimeter. Works on creation of high-power optics for high-power lasers involve introduction of a wide range of technical, technological and design solutions from other areas of science.

1.6.2.3. Temperature Field of Mirrors

Consider a strongly absorbing elastically isotropic body, which at the initial moment of time is kept at zero temperature in a relaxed state and on whose surface an axisymmetric radiative flux is incident. The energy distribution in the laser beam cross section obeys the Gaussian $I(r) = I_0 \exp(-K_0 r^2)$, where $K_0 = 2/w^2$; and w is the radius of the irradiation region on the body surface. We assume that all the thermal and mechanical coefficients of the material do not depend on temperature, and the laser beam is absorbed directly on the irradiated surface. The nonlinear effects, as well as the energy loss by radiation and convection are neglected. Provided that the characteristic size of the beam is $w \ll L$, and the time of energy input is $t_0 \ll L^2/\alpha$ (L is the characteristic size of the body, and a is the thermal conductivity of the material), the problem can be solved by using the half-space model (Figure 6.1). In this formulation, the problem of determining the temperature field is reduced to the solution of the differential equation of heat conduction

$$\frac{\partial T}{\partial t} = \alpha \nabla^2 T \tag{1}$$

for the following initial and boundary conditions:

$$\frac{\partial T}{\partial z}\Big|_{z=0} = -\frac{I_0(1-\tau)}{K}\exp(-K_0 r^2), \tag{2}$$

$$T(r,z,0) = 0, \lim_{z\to\infty} T\,(r,z,t) < N, \lim_{r\to\infty} T\,(r,z,t) < M. \tag{3}$$

Here, K is the thermal conductivity and # is the reflectivity of the material surface.

By applying to (1) the Hankel transform over r and the Laplace transform over t and using the initial and boundary conditions (3), we obtain [4]

$$T(r,z,t) = \frac{I_0(1-\tau)}{4KK_0}\int_0^\infty \mathcal{J}_0(\xi r)\, e^{-\xi^2/4K_0} \times \left[e^{-\xi z\Phi^*}\left(\frac{z}{2\sqrt{at}} - \sqrt{at\xi}\right) - e^{\xi z\Phi^*}\left(\frac{z}{2\sqrt{at}} + \sqrt{at\xi}\right)\right] d\xi, \tag{4}$$

where $\mathcal{J}_0(\xi r)$ is the zero-order Bessel function and $\Phi^*(x) = 1 - \Phi(x)$; $\Phi(x)$ is the error function.

The behavior of the temperature field on the surface of the body exposed to a laser beam is shown by the example of a quartz sample with a gold coating ($h \approx 10^2$nm). Figure 6.2 shows the surface temperature profiles calculated for several time points. However, if the exposure time satisfies the inequality $w^2/a \ll t_0 \ll L^2/a$, then a stationary temperature fiel can establish. In this case, the expression for the temperature has the form

$$T(r,z,\infty) = \frac{I_0(1-\tau)}{2KK_0}\int_0^\infty \mathcal{J}_0(\xi r)\exp(-\xi^2/4K_0)\exp(-\xi z)d\xi, \tag{5}$$

at the point r=z=0, where the maximum temperature is reached. Then, expression (5) is simplified:

$$T(0,0,\infty) = \frac{\sqrt{\pi}}{2\sqrt{2}}\frac{I_0(1-\tau)w}{K}. \tag{6}$$

For comparison Figures 6.3 and 6.4 show the calculated distribution of the stationary temperature fields in gold-coated aluminum and quartz samples for different values of K_0 at a given power level.

The analysis of expression (5) confirms the localization of the temperature field, the intensity of which decreases sharply with increasing distance from the center of the irradiated region on the surface and deep inside the material. The half-width of the intensity distribution of the laser beam is several times narrower than the half-width of the corresponding temperature profile on the irradiated surface, and the broadening of the latter is determined solely by the thermal conductivity of the material:

$$\Delta r \sim 1/K.$$

If, during the laser exposure the melting point of the material is reached, then the optical surface is irreversibly deformed. By setting $r=z=0$ in (4) and equating the obtained expression for $T(0,0,t)$ to the melting temperature T_m of the material, we can estimate the value of the maximum laser intensity, which during the exposure time t_0 leads to the melting of the surface layer:

$$I_{T=Tmelt} = \frac{2\sqrt{2\pi T_{melt} K}}{(1-\tau)w} f_T(F_0), \tag{7}$$

where $f_T(F_0) = (\pi - 2arctg\frac{1}{2\sqrt{2F_0}})^{-1}$ is a temperature function, which depends on the dimensionless parameter (the Fourier number) $F_0 = \frac{at_0}{w^2}$.

Thus, knowing the value of the temperature function at a given F_0 (Figure 6.5), we can determine the maximum intensity of the laser radiation, which leads to the melting of the optical surface, for any material and any beam geometry at known values of the parameters T_{melt}, K.

From this discussion it follows that for$F_0 < 10^{-2}$ the value of $I_{T=T_{melt}}$increases linearly with decreasing F_0 and remains constant for $F_0 \geq 3 \dots 5$. In particular, for an aluminum sample with a reflectance $\tau = 98{,}2\%$, the maximum intensity, leading to the melting of the surface, is 8.6×10^7W/cm^2 with a pulse duration $t_0=10^{-6}$s and is virtually independent of the parameter w. If the duration is $t_0 = 1$ s, the values of the maximum intensity differ for different values of the parameter $w(I_{T=T_{melt}}, = 1{,}6 \cdot 10^5;\ 2{,}8 \cdot 10^5;\ 6{,}5 \cdot 10^5 W/cm^2 and\ 0.2cm$, respectively $\mathrm{w} = 1; 0{,}5; 0{,}2cm)$, which is due to the mechanism of the material conductivity in the interaction process, resulting in a significant energy exchange between neighboring areas of the irradiated body. Note that this solution is valid for describing the processes preceding the thermal breakdown of the material.

The nonstationary temperature field $T(r,z,t)$ causes a stress in the irradiated body, which also changes with time.

Let us now define thermal stresses in a solid body.

1.6.2.4. Thermoelastic Stresses

In the quasi-static formulation, the problem of determining stresses in the half-space irradiated by an axisymmetric flux leads to the equation [5]

$$\nabla^2 \varphi = mT \ (m = \alpha(1+\nu)/(1-\nu)) \tag{8}$$

with the initial conditions $\varphi|_{t=0} = \frac{\partial \varphi}{\partial t}|_{t=0} = 0$ and boundary conditions at which $\sigma_{zz}(r,0,t) = \sigma_{rz}(r,0,t) = 0$. Here φ is the thermoelastic displacement potential $\left(\frac{\partial \varphi}{\partial i} = V_i\right)$; α is the coefficient of linear expansion, and ν is Poisson's ratio.

The solution of the problem is the sum of two fields of thermoelastic stresses: the fields $\overline{\sigma}_{ij}$, whose components of the stress tensor satisfy thermoelasticity equations, but do not satisfy the boundary conditions, and the fields $\overline{\sigma}_{ij}$, whose stress components characterize the temperature-free solution of the corresponding classical theory of elasticity and satisfy the boundary conditions of the thermoelastic problem. After necessary transformations, we obtain the desired stress field:

$$\sigma_{rr} = C_1 \int_0^\infty e^{-\xi^2/4K_0} \left\{ \mathcal{J}_1(\xi r)\frac{\xi}{r} P - \mathcal{J}_0(\xi r) Q + \left[\mathcal{J}_1(\xi r)\frac{1-\xi z-2\nu}{r} - \mathcal{J}_0(\xi r)\xi(1-\xi z)\right]\xi e^{-\xi z} R \right.$$
$$\left. + \left[\mathcal{J}_0(\xi r)\xi(2-\xi z) - \mathcal{J}_1(\xi r)\frac{2-\xi z-2\nu}{r}\right] e^{-\xi z_t} \right\} \xi d\xi;$$

$$\sigma_{\varphi\varphi} = C_1 \int_0^\infty e^{-\xi^2/4K_0} \left\{ \left[\mathcal{J}_0(\xi r)\frac{1}{r}\right]\xi P - \mathcal{J}_0(\xi r) Q - \left[\mathcal{J}_0(\xi r)2\nu\xi + \mathcal{J}_1(\xi r)\frac{1-\xi z-2\nu}{r}\right]\xi e^{-\xi z} R \right.$$
$$\left. + \left[\mathcal{J}_0(\xi r)2\nu\xi + \mathcal{J}_1(\xi r)\frac{1-\xi z-2\nu}{r}\right] t e^{-\xi z} \right\} \xi d\xi; \tag{9}$$

$$\sigma_{zz} = C_1 \int_0^\infty e^{-\xi^{\frac{2}{4K_0}}} \mathcal{J}_0(\xi r)\left[P - (1+\xi z)e^{-\xi z} R + z t e^{-\xi z}\right]\xi^3 d\xi;$$

$$\sigma_{rz} = C_1 \int_0^\infty e^{-\xi^{\frac{2}{4K_0}}} \mathcal{J}_1(\xi r)\left[e^{-\xi z}(\xi z - 1)t - \xi^2 e^{-\xi z} z R + S\right]\xi^2 d\xi.$$

Here

$C_1 = \frac{\alpha a I_0(1-\tau)G}{KK_0}\frac{1+\nu}{1-\nu}$ (G is the displacement modulus)

$$P = \frac{1}{2\xi}\int_0^t \left[e^{-\xi z}\Phi^*\left(\frac{z-2a\xi t`}{2\sqrt{at`}}\right) - e^{\xi z}\Phi^*\left(\frac{z+2a\xi t`}{2\sqrt{at`}}\right)\right] dt`;$$

$$Q = -\frac{1+\xi z+2at\xi^2}{4a\xi} e^{\xi z}\Phi^*\left(\frac{z+2a\xi t}{2\sqrt{at}}\right) + \frac{1-\xi z+2at\xi^2}{4a\xi} e^{-\xi z}\Phi^* \times \left(\frac{z-2a\xi t}{2\sqrt{at}}\right)$$

$$+\frac{\sqrt{t}}{\sqrt{a\pi}}e^{-\left(at\xi^2+\frac{z^2}{4at}\right)};$$

$$R=\left(\frac{t}{\xi}-\frac{1}{2a\xi^3}\right)\Phi(\xi\sqrt{at})+\frac{\sqrt{t}}{\xi^2\sqrt{a\pi}}e^{-a\xi^2 t};$$

$$S=\frac{z+2at\xi}{4a\xi}e^{\xi z}\Phi^*\left(\frac{z+2at\xi}{2\sqrt{at}}\right)-\frac{z-2at\xi}{4a\xi}e^{-\xi z}\Phi^*\left(\frac{z-2at\xi}{2at}\right).$$

The expressions obtained show the variation of the stress field of the irradiated body during exposure to laser radiation. The maximum values of the radial and circumferential stresses are at a point r=0 on the surface, where they are equal to each other, and can be written in the form:

$$\sigma_{ii}(0,0,t)=C_1\left\{\frac{1+\nu}{3a\sqrt{\pi at}}{}_2F_1\left(\frac{3}{2};2;\frac{5}{2};-\frac{1}{4K_0at}\right)-\frac{\sqrt{t}}{\sqrt{a\pi}}(1+r)\frac{2K_0}{1+4K_0at}\right.$$
$$\left.-\frac{1-\nu}{2a}\left[\sqrt{\pi K_0}-\frac{1}{\sqrt{\pi at^2}}F_1\left(\frac{1}{2};1;\frac{3}{2};-\frac{1}{4K_0at}\right)\right]\right\} \tag{10}$$

[${}_2F_1(\alpha,\beta,\gamma,\delta)$ is the hypergeometric function].

Expression (10) allows one to evaluate the maximum laser intensity $I_{\sigma=\sigma_T}$, leading during the irradiation time t_0 to the achievement of the yield strength of the material, and to determine the maximum permissible duration of the laser pulse, corresponding to the given intensity. It is known that at intensities greater than the limit, plastic deformations arise in the surface layer, and after the termination of the laser pulse inevitable there appear residual stresses. Multiple repetition of the described process leads to a rapid destruction of the optical surface.

Figure 6.6 shows the dependence of the maximum laser intensity on the pulse duration for three values of the parameter K_0 for a polished aluminum sample. Comparing the dependence obtained with the dependence $I_{T=Tmelt}=f(t_0)$, one can see that the maximum intensities, leading to a plastic yield of the material, are by 1.5 ... 2 orders of magnitude lower than the values corresponding to the melting of the material. In the case of CW exposure a stationary stress field can arise, which for the half-space has the form

$$\sigma_{rr}(r,0,\infty)=-\frac{\alpha I_0(1-\tau)E}{2KK_0}\frac{\sqrt{\pi K_0}}{2}{}_1F_1\left(\frac{1}{2};2;-K_0r^2\right);$$

$$\sigma_{\varphi\varphi}(r,0,\infty)=-\frac{\alpha I_0(1-\tau)E}{2KK_0}\left[\sqrt{\pi K_0}\exp\left(-\frac{K_0r^2}{2}\right)\times\sum_{m=0}^{\infty}\frac{(K_0r^2/4)^{2m}}{m!\,\Gamma(m+1)}\right.$$
$$\left.-\frac{r}{2}\sqrt{\pi K_0}\quad {}^1F^1\left(\frac{1}{2};2;-K_0r^2\right)\right] \tag{11}$$

Here ${}_1F_1(\alpha,\beta,\gamma)$ is the degenerate hypergeometric function and E is Young's modulus.

The components of the stress tensor σ_{rr} and $\sigma_{\varphi\varphi}$on the metal surface as a function of distance from the center of the irradiated region are shown in Figure 6.7. The change in the sign of the component $\sigma_{\varphi\varphi}$ at a distance r_0>0 seems interesting. This behavior of the stress field can lead to cracking of the material in the irradiation region and to its subsequent chipping.

Let us analyze now the other two components of the stress field in a solid body, arising during its exposure to a high-power light of finite duration, namely, σ_{zz} and $\sigma_{\gamma z}$. Of most interest is the consideration of the behavior of the component σ_{zz}, because the component $\sigma_{\gamma z}$ due to the axial symmetry of the problem everywhere on the axis is zero and σ_{zz} has a maximum at point z_0>0.

Figure 6.8 shows the dependence of the stress field component σ_{zz} at the axis (r=0) of the aluminum sample during the exposure to laser radiation with intensity $I_0\exp(-K_0r^2)$ on the pulse duration t_0. The feature of the behavior in the component σ_{zz} is the fact that as the exposure time decreases the maximum stress at point z_0 passes through a maximum, and the coordinates decreases monotonically, approaching the surface. By equating the expression for the component σ_{zz} at point z_0 to the ultimate tensile strength of the metal, one can estimate the intensities at which the coating starts sagging during the exposure.

Thus, the destruction of the optical surface by laser radiation can occur at intensities much lower than those required for its melting.

1.6.2.5. Surface Deformation

Let us determine the value of the displacement of the solid-body surface due to the impact of laser radiation. Displacement of the body surface can be represented as the sum of the derivatives in the coordinates of the thermoelastic displacement potential and Love' function. The expression for the normal surface displacement that matches the temperature distribution has the form

$$V(r,0,t) = -(1+\nu)\frac{\alpha a I_0(1-\tau)}{KK_0}\int_0^\infty e^{-\xi^2/4K_0}J_0(\xi r)\times\left\{t+\Phi(\xi\sqrt{at})\left[\frac{1}{2a\xi^2}-t\right]-\frac{\sqrt{t}}{\xi\sqrt{a\pi}}exp(-a\xi^2 t)\right\}\xi d\xi \quad (12)$$

The displacement is negative, because the Z-axis is directed into the solid body. Profiles of the optical surfaces of various materials, deformed by the laser beam, are calculated at t_0=1s, K_0=2 and P=1kW and are shown in Figure 6.9.

The graph shows that the minimum deformation corresponds to a gold-coated quartz sample. This is explained by a slow increase in the material heating region due to the low thermal conductivity of quartz, and by the low value of its thermal expansion coefficient. However, the temperature of the surface of the quartz – gold sample is two orders of magnitude higher than the temperature of the copper sample.

Equally significant is the dependence of the deformation of the irradiated surface on the energy distribution in the laser beam cross section.

Figure 6.10 shows the deformation profiles of the surface of the nickel sample at different K_0.

Consider the dependence of the surface deformation on the time of its irradiation. To do this, if r=0 in (12) and the previously adopted notation is used, we obtain

$$V(0,0,F_0) = -(1+\nu)\frac{\alpha I_0(1-\tau)}{KK_0}\left[4F_0(1-2R) + 0{,}251n\frac{1+2R}{1-2R} - R\right], \quad (13)$$

where $R = \frac{\sqrt{2F_0}}{\sqrt{1+8F_0}}$.

Thus, knowing the necessary parameters (ν, τ, K, K_0, a) one can determine the maximum deformation of the surface in the center of the irradiated region for any material and any exposure durations.

Note the following characteristics of the deformation process resulting from the consideration of the expressions obtained:

1. The process is local, and with decreasing exposure time the half-width of the deformation profile, exceeding the corresponding value of the laser beam for large t_0, approaches the half-width of the energy distribution in the beam cross section, which causes deformation of the irradiated surface.
2. The value of the deformation increases rapidly at small values of the Fourier transform F_0 and the rate of its increase is proportional to the intensity of the incident flux. When $F_0 \geq 3 \ldots 4$, the rate of change in the deformation decreases and the displacement of the surface continues to vary logarithmically slowly. The stage of the elastic behavior of the material is characterized by complete recovery of the initial state of the surface when terminating the laser action. However, in some cases, the deformations of the optical surface become inadmissible. Accounting for deformation is crucial for such laser systems as the quasi-optical transmission lines, telescopes, unstable resonators, and focusing systems.

The maximum of laser radiation can be estimated by equating V in (13) to the V_{ad}, corresponding to the maximum admissible deformation of the surface, at which additional losses exceed some value fixed for each system [6]

$$I_{V=V_{ad}} = \frac{KK_0V_{ad}}{(1+\nu)\alpha(1-\tau)} f_\nu(F_0), \quad (14)$$

where $I_{V=V_{ad}} = \frac{KK_0V_{ad}}{(1+\nu)\alpha(1-\tau)} f_V(F_0)$, is the the displacement function.

Thus, knowing the values of the function $f_V(F_0)$, one can estimate the maximum intensity, resulting in admissible deformations of the optical surface of any material that obeys Hooke's law. The dependence of the displacement on the dimensionless parameter F_0 is shown in Figure 6.11. If $F_0 \approx 1$ the rate of a decrease in the function $f_V(F_0)$ varies significantly, due to the saturation of the process of heat exchange by the material conductivity. The further slow drop in the displacement function is due to the increase in the value of the surface displacement from the effective heating region, the volume of which increases with F_0.

Consider the behavior of the maximum intensities $I_{V=V_{ad}}, I_{\sigma=\sigma T} I_{T=T_{melt}}$ as functions of the laser pulse duration t_0 for an aluminum sample (Figure 6.12), calculated for the beam with $w = 1$cm at $T_{melt} = 660°C$, $\sigma_T = 5 \cdot 10^2$kg/cm^2 (the wavelength of the incident radiation is $\lambda = 10{,}6$μm). For pulse durationst $t_0 \geq 2 \cdot 10^{-3}$ and the given parameters of σ_T and V_{ad}, the main process, which limits the maximum value of the laser intensity, is the elastic deformation of the reflecting surface. With a pulse duration less than this, the transition of the material to the inelastic region becomes crucial, which is due the stresses (caused by irradiation) above its yield strength. It is however important to note that the pulse duration $t_0^* = 2 \cdot 10^{-3}$s for aluminum depends on the parameter K_0 of the incident beam and significantly decreases with its increase. Thus, for example, at $w = 0{,}2$cm the value of the pulse duration t_0^*, for which the surface deformations should be taken into account, is $3 \cdot 10^{-5}$ s. Insignificance of the process of the reflecting surface deformation at low laser pulse durations is explained by the smallness of the effective heating of the material.

However, it should be noted that these estimates are valid for ideal surfaces in a vacuum, because the presence of air and slight contamination of the irradiated surface can lead to the breakdown of the gas near the surface [7] and, consequently, to its destruction at laser intensities that are less than the estimated ones.

1.6.3. Porous Structures in Power Optics

The feasibility of using structures with open porosity for cooling heat-stressed laser mirrors was justified for the first time theoretically and experimentally in [1, 2]. An increase in the optical damage threshold of laser reflectors based on porous structures was provided by a 'minimum' thickness of the separating layer (tens of microns), by the heat transfer growth via an increase in the coolant circulation velocity pumped through the structure and by the effect of the significant development of the surface [3]. The authors of papers [4, 5] have shown experimentally that the placement of special porous structures in made it possible to increase (by ~9 times) the effective coefficient of convective heat exchange, which according to [4] reached ~22W/cm^2 × deg.

The test results of water-cooled laser mirrors that are based on the structures with open porosity first indicated the possibility of removal of high heat flows at low values of the mirror surface deformations. The maximum density of the heat flow being removed, which does not lead to destruction, is $q_{max} = 8kW/cm^2$, and at q=2kW/cm^2 the value of thermal deformation Was ~λ/20, where λ=10,6 μm [3].

A further increase in the optical damage threshold of cooled mirror surfaces can be achieved by optimizing the parameters of the porous structure, the appropriate choice of the coolant and the rational design of its inlet and outlet. The development of such 'optimal' laser reflectors requires a detailed study of heat and mass transfer in porous structures, which are currently insufficiently studied. Let us consider some aspects of this problem.

1.6.3.1. Temperaturnoe Field in the Porous Structure for Convective Cooling

The temperature fields in the porous structures are calculated in the one-dimensional formulation under the following assumptions: the incident radiation is uniformly distributed over the irradiated surface; the thickness of the porous layer Δ is much greater than the depth of heating, which makes it infinitely large ($\Delta \to \infty$) and allows consideration of the half-space model; and the temperature and velocity of the flow through the thickness of the porous layer is constant.

Then the heat transfer equation, which describes the temperature distribution over the thickness of the porous layer, can be written as

$$\frac{d^2t}{dx^2} = \frac{h_s}{\tilde{\lambda}} S_v (t - t_T) \tag{15}$$

where x is the coordinate; t, t_T is the temperature material (skeleton) of the porous structure and the coolant, respectively; h_s is the heat transfer coefficient between the material structure and coolant; $\tilde{\lambda}$ is the skeleton thermal conductivity along the axis X; and Sv is heat transfer surface per unit volume. Equation (15) can be written in the dimensionless form:

$$\frac{d^2\Theta}{d\bar{x}^2} = N(\Theta - 1) \tag{16}$$

where $\Theta = \frac{t}{t_T}$; $\bar{x} = x/d_s$ and $N = \tilde{N}uN`$ are the dimensionless temperature, coordinates and parameters, respectively; d_s is the average particle diameter in the structure; $\tilde{N}u = h_s d_s/\tilde{\lambda}$ is the modified Nusselt number, which characterizes the ratio of convective to conductive heat transfer across the skeleton boundary; and $N` = S_v d_s$ is the dimensionless parameter characterizing the structure.

The boundary conditions of equation (16) can be written in the form:

$$\bar{x} = 0; \frac{d\Theta}{d\bar{x}} = -\tilde{N}u\bar{q}\bar{x} \to \infty;\ \Theta \to 1, \tag{17}$$

where $\bar{q} = q/(h_s t_T)$ is the dimensionless heat flux density; q is the heat flux density transmitted through the separating layer.

Solution to equation (16) with the boundary conditions (17) has the form

$$\Theta(\bar{x}) = 1 + \bar{q}\sqrt{\tilde{N}u/N`}\exp(-\sqrt{N}\bar{x}). \tag{18}$$

It follows from (18) that the rate of temperature decrease over the thickness of the porous structure is determined by the parameter $\sqrt{N}$.

We write down some relations required to calculate the thermal characteristics of laser reflectors. The temperature of the separation layer at the interface with a porous substrate ($\bar{x} = 0$) has the form:

$$\Theta_p = 1 + \bar{q}\sqrt{\tilde{N}u/N`} \quad (19)$$

The maximum heat flux density, removed from the reflector due to convective cooling, follows from the condition of equality of the coolant Θ_p and the boiling point Θ_s at appropriate pressure ($\Theta_p \approx \Theta_s$) and has the form:

$$\bar{q}_{max} = (\Theta_s - 1)\sqrt{N`}/\tilde{N}u \quad (20)$$

The degree of heat transfer in a porous structure as a result of the flow circulation and the surface development is determined by the coefficient K_{in} which characterizes the ratio of the amount of heat removed by the coolant in the structure under consideration, to the amount of heat that would be removed directly from the cooling surface of the separation layer by the coolant when it flows in a Δ channel of depth Δ :

$$K_{in} = q/h_\Delta(t_{T=0} - t_T)$$

where h_Δ is the coefficient of convective heat transfer in the coolant flow in a slot gap of size Δ [20]. For example, for the turbulent regime of the coolant flow the Nusselt number Nu_Δ has the form

$$Nu_\Delta = 0{,}023Re^{0{,}8}Pr^{0{,}4}$$

where Re, Pr are the Reynolds and Prandtl numbers.

Let us introduce the coefficient of heat transfer due to the flow turbulence $K_T = h_s/h_\Delta$, where h_s is calculated from the condition that the flow of coolant circulating through the reflector is constant. Then, K_{in} depending on the structure type and parameters and on the thermophysical properties of the coolant is calculated as follows:

$$K_{in} = K_0K_T \quad (21)$$

where $K_0 = q/h_s(t_{T=0} - t_T) = (N`/Nu)^{1/2}$ is the coefficient of ribbing of the cooled surface of the separation layer.

Equation (18) allows one to find the depth of the structure heating that defines the minimum possible thickness of the porous layer, conventionally assuming that the heating ends at the coordinates $\bar{\Delta} = x/d_s$ where the skeleton and the coolant temperatures differ by 1%:

$$\bar{\Delta} = N^{-1/2}ln10^2\bar{q}(\tilde{N}u/N`)^{1/2} \quad (22)$$

In the case of removal of the maximum heat flux densities, the depth of heating is

$$\bar{\Delta}_{max} = N^{-\frac{1}{2}}ln10^2(\Theta_s - 1) \quad (23)$$

Thus, relations (18) – (23) allow one to calculate the temperature fields and thermal characteristics of the cooled laser reflectors. In combination with the expressions describing the flow hydrodynamics, they are the basis for optimizing the parameters of porous structures, ensuring minimal thermal deformations of mirror surfaces or, if necessary, the maximum heat flux, removed in the case of convective cooling.

1.6.3.2. Convective Heat Transfer in a Porous Structure

The regime of the coolant flow in porous materials, which are of interest for high-power optics, is a transition between laminar and turbulent regimes; in this case, the calculation of the heat transfer coefficient between the skeleton and the coolant is very difficult despite numerous experimental data. The most comprehensive review of experimental studies of heat transfer in porous materials, mainly for gaseous coolants, is given in [7], which in addition to criterion equations for calculating the heat transfer coefficient presents the method of processing the experimental data and the defining criteria.

However, the lack of a unified approach to processing of the experimental results and the inability to accurately determine such important characteristics as the S_V, d_s etc. make it difficult to use the equations [7] for the calculation of heat transfer of liquid drops in various types of porous structures with geometrical parameters that vary widely. Therefore, these equations can only be recommended as a first approximation to calculate the volumetric heat transfer coefficient in structures that are identical to the tested ones, and to gas coolants.

It is generally accepted that the criterion equation of the interporous convective heat transfer for gases and droplets has the form [7]

$$Nu = c(RePr)^n \tag{24}$$

where ***c*** and ***n*** are the constants depending only on the structural characteristics of the porous material.

Based on the experimental data of papers [7–10] we analyzed the dependences of ***c*** and ***n*** on the structural characteristics of porous materials for which the latter are quite authentically known; as a result, we found that ***c*** and ***n*** depend mainly on the bulk porosity Π_V.

Thus, relation (24) for the dimensionless Nusselt number with account for correlation expressions ***c*** and ***n*** allows one to calculate the coefficient of convective heat transfer in a porous material.

1.6.3.3. Hydrodynamics of a Single-Phase Flow in a Porous Structure

The temperature field and thermal deformation of the reflector are largely determined by the discharge of the coolant pumped through a porous layer, which depends on the hydrodynamic characteristics and conditions of the coolant inlet and outlet. Hydrodynamic characteristics of a single-phase fluid flow in porous materials, mainly in the region $\Pi_V \leq 0{,}5$ were studied in many experimental works, whose review is given in [7, 8]. The expressions proposed by different authors can be used to calculate the hydrodynamics of the flow in materials that has an identical structure to the samples studied.

In general, the hydrodynamics of the flow in porous materials is described by the modified Darcy's equation (Dupuit – Reynolds – Forchheimer Eq. [11 – 13]):

$$-\frac{dp}{dx} = \alpha\mu u + \beta\rho u^2 \qquad (25)$$

where p is the flow pressure; u is the filtration rate, equal to the ratio of the specific coolant discharge G to the density ρ; α and β are the viscous and inertial is coefficient of dynamic μ resistance coefficient, respectively and viscosity of the coolant.

From equation (25) we obtain the equation for the coefficient of friction, C_f, in the form

$$C_f = 2/Re + 2 \qquad (26)$$

where $C_f = -2(dp/dx)\rho/G^2\beta$; $Re = G\beta/\mu\alpha$ (the characteristic size β / α).

Practical use of relation (26) for the calculation of the hydrodynamics of the flow in different types of porous materials, whose structural characteristics vary widely, is hampered by lack of information about the coefficients β and α, which are usually determined experimentally.

The authors of paper [14] gave a slightly different approach to the calculation of C_f: as a characteristic size they used $\sqrt{K}$ where K is the permeability coefficient, characterizing the hydrodynamics of the flow according to Darcy's law ($Re_{\sqrt{K}} = G\sqrt{K}/\mu$), then

$$C_f = 2(1/Re_{\sqrt{K}} + C)/C \qquad (27)$$

The relationship between the coefficients β and α parameters c and K can be represented as $\alpha = 1/K$ and $\beta = c/\sqrt{K}$.

As shown in papers [14, 15], the parameter c is a universal constant for identical porous structures. For example, for all the materials made of metal powders with spherical or close-to-spherical particles we have $c \approx 0{,}55$, and for materials made of powders of arbitrary shape we have $0.45<c<0.566$. Thus, in the future when calculating the hydraulic characteristics of the structures we assume c=0.55, although in our case this provides a somewhat higher value of the coefficient of friction.

Research conducted with fibrous materials, foams and structures of the layers of particles of arbitrary shape [14, 15] has shown that the parameter c for these materials is relatively small. This is explained by the fact that c, characterizing the fraction of inertial losses, depends largely on the nature of the wake formed behind solid particles due to the flow, and the wake depends on the type of porous structure.

The presence of the free ends (further turbulizing the flow) in metal-fibrous structures increases **c**; therefore, for a relatively short fibers, l/d = 50 - 100, c=0,132 [14].

The permeability coefficient K, which is a structural characteristic of a porous material, does not depend on the flow regime and is determined experimentally from Darcy's law. In connection with the development of works in the field of heat pipes, many experimental data are currently available to determine K for powder and metal fibrous structures [8, 16].

According to experimental data [17], the relation of the permeability coefficient for metal fibrous structures with bulk porosity has the form:

$$K = A\Pi_V^m \tag{28}$$

where A and m are the coefficients depending on the relative length of the fibers l/d [16]. Similar expressions can be obtained for powder materials. In addition, the permeability coefficient is calculated from the known Carman – Kozeny relation:

$$K = \varphi \Pi_V^3 d_s^2/(1-\Pi_V)^2 \approx \Pi_V^3/5S_V \tag{29}$$

where φ is a constant depending on the structure.

Expressions (27) - (29) were used to determine the hydraulic characteristics of the power optics elements, made of porous materials based on metal powders and metal fibrous structures.

1.6.3.4. Effect of the Coolant Inlet and Outlet Conditions on the Hydraulic Characteristics of the Reflector

Usually, in cooled laser reflectors the coolant is supplied to and removed from the porous structure through evenly distributed alternating channels, which are made in the form of slots or holes on the cooling surface. With a slotted inlet, a more uniform velocity field of the coolant and a minimum differential pressure required for its pumping are provided. In the case of inlets and outlets in the form of alternating holes we may deal with a significant nonuniformity of the velocity field in calculating the flow in radial directions. This leads to additional pressure drops in the circulation of the coolant, which are considered by the coefficient K_r; in this case, the total pressure drop in a porous layer has the form

$$\Delta P = \Delta p K_r \tag{30}$$

where Δp is the pressure in the case of a uniform velocity field.

Under the assumption of uniform injection (outflow) of the coolant in the radial directions in the regions of supply (outlet) restricted by the region $r_0 \le r \le s/2$ (s is a step between alternating holes, r_0 is the radius of the hole), the coefficient K_r, characterizing the influence of collector effects on hydraulic resistance during the motion of the fluid in a porous layer, can be written as:

$$K_r = \frac{F}{\pi s^2 n(1-a)} \frac{cG(1/a-1)/\pi n\rho s\Delta - (\nu/\sqrt{K})\ln a}{\nu/\sqrt{K} + c\nu} \tag{31}$$

Here G is the mass flow rate of the coolant, F is the area of the exposed surface, n is the number of channels for the supply (outlet) of the coolant, and $a = 2r_0/s$ is a relative spacing between the holes. One can see from equation (31) that K_r depends both on the geometric characteristics of the supply and removal of the coolant (on a) and on G; with increasing a

and G, the coefficient $K_{\S}$ increases. Thus, the coefficient K_r is characterized by design excellence of the inlet and outlet system of the mirror coolant.

With the known K_r the total pressure drop in the porous layer is calculated by the formula (30), taking into account the expression for calculating Δp:

$$\Delta p = \nu \rho s(1-a)(v/\sqrt{K} + cv)/\sqrt{K} \tag{32}$$

where $v{=}Gs/\rho F\Delta$ is the coolant filtration rate.

1.6.3.5. Thermal Conductivity of Porous Materials

As for the problems of cooling of power optics elements, of interest is to study the thermal conductivity of a porous material skeleton without account for the thermal conductivity of the coolant. The authors of papers [8, 17] summarized the results of numerous experimental studies of thermal conductivity of materials of various structures and showed that the effective thermal conductivity depends not only on the volume porosity, but also on such factors as the sintering behavior of the material, the size and shape of the primary particles, pressing and sintering technologies, etc., which leads to a large scatter in the data and makes it difficult to generalize the results.

Practical data are summarized in the form of the dependence $\tilde{\lambda}$ (Π_V) for the samples, manufactured using the single technology and the same type of material. Analysis of experimental results shows that for the calculation of a wide range of porous powder materials can be based on the Odolevsky equation [8]:

$$\tilde{\lambda} = \frac{\lambda_K(1-\Pi_V)}{1+\Pi_V} \tag{33}$$

where λ_K is the thermal conductivity of a compact material.

Effective thermal conductivity of metal fibrous felt structures may have a considerable anisotropy depending on the direction of the fibers in the felt. Thus, on the basis of [17] $\tilde{\lambda}$ is generalized by the relations:

$$\tilde{\lambda} = \lambda_K(1-\Pi_V)\exp(-\Pi_V) \tag{34a}$$

$$\tilde{\lambda} = \lambda_K(1-\Pi_V)^2 \tag{34b}$$

where $\tilde{\lambda}$ is the effective thermal conductivity in the direction parallel to (34a) and perpendicular (34b) to the felt-making plane. In the latter case one can also use the expression

$$\tilde{\lambda} = \lambda_K(1-\Pi_V)^3 \tag{35}$$

Expressions (34b) and (35) satisfactorily approximate the experimental data and can be used in the determination of the thermal characteristics of cooled laser mirrors, made of metal fibrous structures. In this case, expression (34b) describes the upper limit of the experimental data (an optimistic estimate), and (35) - the lower (a pessimistic estimate).

1.6.3.6. Estimate of the Thermal Deformation of the Optical Surface

In general, thermal deformations of a mirror are calculated by using the temperature field of the separating layer and the porous base and the known conditions of their attachment. The exact solution of this problem of thermal strength under real conditions of the reflector exposure is very difficult. To estimate the low distortions of the optical surface, which are characteristic of deformation of power-optics elements, one can make the assumption of free expansion of the porous base and the separating layer according to the temperature fields. Then, the thermal deformation of the mirror surface W is the sum of extensions of the separating and porous layers:

$$W = \alpha_p \Delta_p t_T [1/2(\Theta_1 + \Theta_2) - \Theta_0] + \alpha_n \Delta t_T [(1 - \Theta_0) + \bar{q} d_s / (N`\Delta)] \quad (36)$$

where $\Theta_i = t_i / t_T$ is the dimensionless temperature; t_0 is the coolant temperature at the inlet of the reflector; and t_1 and t_2 are the temperatures of the outer and inner surfaces of the separating layer, respectively.

It follows from equation (36) that the higher the coolant flow rate, the lower the W. The maximum deformations of the optical surface are realized in the region of the outflow of the coolant from a porous layer, the heating value of the coolant through the slot channels having the form

$$\Delta t_T = qF / GC_T$$

where G_T is the heat capacity of the coolant.

1.6.3.7. Results of Calculations of Thermal Characteristics of Reflectors

The above-derived expressions describing the processes of heat and mass transfer in porous structures, were used to calculate the characteristics of cooled laser mirrors with the coolant inlet and outlet system in the form of uniformly (with a step s) alternating holes; the distribution of the heat load on the surface of the reflector was assumed uniform.

Figure 6.13 shows the qualitative dependence of the deformation on the maximum density of the heat flow being removed (the constant Π_V at a constant d_s) for two selected regions of the reflector, corresponding to the regions of injection $W_1(q_{max})$ and outflow $W_2(q_{max})$ of the coolant. Varying the average grain size d_s (or the diameter of the fiber), we can construct a family of curves, characterized by a constant value of d_s and variable porosity Π_V, for reflectors with the same type of capillary structure. The curves are plotted at a constant pressure drop and by taking into account the coolant heating in the porous layer; the temperature of the coolant at the inlet temperature was assumed equal to the temperature of the final finishing of the mirror.

The deformation of the optical surface in the coolant outflow region obeys $W_2>W_1$. Crucial to the selection of the structural characteristics of the porous material is curve *2*, and the difference between curves *1* and *2* characterizes the degree of the cooling system perfection. The curves are the envelopes of the working thermal deformation characteristics of a family of reflectors with this type of structure.

Performance characteristic of the reflector with a given porosity structure Π_V is obtained by connecting the straight line from point *C* with the origin of the coordinates. Point *C* corresponds to the maximum heat flow, removed due to convective cooling, and the line segment *OC* is a dependence of thermal deformations of the mirror surface on the heat load.

In general, curve *2* has two main points: point A corresponding to optimal porosity Π_{V1}^{opt} which facilitates removal of the maximum heat flux density for the selected grain size, coolant pressure drop, and coolant supply and removal conditions; and point B (the point of tangency of curve *2* with the straight line from the origin of the coordinates) corresponding to the porosity Π_{V2}^{opt}, for which in the porous structure the optimal thermal distortions of the mirror surface are realized.

The choice of material and the basic parameters of the structure (d_s, Π_V) must be based on a comparison of a family of curves *2* with possibilities of obtaining the desired porous structures and separating layers.

Figures 6.14 and 6.15 present the results of numerical calculations of thermal deformation characteristics of a family of water-cooled reflectors with porous structures made of the widely used copper powders (Figs. 6.14a, b) and molybdenum powders (Figs. 6.14c, d), illustrating the capabilities of the method. The average grain size and bulk porosity varied between $10<d_s<200\mu m$ and $0,1<\Pi_V<0,9$.

One can see from Figure 6.14 that the maximum heat flux densities are as follows: in the region of the coolant inlet q_{dt1}=12,8kW/cm^2, in the region of the coolant outlet q_{dt2}=3,75kW/cm^2, with W_1=0.1–0.4μm, and W_2=0.4–3.5μm. In the region of the maximum heat flux density W_2=1.5–2μm, which is in most cases much higher than the maximum allowable deformation of the reflectors for CO_2 lasers.

With the increase in coolant flow (Figure 6.14b) W_2 can be dramatically reduced by using the materials with high bulk porosity (0.7–0.8) and the transition to a larger grain size. In this case, q_{max}=2kW/cm^2 at W=0.5μm. Powder-material structures with $\Pi_V>0,75$ for the power optics are very difficult to produce. With a decrease in porosity to the real values (~0.6) the heat flux densities being removed will be ~ 1.4kW/cm^2 at a 0.5-μm deformation. Note that increasing the coolant flow rate is also achieved through rational design of the coolant inlet and outlet system, providing a more uniform distribution of the coolant along the cooling surface.

One can see from Figures 6.14c and d that the use of porous molybdenum allows an approximately four-fold decrease in the level of thermal deformations of mirror surfaces in both the first and second regions; in this case q_{dt} somewhat decreases: q_{dt1} = 7,8kW/cm^2, (q_{dt2}=2,8kW/cm^2). For this structure (Figure 6.14c) in the implementation of the maximum removed flows the arising thermal deformations are below the characteristic optical damage thresholds of the reflectors for CO_2 lasers. For example, a structure with Π_V=83% и d_s=20μm facilitates the removal of q_{max}=2.8kW/cm^2 at W=0.45μm. With increasing the coolant flow $W_2 \rightarrow W_1$ (Figure 6.14b), q_{max} decreases. In this region the level of heat flows to be removed can be increased by the use of materials with the larger size of the particles; for example, the molybdenum powder structure with Π_V=83% и d_s=20μm facilitates the removal of q_{max}=2.1kW/cm^2 at W=0.23μm.

Figure 6.15 shows the thermal deformation characteristics of the reflectors with porous structures made of copper (Figures 6.15a, b) and molybdenum (Figures 6.15c, d) felt. One can see from Figures 6.15a and c that the transition to metal fibrous structures leads to a significant (~ 1.7) decrease in q_{dt}, which is associated with a decrease in the effective thermal conductivity of these structures compared to powder structures.

In regions of the coolant outflow (Figures 6.15b and d) the maximum heat flux densities are on the same level as for the mirrors with powder structures; however, an increase in the coolant flow leads to a strong dependence of W on Π_V and to less sensitivity to the diameter of the initial fibers. Deformations corresponding q_{dt} are approximately two times lower compared to the powder structures: for copper and molybdenum reflectors W=1.2 and 0.3μm at q_{dt}=3,9kW/cm^2 and q_{dt}=2.6kW/cm^2, respectively.

For mirrors with a copper fiber structure the heat flux densities, which can be realized with $W=\lambda/20$ (where λ=10.6μm), of 2.4kW/cm^2, are 2.4kW/cm^2, which exceeds the corresponding level of q_{max} for powder materials. Thus, the use of metal fibrous structures for producing power optics elements is very promising; for example, the structure of molybdenum fibers with a diameter of 100–200μm and a porosity of 50%-60% q_{max}=2.2kW/cm^2 at W= 0.15μm. In addition, the possibility of obtaining materials with high Π_V (70%-80%) makes it possible to manufacture optimal reflectors.

1.6.3.8. Comparison of Theoretical and Experimental Results

We have compared the calculation results of thermal mirror-surface deformations of the laser reflector based on a porous copper-powder material with the experimental data [3].

Figure 6.4 presents the results of experimental studies of thermal deformations of a water-cooled mirror as a function of the density of the heat flow removed, and the results of calculation of the maximum thermal deformation characteristics in the region of the coolant removal for a family of reflectors with a 55%-70% porosity. Straight lines are the calculated characteristics of the tested reflector for the maximum porosity of 60% and 62%. It is seen that there is good agreement between theoretical and experimental data.

Figure 6.16b shows the calculation of the thermal deformation characteristic of a family of reflectors in the region of the coolant inlet, which, like the curve in Figure 6.16a, is obtained for the cooling surface temperature of the separating layer equal to 100°C. It follows from Figure 6.16b that q_{max} =8kW/cm^2, in good agreement with the experimental data.

Note in conclusion that this method of predicting the thermal and deformation characteristics of porous laser reflectors offers optimized type and parameters of the structure to ensure the necessary removal of heat fluxes at the admissible values of the distortion of the mirror surface. The results of theoretical and experimental studies suggest that water-cooled laser reflectors with porous structure have a high optical damage threshold. With the accumulation of new experimental data on convective heat transfer and hydrodynamics of flows in porous structures the latter can be quite easily introduced into the calculation algorithm.

1.6.4. PURSUIT OF POWER OF SEMICONDUCTOR LASER STRUCTURES

One of the brightest and most promising implementations of the ideas of cooled power optics is now the introduction of forced heat transfer in high-power semiconductor lasers. Currently, this type of laser has the highest efficiency reaching 80%-90%. Modern manufacturing technologies of semiconductor structures made it possible to significantly increase the laser lifetime, which is tens of thousands of hours of continuous operation. The variation of the semiconductor material composition can change the wavelength range of the radiation from the near-IR to the UV. They are very compact, reliable and easy to operate. The power output can be increased by the simultaneous use of a large number of laser diodes, which are formed in one-dimensional or two-dimensional effectively cooled structures.

Cooled laser diode assemblies possess almost all the remarkable properties of single semiconductor lasers: high-density, high reliability, long lifetime (up to 10^{12}cycles). These lasers have much smaller weight and size dimensions in comparison with other types of lasers, can easily be fed from independent low-voltage power supplies (solar, nuclear energy) without bulky transformers. Equipment based on laser diode assemblies really becomes a reliable high-performance instrument that can be used in industry, medicine, research, and military applications.

The stability of operation of laser diode assemblies and the value of their output are completely determined by the heat transfer efficiency. Laser diode arrays are soldered with a low-temperature solder to the surface of the heat exchanger, which is produced by high-power optics technology. It should be noted that the levels of heat fluxes which are to be removed from the contact region of the array with the heat exchanger have already approached the characteristic values of the power optics. To date, there have been three main areas of use of cooled high-power laser diode assemblies.

First, pumping of solid-state crystal lasers. These lasers, having high-quality output radiation, work as light transformers. The efficiency of such lasers is much higher that the efficiency of solid-state lamp-pumped lasers; it is pumping by semiconductor lasers that makes it maximal and has already exceeded the value of a few tens of percent for kilowatt power levels. Radiation of semiconductor lasers does not deteriorate the gain medium of crystal lasers, thereby extending significantly the lifetime of the entire laser system. Weight and size dimensions of such systems are small, which makes them transportable.

The main purpose is the development and creation of diode arrays with the highest possible power density. The record of 200W from a 1-cm-long diode array has been demonstrated recently thanks, above all, to a highly efficient heat transfer. The maximum output power from one array, according to calculations, can reach 500W. High-power solid-state diode-pumped lasers with an output power of 100kW and higher are being developed in the U.S.

Second, an important task is the coupling of the diode array radiation into the fiber for further transport to the processing facility. Radiation losses in the laser diode assembly – optical fiber system is about 20% and are mainly determined by the same heat transfer, which manifests itself in the form of a smiley, which does not allow one to effectively combine the lasing zones of separate diodes of the array into a minimal spot with the help of cylindrical optics. It is now known about such systems with an output power of a few kW, manufactured commercially in the U.S., Japan and Germany.

Third, and in our opinion, the most promising line of laser diode assemblies is the creation of a phase-locked diode array. The output radiation of such a laser is a set of narrowly focused interference peaks (in particular one peak) with a small divergence. Based on theoretical estimates, the output power of such systems can be as high of the power of a conventional out-of-phase diode array. Therefore, this laser is a light source with a high efficiency inherent in semiconductor lasers and with low divergence characteristic of solid-state crystal lasers. By analogy with the technique of phase-locked antennas in the radio range, the radiation divergence in the case of phase-locked array diode arrays can be increased to 10^{-6}rad without expensive telescopic systems, which provides ultra-long communication and transmission of energy over long distances.

The General Physics Institute has already shown the possibility of scaling the process of phase-locking of the arrays and bars of laser diodes. Besides, 500-W laser diodes have been produced and phase-locking of laser diode structures at high power levels is being investigated. Thus, water-cooled high-power laser diode bars and arrays are promising sources of radiation.

Among the effective implementations are the diamond turning and creation of unique machining tools ensuring high accuracy and high purity of finishing. They are currently widely used to manufacture memory disks, copying machine drums, etc. From 'non-mirror' practice, of significance are the technology of the production of a wide range of capillary-porous structures based on a variety of metals and composites, the technology of deposition of resistant and hard metallic and intermetallic coatings, etc.

1.6.5. Optics Based on Silicon Carbide

Currently, the development of high-power optics stimulates three trends of efficient use of its technical and technological solutions:

1. Lightweight large ground- and space-based telescopes for studying the universe;
2. Astronomical optical remote sensing instruments for small satellites;
3. Mirrors for high-power lasers for special purposes.

All the three trends are based on cutting-edge technology for the conservation and development of scientific and industrial potential of the country and its defense industry. The main element of the basic objects of all the three trends is large lightweight mirrors with a diameter of up to 1000mm. The choice of mirror material is a key issue in production of a new generation of optical objects. Comparative evaluation of materials with the help of optical quality criteria developed by us in the early 1970s showed that for a wide range of mirror applications (including heavy weight ones) silicon carbide has a distinct advantage over traditional materials, such as aluminum, glass, pyroceram, beryllium, copper, etc. This conclusion is consistent with more recent conclusions of foreign experts, who are conducting research in this area. As an example, we can single out the Advanced Mirror System Demonstrator (AMSD) program of NASA and the U.S. Department of Defense, within the framework of which large-scale works were carried out to choose the necessary material and develop basic technologies for the production of power and large cooled lightweight mirrors.

The program provided the basis for large-scale studies of this problem in Germany, France, Japan, and the USA. Today the technology of silicon carbide is successfully being developed in China. Based on the studies performed, it was concluded that silicon carbide has an indisputable advantages when it comes to production of large, lightweight and cooled mirrors for space-based telescopes. This conclusion is implemented by industrialized countries in dozens of new generation telescopes. Thus, a bulky silicon carbide mirror with a diameter of more than 500mm has a weight that is 5-7 times less than the weight of a similar mirror made of glassceramic, the best quality in terms of surface scattering, high thermal stability and a minimum time constant when changing the external temperature. Obviously, the transition to a new generation of space-based telescopes, high-power large mirrors of high-power lasers is not possible without the introduction of silicon carbide in practice.

1.6.6. Conclusion

Thus, we have conducted a series of studies and developed the technology of manufacturing highly reflective multilayer mirrors. In the experiments, it was found that at a level of rear mirror deformations corresponding to the optical power of $1m^{-1}$, the combined effects of mechanical stresses and exposure to laser radiation lead to failure. To reduce the stress, a multilayer combined metallic dielectric coating consisting of 3-5 layers was developed (in a traditional coating there are 20-25 layers). To ensure the radiation resistance of coatings use was made of the technology of ion bombardment and implantation, developed in St. Petersburg and Podolsk. The coatings were tested experimentally in lasers with an output power up to 1.5kW. These tests showed high radiation resistance in the high-frequency pulse-periodic regime. Ion treatment significantly improves the coating deposition technology and the optical and operational characteristics, such as radiation resistance, scattering and absorption coefficients. The most important areas of ion treatment when applying optical coatings are as follows:

- ion polishing of wafers in vacuum chamber for coating deposition;
- ion polishing of coatings (ion treatment of the produced film surface);
- ion sub-polishing of coatings (layer-to-layer ion polishing of the film during its deposition);
- ion bombardment of the growing film during its deposition;
- deposition of coatings by ion beam sputtering;
- deposition of coatings by ion beam sputtering with ion bombardment of the growing film;
- deposition of coatings from an ion beam incident on the substrate.

Ion treatment of optical materials for production of mirrors for high-power lasers is an effective tool for advanced optical technology. Ion treatment has already made it possible to elaborate a number of innovative technological processes that can significantly enhance the precision and performance capabilities of high-power laser industry, which is especially important for mirrors of high-frequency pulse-periodic lasers. The value of the scattering coefficient of 0.1–0.2-μm-thick Si films is 0.015%–0.02%. Microphotographs show that the surface of conventional films is significantly non-uniform: against the general background of particles measuring 7–10nm there is a significant amount of clusters measuring 25–60nm. After ion polishing the surface of films becomes uniform, the crystallite size being 5 to 7nm. The coefficient of light scattering after ion polishing with the help of oxygen ions with an energy of 3keV for 5min decreased to 0.005%–0.008%. Radiation resistance of optical coatings of the mirrors increases significantly because due to the appropriate choice of the regimes of ion bombardment, the damage threshold of optical coatings increases by 65%, and the lifetime of the mirror increases by an order of magnitude and reaches the desired value of 3000h, which corresponds to today's serviced life of laser diodes used to pump the gain medium. These results are of practical interest in designing optical interference systems with small loss of light, elements of integrated optics and mirrors of high-power high-frequency pulse-periodic laser systems.

The developed coatings made it possible to use membrane mirrors in solid-state laser resonators. With these mirrors the feasibility of *Q*-switching with a deformable rear mirror has been experimentally.

1.6.7. FIGURES

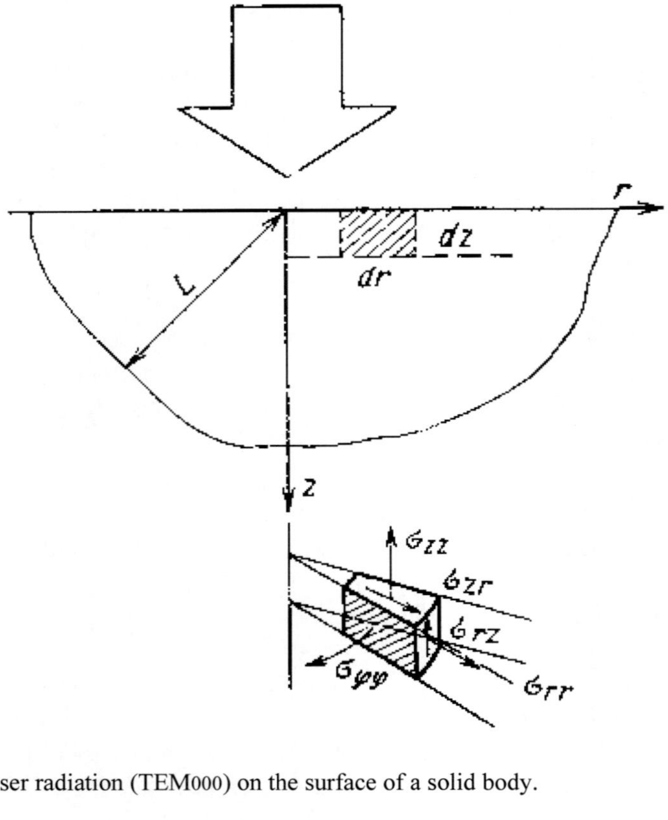

Figure 6.1. CO_2-laser radiation (TEM_{000}) on the surface of a solid body.

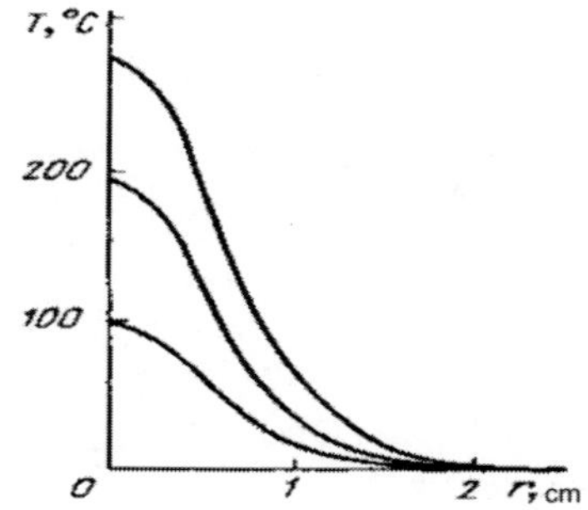

Figure 6.2. Time dependence of the temperature profile of the surface of the gold-coated quartz sample ($\tau = 98.6\%$; P=1kW, $t_0 = 1$cm); t0=2, 10 and 30s.

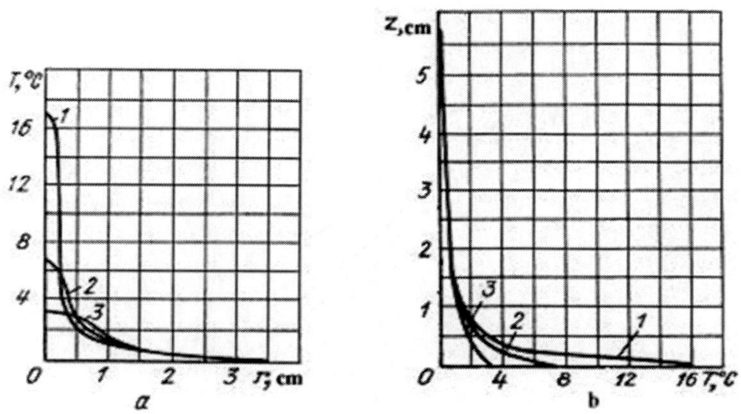

Figure 6.3. (a) Calculated radial temperature field distribution and (b) calculated temperature field distribution along the coordinate z in an aluminum sample [$\tau = 98.2\%$; P=1kW, $w =$ (1) 0.2, 0.5 (2) and (3) 1cm].

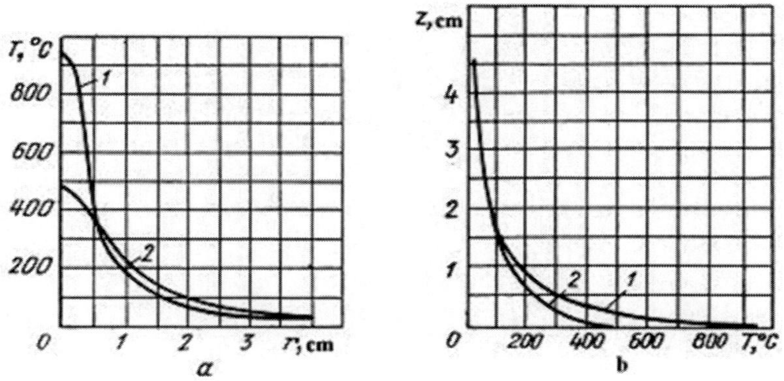

Figure 6.4. (a) Calculated radial temperature field distribution and (b) calculated temperature field distribution along the coordinate z in a gold-coated quartz sample [$\tau = 98.6\%$; P =1kW, $w =$ (1) 0.5 and (2) 1cm].

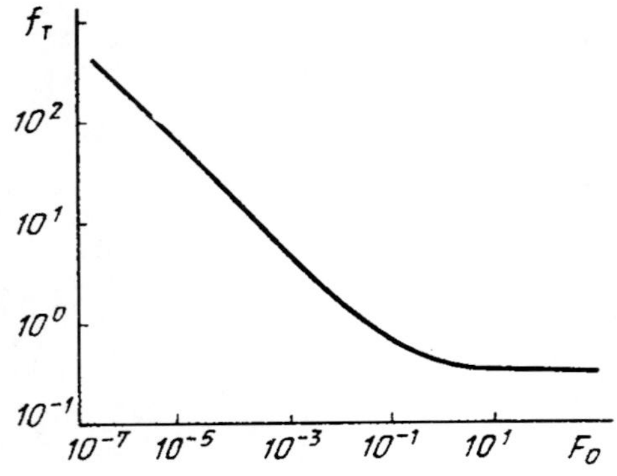

Figure 6.5. Dependence of the temperature function f_T on the parameter F_0.

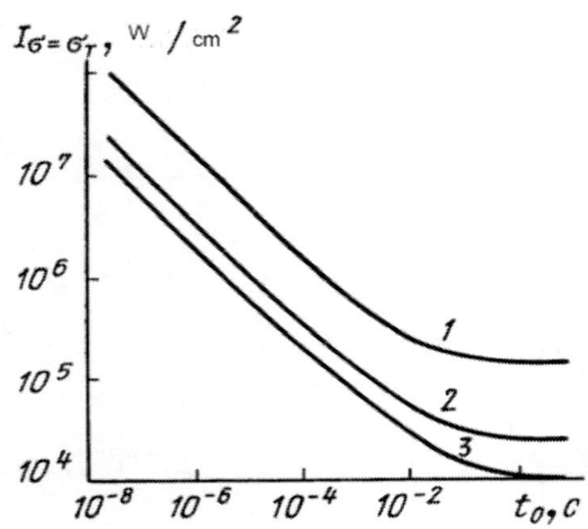

Figure 6.6. Dependences of the limiting radiation intensity on the pulse duration at K_0=(1) 50, (2) 8, and (3) 2 for an aluminum sample.

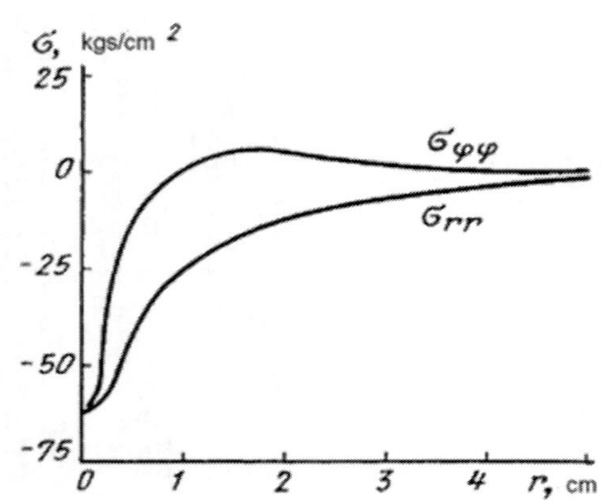

Figure 6.7. Stress field at the aluminum surface ($\tau = 98.2\%$; P=1kW, $w = 0.5$cm).

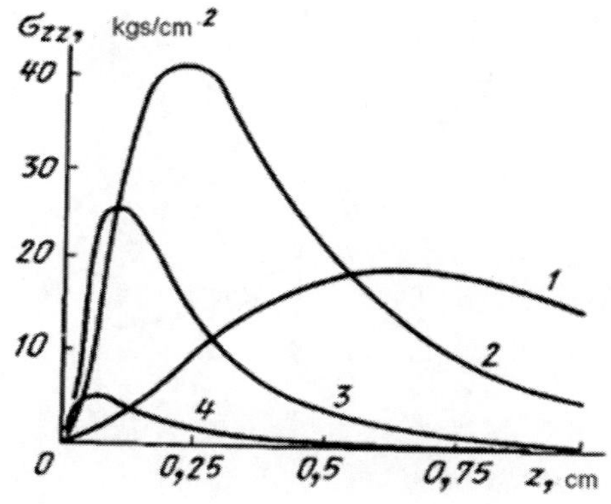

Figure 6.8. Dependence of the axial stress for the z coordinate for an aluminum sample at the laser pulse duration t_0=(1) 0.1, (2) 0.01, (3) 0.001, and (4) 0.0001 s (P=1kW; K_0=50).

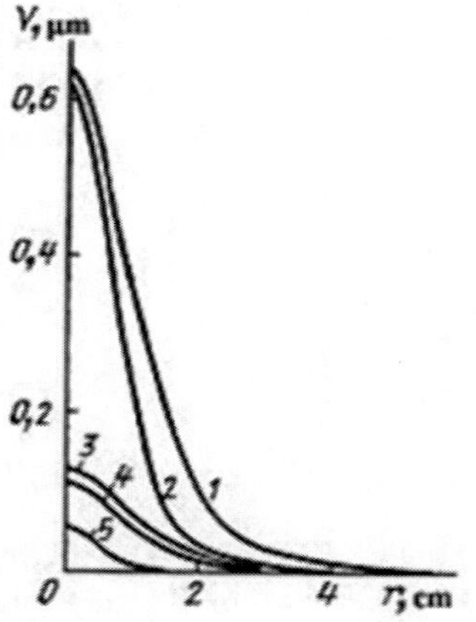

Figure 6.9. Deformation profiles of surfaces of various materials: 1 – aluminum, 2 – nickel, 3 – copper, 4 – tungsten, 5 – gold-coated quartz.

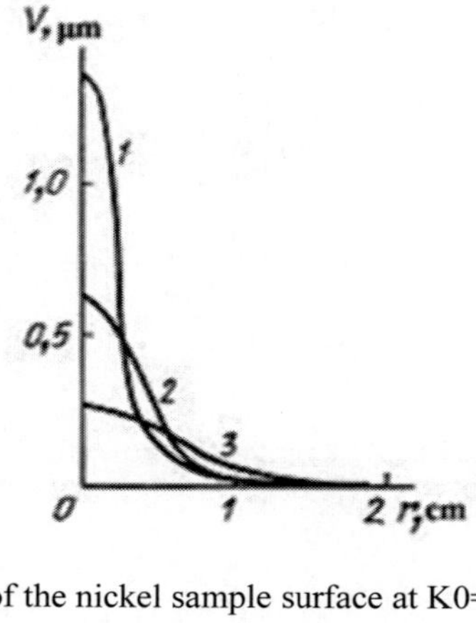

Figure 6.10. Deformation profiles of the nickel sample surface at K0=(1) 50, (2) 8, and (3) 2 at P=1kW, t0=0.25s.

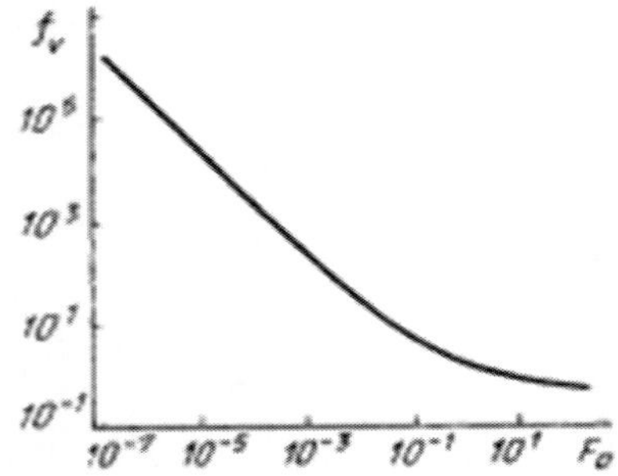

Figure 6.11. Dependence of the shift function f_v on the parameter F_0.

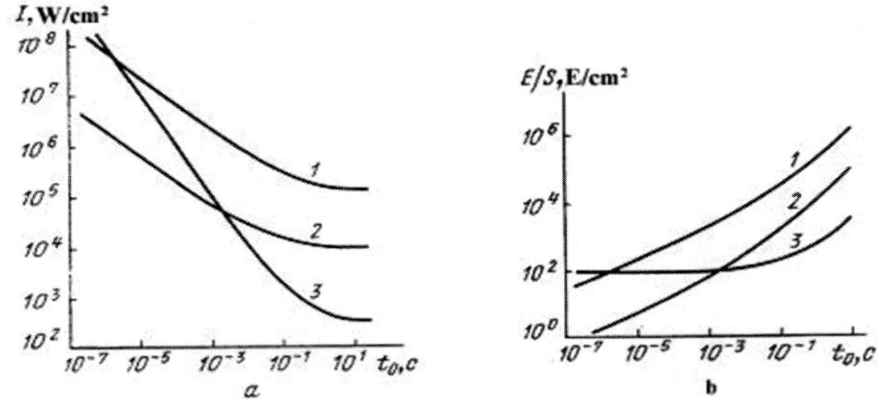

Figure 6.12. Dependences of (a) the limiting intensity and (b) the energy density of laser radiation on the light pulse duration at $K_0=2$ for the three stages of thermal treatment: (1) $T=T_m$, (2) $\sigma=\sigma_т$, and (3) $V_{доп}=\lambda/20$.

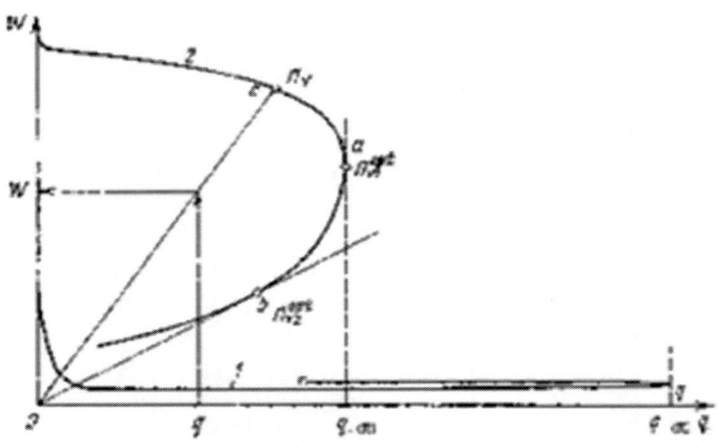

Figure 6.13. Qualitative dependence of the thermal deformation on the maximum heat flux density removed from two regions of the reflector, corresponding to the region of the coolant injection (*1*) and outflow (*2*).

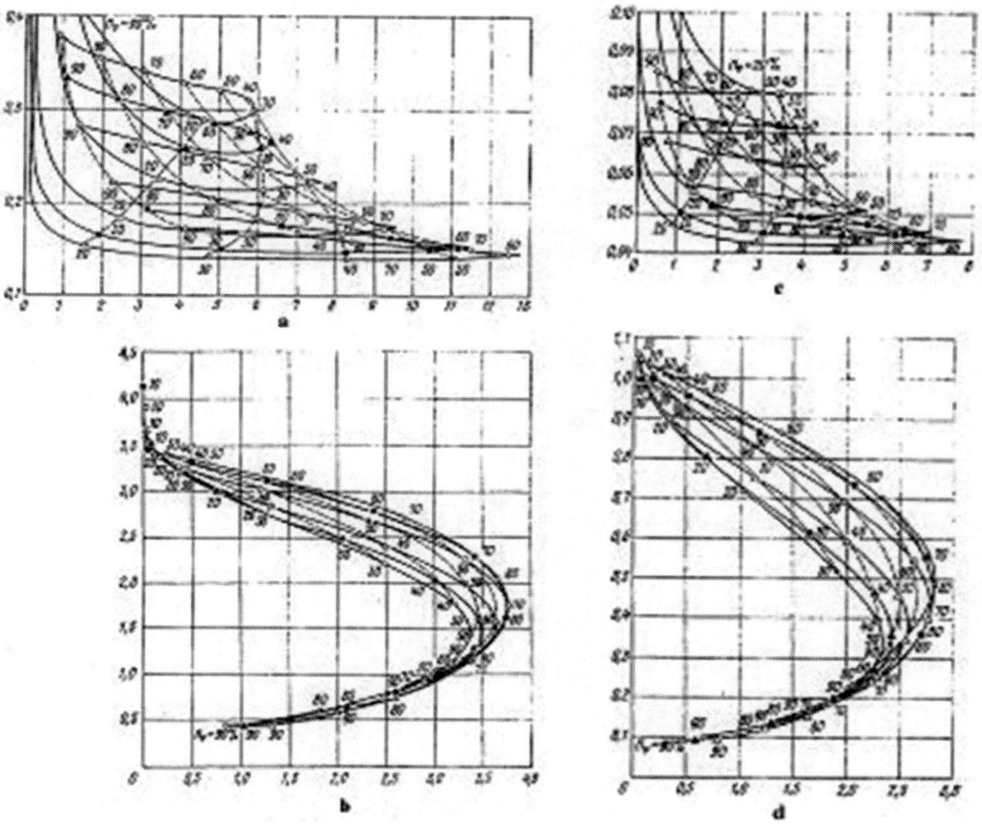

Figure 6.14. Nomograms of thermal deformation characteristics of a family of water-cooled reflectors made of copper (a, b) and molybdenum (c, d) powders for the injection region (a, c) and the outflow region (b, d); points in this and other figures correspond to the following particle diameters or felt : ○−20, ■−30, □−50, Δ−150 и ○−200μm.

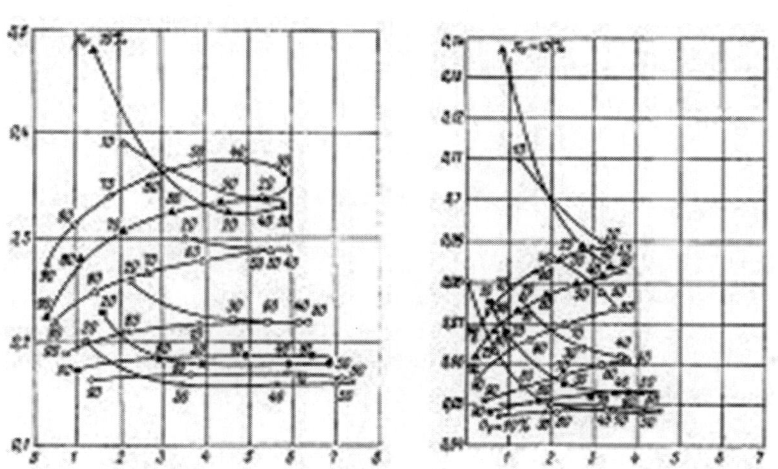

Figure 6.15. (Continued)

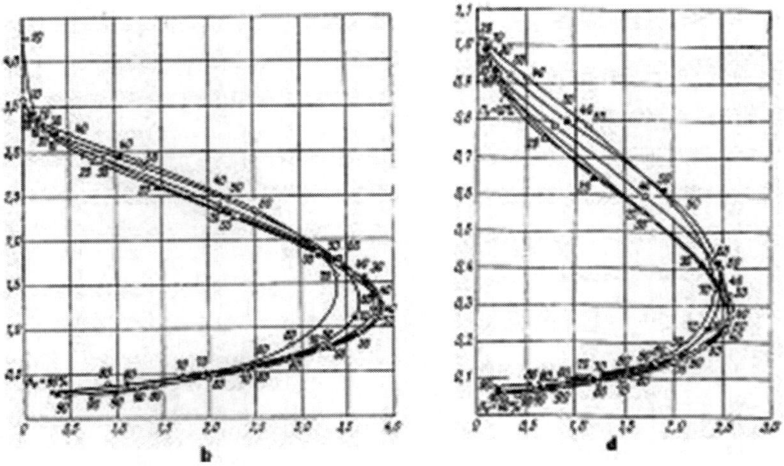

Figure 6.15. Nomograms of thermal deformation characteristics of a family of water-cooled reflectors based on porous structures made of copper (a, b) and molybdenum (c, d) felt: the region of the coolant injection (a, c) and region of the coolant outflow (b, d).

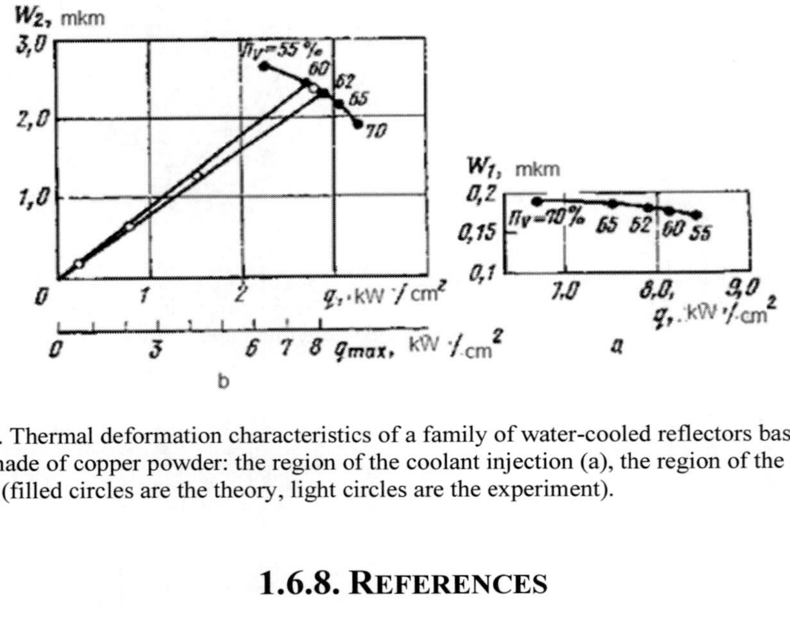

Figure 6.16. Thermal deformation characteristics of a family of water-cooled reflectors based on porous structures made of copper powder: the region of the coolant injection (a), the region of the coolant outflow (b) (filled circles are the theory, light circles are the experiment).

1.6.8. References

[1] Apollonov V.V., Barchukov A.I., Prokhorov A.M. *Proc. IV All-USSR Meeting On Nonresonance Interaction of Optical Radiation with Matter*. – L.: GOI, 1978, p. 31.

[2] Apollonov V.V., Barchukov A.I., Borodin V.I., Bystrov P.I., Goncharov V.F., Ostrovskaya L.M., Prokhorov A.M., Rodin V.N., Trushin E.V., Tsypin M.I., Shevakin Yu.F., Shur Ya. Sh. Sov. *J. Quantum Electron*., 8, 672 (1978).

[3] Apollonov V.V., Barchukov A.I., Borodin V.I., Bystrov P.I., Goncharov V.F., Ostanin V.V., Ostrovskaya L.M., Prokhorov A.M., Rodin V.N., Trushin E.V., Tsypin M.I., Shevakin Yu.F., Shur Ya. Sh. *Pis'ma Zh. Tekh. Fiz*., 4, 1193 (1978).

[4] Megerlin F.E., Murphy R.V., Berles A.E. Teploperedacha, Ser. C, No. 2, 30 (1974).

[5] Polyaev V.M., Morozova L.L., Kharybin E.V., Avraamov N.I. Izv. Vyssh. *Uchebn. Zaved., Ser. Mashinostroenie,* № 2, 86 (1976).

[6] Isachenko V.P., Osipova V.A., Sukomel A. Teploperedacha (Heat Transfer) [In Russian]. – *Moscow: Energiya*, 1969.

[7] Mayorov V.A. *Teploenergetika*, № 1, 64 (1978).

[8] Belov S.V. *Poristye metally v mashinostroenii* (Porous Metals in Engineering) [In Russian]. – Moscow: Mashinostroenie, 1976.

[9] Kays W.M., London A.L. *Compact Heat Exchangers*. – New York: McGraw-Hill, 1984.

[10] Lykov A.V. *Teplomassoobmen* (Heat and Mass Exchange) [In Russian]. – Moscow: Energiya, 1978.

[11] Duruit J. *Etudes theoriques et pratiques sur le mouvement des eaux*. Paris, 1863.

[12] Reynolds O. *Papers on Mechanical and Physical Subjects*. Cambridge University Press, 1900.

[13] Forchheimer P. *Wasserbewegung durch Boden*. Z. Vereines deutscher Ingenieure, 45, (1901).

[14] Beavers G.S., Sparrow E.M. *Journal of Applied Mechanics*, 36, 711 (1969).

[15] Beavers G.S., Sparrow E.M., Rodenz D.E. *Journal of Applied Mechanics*, 40, 655 (1973).

[16] Semena M.G., Kostornov A.G., Gershuni A.N., Moroz A.L., Shevchuk M.S. *Teplofiz. Vysok. Temp.,* 13, 162 (1975).

[17] Semena M.G., Kostornov A.G., Zaripov V.K., Moroz A.L., Shevchuk M.S. *Journal of Engineering Physics*, 31, 1121 (1976).

[18] Witte H.H. *Opt. Laser. Technol.*, 3, 31 (1971).

[19] Apollonov V.V., Barchukov A.I., Konyukhov V.K., Prokhorov A.M. *Pis'ma Zh. Eksp. Teor. Fiz.,* 15, 248 (1972).

[20] Apollonov V.V., Barchukov A.I., Prokhorov A.M. IEEE J. Quantum Electronics, QE-10, 505 (1974).

[21] Apollonov V.V., Shefter E.M. *Preprint FIAN*, No. 105 (1974).

[22] Melan E., Parkus G. Wärmespannungen *infolge* stationärer Temperaturfelder. – Wien: Springer, 1953.

[23] Barchukov A.I., Konev Yu.B. *Preprint FIAN*, No. 158 (1973).

[24] Barchukov A.I., Bunkin F.V., Konov V.I., Lyubin A.A. Zh. *Eksp. Teor. Fiz.*, 66, 965 (1974).

PART 2. MODELS OF PHYSICAL PROCESSES OF LASER PLASMA FORMATION

Chapter 7

2.1. IONIZATION

Many theoretical and experimental works are devoted to physical models of laser plasma formation and heating [1−7]. All the theoretical calculations are mainly based on the assumption about the existence of a thermodynamic equilibrium of any type in a plasma torch. The solutions of such problems are represented as dependences of the average ion charge on the radiation flux density, electron temperature and particle density during the action of a laser pulse. As a result of the calculations, the charge distribution is given in the form of narrow peaks, charge value of which is a function of the electrons temperature and radiation parameters.

Experimental measurements of the electrons temperature at an early stage of the torch development (during the time interval of the radiation pulse action) indicate the essential differences of the velocity distribution of electrons from the Maxwell function. In addition, the data on the charge, velocity and energy characteristics of the ions in the laser plasma at the stage of their expansion do not coincide with the results of theoretical calculations based on the assumption about the existence of a thermo-dynamic balance in the plasma.

Detail characteristics of the laser plasma expansion products can be obtained only at large distances from the place of plasma formation, in a zone, where the ionization processes have been already terminated.

In this chapter we have made an attempt to describe the ionization processes proceeding in the hot part of the torch. In this case we rely on the information about the laser plasma characteristics obtained at the stage of the laser plasma expansion.

After the laser radiation intensity reaches a particular value for the given sample and takes a rise above it, the material surface is evaporated. The values of such boundary values for different materials are listed in the Table 2.1. The laser radiation intensities marking the beginning of the sample surface destruction are several times lower of the similar values measured earlier with the ionic plasma component taken into account, which indicates the existence of the evaporation stage and the presence of uncharged particles in the course of plasma development at the initial stage.

The energy distributions of atoms obtained during the evaporation of carbon, aluminum, cobalt, cadmium and plumb samples at the boundary values of the flux density are represented in Figure 2.1.

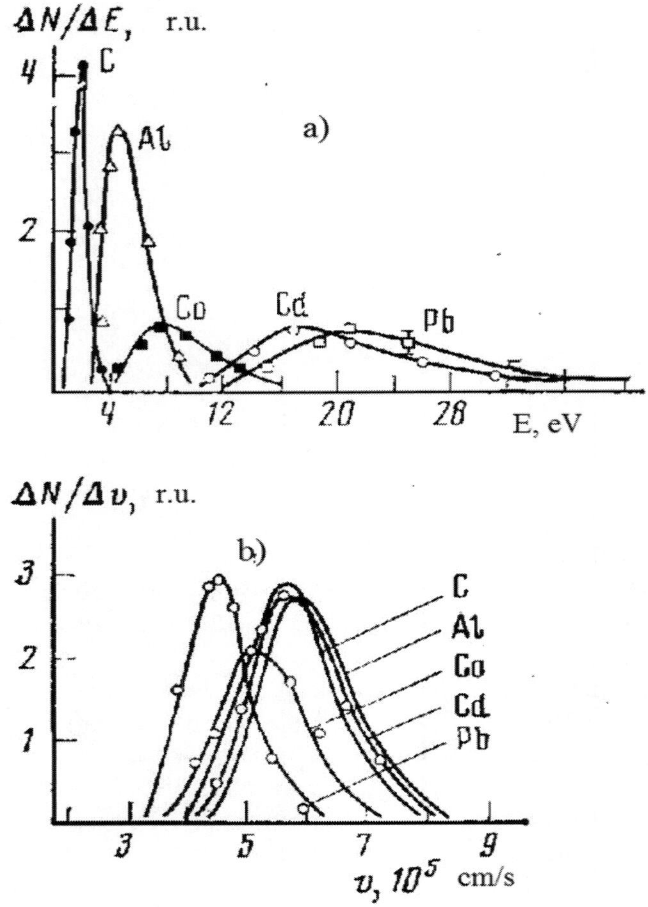

Figure 2.1. The energy distributions of atoms obtained during the evaporation of carbon, aluminum, cobalt, cadmium and plumb samples at the boundary values of the flux density.

The shape of the distributions indicates that emergence of such atom fluxes is explained by thermal evaporation. To evaluate the possible temperature, energy and velocity of the atoms, we studied the shape of the velocity distributions of the particles by comparing the experimentally obtained atomic spectra with the Maxwell distribution function.

The results of the comparison of the Maxwell distribution function (light-colored circles) with the experimental data are presented in Figure 2.1 and shows satisfactory coincidence of these functions, which undoubtedly confirms the thermal nature of the atom formation. The experimentally obtained energy and velocity parameters of the atom fluxes and the parameters evaluated using the Maxwell function at a threshold intensity of the radiation are listed in Table 2.1.

With a further increase in the radiation flux density on the sample, the atom distributions of all the elements exhibit the atom fluxes that differ in velocities and expansion energy. Thus, the distribution smooth shape is destroyed and the second intensity maximum emerges. The values of the 'second boundary' radiation intensities are represented in the Table 2.2.

Table 2.1. The experimentally obtained energy and velocity parameters of the atom fluxes and the parameters evaluated using the Maxwell function at a threshold intensity of the radiation

Target	q_1, 10^7 W/cm²	N_a, 10^{12}	E_m^{exp}, eV	E_0^t, eV	v_m^{exp}, 10^5 cm/s	v_0^t, 10^5 cm/s	u_m, 10^5 cm/s	kT_a,
C	8	3,2	2	0,3	5.8	2,4	4.3	0,25
AI	10	1,6	5	0.7	5.8	2.3	4.8	0,68
Co	8	1,6	8	1,1	5,0	1,9	4.8	1,5
Cd	2	2,1	18	2,8	5,9	2.2	4.5	2,5
Pb	1	3,3	21	3,8	4,5	1,9	3.5	2,9

Table 2.2. The values of the 'second boundary' radiation intensities

Target	q_2, W/cm²	I, eV	E +I, eV
Pb	I·10⁸	7,4	9
Al	I,5·10⁸	5,9	9
Cu	2·10⁸	7,7	II
Co	2·10⁸	7,9	I2
C	5·10⁸	II,3	I4,9

The emergence of the high-energy particles (*E*>40eV) in the beam of neutral atoms is due to the process of the ion recombination in neutral atoms, which undoubtedly follows the ionization process. This fact points to the presence of the ionization process in the torch and the laser plasma formation, i.e., these values of the flux densities (the emergence of the high-energy atoms) indicate the initial moment of the ionization processes. Therefore, we have established experimentally that after reaching certain (threshold) radiation energy on a given sample, the evaporation process starts and afterwards as the radiation flux density rises up to the 'second boundary' value, the ionization process occurs in the torch, which indicates the moment of the laser plasma formation.

2.1.1. Influence of the Radiation Flux Density on the Ion Formation Processes

The most essential parameter affecting the plasma torch heating process is undoubtedly the value of the flux density on a sample or in plasma. If we consider a limited enough range of laser pulse durations (from 10^{-8} to 10^{-9} s), it is possible to make use of the radiation energy as well. With the radiation flux densities from 10^7 to 10^8 W/cm², there are few charged particles in the torch formed, whereas neutral atoms are effectively emitted from a solid surface. Starting with some value of the radiation density in the range from 10^8 to 10^9W/cm^2 (for a certain sample the exact threshold value for the ion emergence is dependent on the ionization potential, evaporated material mass and thermal and physical characteristics of the sample), the singly charged ions are registered in the plasma torch. With a further increase in the laser energy, the number of singly charged ions drastically increases. And only after reaching the next 'threshold' value of the radiation flux density in the torch, the doubly charged ions are registered. A gradual increase in the radiation flux density will lead to an increase in the number of both singly charged and doubly charged ions; however, the number of ions with the charge Z=2 increases faster. After reaching the next 'threshold' density, the ions with the charge Z=3 will be registered. This is a common property, which is observed during the ion emergence with the different charges Z for all elements in the radiation flux density range from 10^8 to 10^9W/cm^2, which is presented quantitatively in Figure 2.2.

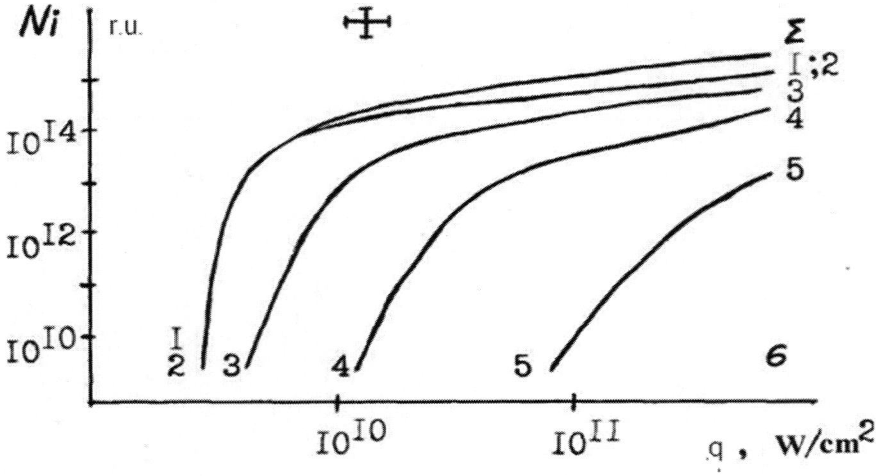

Figure 2.2. Influence of the radiation flux density on the carbon ion formation: 1, 2, 3, 4, 5, 6 are the ion charge values and Ni is the total number of charged particles.

The dependence of the number of ions on the radiation flux density is identical for ions of all charges Z and consists of two parts: an initial drastic increase in the number of particles in the torch and a smoother and lower growth of their number after reaching some value of the radiation flux density. The region of the kink in the curve showing the dependence of the number of ions on q corresponds to the 'threshold' flux density leading to the appearance of ions with a higher charge. The fall in the growth rate of the particles with increasing ion charge is explained by the fact that to form ions with a greater charge (Z) requires more radiation energy for every particle in connection with an increase in the ionization potential for the ions with a higher charge. The presence of the kink in the dependence $N_Z=f(q)$ is also related to the radiation energy redistribution and the transition of the particles into the ions having a higher charge. Because the radiation flux density on the sample is one of the main parameters affecting the ionization and plasma heating, we represent the dependence of the emergence of the ions with Z_{max} in the torch expansion products as a function of the radiation flux density. As follows from the Figure 2.2, the ion emergence with a particular charge has a threshold nature and depends on q, which can be correlated well with the ionization potential dependence for ions of a particular charge.

Figure2.3 presents the dependences of Z_{max} on the radiation flux density at the sample surface and a scale of changes in the ionization potentials for these elements. An overlap of these dependences allows a conclusion that the emergence of ions of a particular charge is directly related with the emergence of the electrons with the energy higher than the ionization potential $E_e > I(z)$ or equal to it in the plasma. It follows from the experiments that the dependence of the appearance of ions with a maximal charge on the radiation flux density can be written as

$$Z_{max} = k \lg q \tag{2.1}$$

This relationship is important for solving practical problems of producing ion beams of a given charge, because it is possible to calculate and select in advance a suitable type of a laser device.

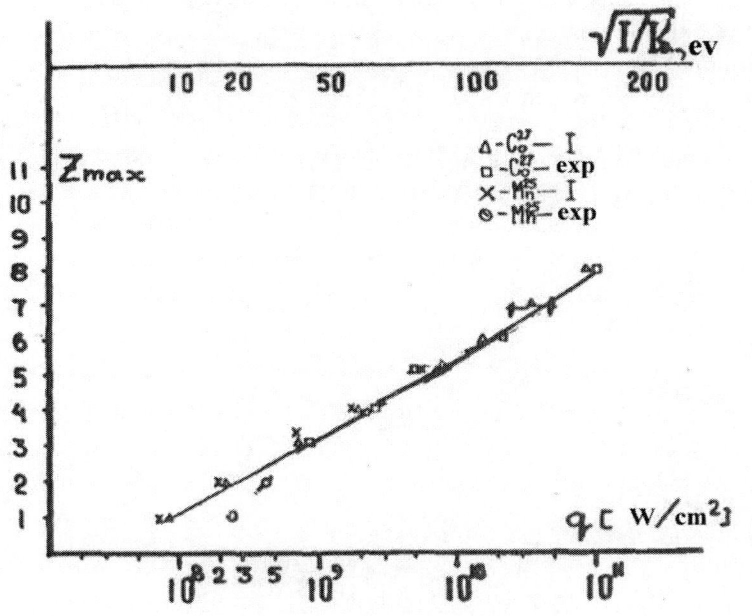

Figure 2.3. Comparison of the dependences of the maximal charge of cobalt and manganese ions on the radiation flux density with different ionization potentials: ***I*** stands for the ionization potentials, and *EXP* stands for the experimental values.

2.1.2. Charge Distribution of Particles in the Plasma Torch

An important characteristic of a plasmoid is the shape of the charge distribution of ions. Depending on the type of such a distribution, we may speak of the presence or the absence of the ionization equilibrium in the plasma.

Thus, for instance, for the long-living or stationary plasmas, the charge distributions can be calculated on the basis of such models as local thermodynamic equilibrium or corona equilibrium, whereas for the short-living plasmas, the density, electron energy and the torch lifetime come to the fore in determining the ion charge distributions. The number of ions of some charge Z can be found using the formula:

$$N_Z = \int_0^{\tau_e} n_e n_0 < \sigma_Z v_Z > dt \tag{2.2}$$

where *no* is the density of neutral atoms.

For the laser plasma, typical are the conditions when at high densities of neutral atoms, ions and electrons, in the torch there appear a remarkable number of high-energy electrons. In this case, the time needed for the ionization process to proceed till the emergence of high-charged ions will be small and equal to 10^{-8} ~10^{-7}s.

Probably, this is one of the key moments leading to the formation of the beams of the ions of different elements with very high charge values, in particular, cobalt ions with the charge Z=25 [8], aluminum nuclei with Z=13 and carbon with Z=6 [9] in the laser plasma.

As an example Figure 2.4 presents the charge distribution of cobalt ions (typical of laser plasma) at the radiation flux density of 10^{13}W/cm^2. With increasing ion charge Z, the number of particles of the given charge monotonically decreases in the plasma. For comparison, we laid on this distribution the dependence of the ionization cross sections on Z of the electron collisions with an electron energy of 10keV.

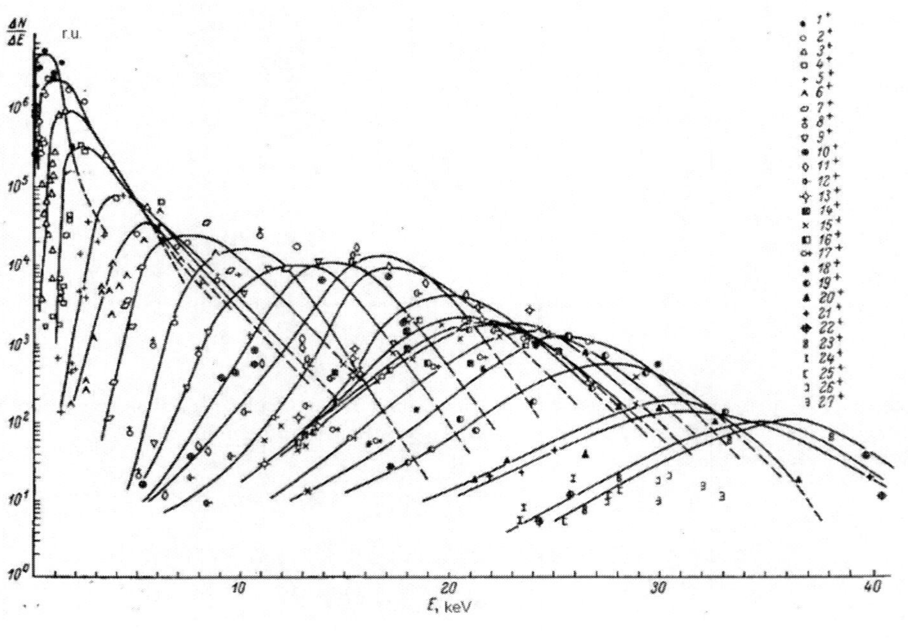

Figure 2.4. Energy distributions of cobalt ions at 10^{13}W/cm^2.

These curves coincide qualitatively but to reach a complete coincidence, it is necessary to construct the exact dependence Nz=f(z) with all the process features taken into account. Such a type of the charge distribution obtained experimentally in our experiments indicates that the ionization processes proceeding in the laser torch are non-stationary and non-equilibrium.

One can assert that the ion distribution in Z is determined mainly (at the given radiation flux density) by the shape of the function of the effective ionization cross section of the atoms in their successive single-electron collisions with electrons. Complete coincidence of these plots is hard to obtain because of the absence of information about the shape of the energy distribution of electrons in the dense part of the plasma torch.

We should also mention the presence of an interesting feature in such distributions: wave-like ascents of the dependence of the number of ions with a specified charge Z, namely in the region of the ions with the charge equal to Z=6−8 and Z=10−11, which can be explained by the dependence of the shape of the total ionization cross section (presented in paper [10]) on the charge value Z.

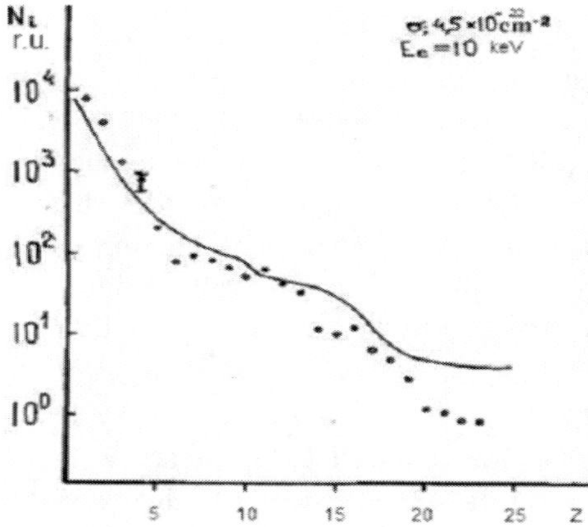

Figure 2.5. Charge distributions of cobalt ions and values of the effective ionization cross sections of electrons. The electron energy is equal to 10keV.

2.1.3. Influence of the Wavelength of Heating Radiation on the Ionization Process

In the course of investigations, three laser types were used: ruby lasers (0.6943μm) with the maximal output energy up to 20J and the generation time 30 ns, neodymium lasers (1.06μm) with the maximal output energy up to 100J and the generation time 10 and 30ns, and CO_2 lasers (10.6μm) with the maximal output energy up to 200J and the generation time of the first spike 50−70ns. In the experiments, the following parameters were varied: the output energy, the focusing spot, the elemental composition of the sample and the laser pulse duration, which changes when use is made of longer wavelength lasers. When comparing the characteristics of the expansion products of the plasma generated by radiation with different wavelengths, the initial conditions of the experiments were the same, and the radiation flux density, incident on the sample surface, was selected as an unalterable parameter, on a background of which it is convenient to make the comparison.

The most essential differences in the properties of the expansion products of the plasma generated by the lasers emitting at different wavelengths are observed in the values of the maximal charge (Z_{max}) of the ions in the plasmoid. Figure 2.5 presents the plots for comparison of Z_{max} of different elements in the plasma generated by the lasers emitting at two wavelengths (1.06 and 10.6μm). When plasma is heated by 10.6-μm radiation, there appear ion beams with the charge Z_{max}, which is 2 or 3 times higher than in the case of the lasers with shorter wavelengths (0.6943 and 1.06μm) [11]. In the plasma heated by long-wavelength radiation, the ionization processes are more effective, which probably stems from the fact that in the course of the radiation action, the electrons are accelerated up to higher energy. It was found in the experiments that the ions with equal values of Z_{max} can be produced in the plasma under the following flux densities and radiation wavelengths:

$$Z_{max} \sim \lambda^2 q \tag{2.3}$$

Thus, to obtain ions obtaining with the same value of Z_{max} as that in the plasma generated by a CO_2 laser (10.6μm), or a ruby laser (0.6943μm) or a neodymium laser (1.06μm), it is required to increase the flux density at least by two orders of magnitude with respect to q.

Let us consider in detail the changes in the quantitative characteristics of the ions with different charges Z under the action of 1.06- and 10.6μm lasers. It was noted in the experiments that at equal radiation flux densities on the sample surface, the number of the particles in the plasma torch generated by the CO_2 laser radiation is two orders of magnitude higher, some differences in other initial conditions being neglected. The first parameter, which should be taken into account in this connection, is the focusing spot on the sample. Because in our experiments it was impossible to obtain a focusing spot less than one millimeter in diameter at 10.6μm, the obtained result of the excess of the number of the ions for the plasma generated under such conditions is improper and should be compared with the results obtained in a similar geometry for a neodymium laser. In order to make a correct comparison of the experiments results, the 1.06μm radiation acting on the sample was intentionally defocused down to the similar 'spots' with the 10.6μm radiation and represented as the dependence of the total number of particles on the incident radiation wavelength and flux density. At the same radiation density, more singly charged particles are produced in the case of 1.06μm radiation, whereas the number of highly charged ions is higher in the torch generated by the radiation of the CO_2 laser (Table 2.3 presents data for an aluminum sample). Under the action of the radiation with the various wavelengths, the total number of the charged particles in the torch differs insignificantly, i.e., within an experimental error it can be considered independent of the incident radiation wavelength. A comparative analysis shows that the radiation energy in the torch re-distributed to generate either the multiply charged ions or the particles having a small charge. Along with this, it is necessary to remember the differences present in the pulse structures of the neodymium and CO_2 lasers used in the experiments. Simple calculations of energetic balances for these cases indicate the absence of any essential differences in processes of energy absorption and re-distribution for the generation of particles with different charges. The energy of the directed motion of the ions in the plasma generated by radiation with various wavelengths was considered almost constant.

From the comparison of the characteristics of the laser plasma ion component at the same charge composition and different flux density (Table 2.3) during the exposure to the long-wavelength radiation, it follows that the number of ions of all the charges with $Z>1$ increases.

Table 2.3. The comparison of the characteristics of the laser plasma ion component at the same charge composition and different flux density

Radiation wavelength (μm)	Ion charge Z				Radiation flux density (W cm^{-2})
	1	2	3	4	
1.06	8.1x10^9	1.6x10^9	1.5x10^8		8x10^8
10.6	4.1x10^9	4.5x10^9	2.6x10^9	7.8x10^8	8x10^8
1.06	9.4x10^9	3.2x10^9	8.2x10^8	3.7x10^7	2x10^9

The energy balance is well fulfilled in this case, too, i.e., all the additional energy of the CO_2-laser radiation having a longer pulse is spent on the generation of the ions with a higher charge.

2.1.4. Dependence of the Ionization Processes on the Plasmoid Initial Size

For the ionization process to be effective in the laser plasma, at least two factors are required: 1) the presence of the electrons with the energy exceeding by 2 or 3 times the ionization potential $I(z)$ in the plasmoid and 2) the necessary duration of the interaction process of such electrons with the atoms and the ions of the plasmoid. Therefore, in our works we noted the necessity for taking into account the interaction time of the fluxes of the ionizing and the ionized particles or the lifetime of the torch, i.e., the condition must be fulfilled:

$$\tau_i \leq \tau_0 \quad (2.4)$$

where τ_i for the ionization time in the case of the successive single-electron ionization;

τ_0 for the torch lifetime.

For the plasma lifetime the relation can be written:

$$\tau_0 \approx \frac{R}{u} \quad (2.5)$$

where R=d/2 is the initial radius with a spherically symmetrical plasma torch; and U is the plasma motion velocity, which can be taken equal to a local sound speed.

Hence it follows that by controlling the lifetime of a high-density plasma, in which high-energy electrons exist, it is possible to obtain pre-programmed ion beams with a specified charge Z. A necessary requirement for ionization is the lifetime exceeding the time needed for ionization of an ion with a specified Z. Thus, the attempt to increase the torch lifetime is quite understandable. Based on formula (2.5), this can be implemented by changing the initial size of the plasma torch. In our experiments, the lifetime was varied either by changing the diameter of the radiation passed through a lens from a single-mode neodymium laser (in this case the target was placed in front of the focal plane of the focusing lens), or by changing the diameter of the focusing spot in the process of the lens motion to the target surface such that the lens focus was placed behind the sample surface. After irradiation, the produced crater was measured and the energy necessary to provide the constant radiation flux density for each focusing spot size was selected with the help of a calorimeter.

Using this method we measured the energy spectra of aluminum, copper and cobalt ions at 2×10^8, 8×10^8 and 2×10^9W/cm^2 and the initial torch size in the range from 0.2 to 2mm. In addition to mass spectrometry the integral characteristics of the ion emission from the plasma were obtained by repeating these measurements with the help of collectors. This allowed excluding the occurrence of any instrumental errors and possible saturation effects emerging during the registration of the increasing particle flux from the torch.

Figure 2.5 shows the changes in the emission of the ions with different charges Z as a function of the initial torch size. When the initial plasmoid size ranges from 0.2 to 1mm, the number of the ions in the plasma grows exponentially and after the torch size becomes equal to 1mm, the increase in the particle emission is not observed. On the contrary, their number decreases somewhat. Thus, if we observed only the ions with Z=2 at a focusing spot diameter of 0.2mm, then at d=0.5 and 0.7mm, aluminum ions with Z=4 emerge in the torch, too. Therefore, the plasma torch ionization effectiveness changes. It follows that the emergence of the ions with a higher charge (at the same radiation flux density on the sample) is due to the change in the initial torch size, i.e., its lifetime. From these plots one can see that the change in the number of the ions as a function of Z obeys the expression:

$$\frac{dI(z)}{dR} = f(z) \tag{2.6}$$

where $I(z)=N(z)/S$

Nz for the ions quantity with the charge equal to Z;

S for the square area of the radiant focusing spot;

$I(z)$ for the “brightness” of the ion beam;

R for the focusing spot initial size

The experimental data made it possible to measure the ionization time of the ions with different charges Z in the laser plasma by using the calculations carried out in paper [12]. The higher the ion charge, the longer the ionization and the less the inclination of the function N(z)=f(R). The ionization times obtained from the experimental results depend on Z and other parameters in the range from 10^{-10} to 10^{-8}s.

It follows from these plots that to obtain the specified intense fluxes of the ions with a high charge, it is necessary to select such experimental parameters, which depend on the initial torch size or the focusing spot. The comparison of the ways of increasing ion charge with increasing radiation flux density and selecting an optimal focusing spot shows that at 1.5×10^8W/cm^2 and d=200µm, the doubly charged aluminum ions are registered in the torch.

To produce ions with $Z = 4$ at a constant d, the laser radiation energy should be increased by no less than 30 times.

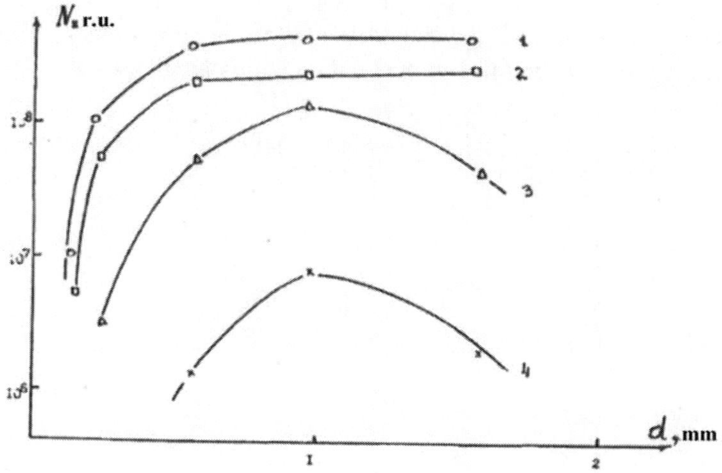

Figure 2.6. Dependence of the number of aluminum ions in the plasma on the diameter focusing spot of radiation incident on the sample: 1, 2, 3, 4 are the ions charges.

The same effect (the ion beam with Z=4) can be achieved by increasing the initial size of the plasma up to 350μm and the laser energy by 8 times. In our experiments we showed for the first time that this method makes it possible to increase the ionization process efficiency by no less than 4 times with decreasing output energy of the laser.

2.1.5. Ionization Processes as Functions of the Average Velocity of the Plasma Torch Expansion

The second parameter, which affects the plasma torch lifetime, as follows from formula (2.5), is the plasma expansion velocity. The plasmoid velocity, taken equal to the sound speed, is the function of the atomic weight of the plasmoid particles. Thus, we obtain that the plasmoid lifetime will be proportional to the square root of the particle atomic weight:

$$\tau_0 \approx \sqrt{M_c} \bullet R \tag{2.7}$$

i.e., the heavier the sample element, the longer the plasma torch lifetime.

Let us observe the changes in the plasma torch characteristics under smooth changing of the average atomic weight of the plasmoid, which can be defined as:

$$M_C = \frac{m_1 n_1 + m_2 n_2 + \cdots + m_K n_K}{n_1 + n_2 + \cdots + n_K}. \tag{2.8}$$

where $m1, m2, ..., mk$ for atomic weights of the sample elements;
$n1, n2, ..., nk$ for values of atoms concentration in the sample.

Such a smooth change in the average velocity of the torch expansion can be achieved, for example, by changing the carbon content in a Sn sample. Figure2.7 presents the ionization efficiency of the carbon ions as a function of the square root of the average plasmoid mass, which is proportional to the plasma torch lifetime. With increasing average plasmoid mass (due to an increase in the number of Sn atoms), the relative number of carbon ions grows in the torch.

With increasing torch lifetime, the emission efficiency of singly charged ions from the plasmoid increases more than by an order of magnitude.

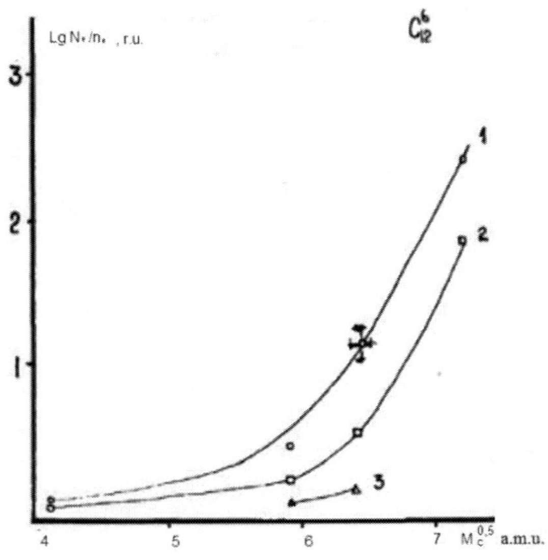

Figure 2.7. Efficiency of carbon ion formation (Z=1, 2 and 3) as a function of the plasmoid expansion time.

Figure 2.8 shows the quantitative dependence of the carbon ion formation on the content of these atoms in the sample (the carbon ions were mixed in into Sn). When the content of the atoms is large, the number of carbon ions increases in the torch with increasing number of these atoms in the sample. As soon as the number of Sn atoms becomes less than half, one can observe a drastic decrease in the number the carbon ions in the torch, despite an increase in the content of carbon atoms in the sample.

It is important to note that these experiments were performed at the same radiation flux density and do not depend on the shape of the intensity distribution along the sample surface.

In the case when it is necessary to obtain the maximal number of particles with a maximal charge, the most appealing approach for pursuing such a goal is an optimal selection of the atomic weight of the whole plasmoid (M_c) by addition of a second element.

To verify the conclusions made, we performed the experiments aimed at changing the yield of carbon nuclei from the plasma generated from such targets as (CH)m–polyethylene, (SF_2)m–teflon, (TiC, CaC, ZrC, Cr_2C_3)–carbides and graphite (C). Figure 2.9 presents the change in the maximal charge of the carbon ions in the plasma torch as a function of the sample composition. The carbon nuclei were observed only in the plasma generated on a teflon target at 4×10^{11}W/cm^2. On the samples made of graphite and titanium carbides, calcium and chrome, the maximal charge was Z=5, and in the polyethylene and zirconium carbide (ZrC) plasmas, only quadri-charged ions were observed. With increasing radiation flux density up to 7×10^{12}W/cm^2, a 5-fold higher carbon nuclei emission was observed in the teflon target in comparison with the graphite target. Taking into account the content of carbon ions in the teflon sample, one can speak of an effective increase in the ionization process with decreasing radiation energy by approximately 15 times due to the optimal selection of the atomic weight of the whole plasmoid (M_c) complying with the ratio of the lifetime and ionization time for carbon nuclei.

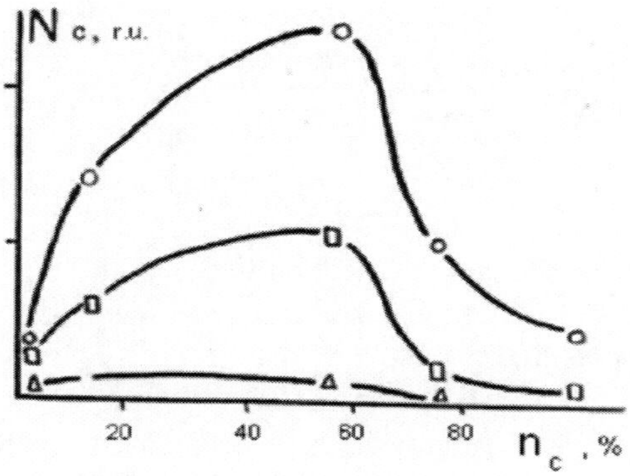

Figure 2.8. Change in the number of carbon ions in the torch as a function of their concentration in the sample (**o**–Z=1; □–Z=2; Δ–Z=3).

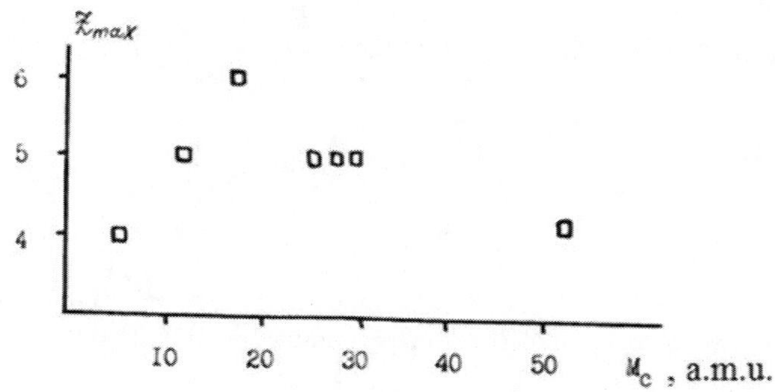

Figure 2.9. Dependence of the maximal charge of carbon ions in the plasma on the sample composition.

2.1.6. Mechanisms of Atom Ionization in the Plasma

Several schemes are known for the ionization process proceeding in the plasma. Those schemes are based on inelastic electron–atom or electron–ion collisions and on ion–atom or ion–ion collisions. Let us mention briefly some of them most probably occurring in the laser plasma.

1) Successive single electron ionization

$$A^{z-1} + e^- \rightarrow A^z + 2e^- \tag{2.9}$$

where Z takes the value from 1 to the nucleus charge value. We can single out several experimental works [13] containing information about the values of effective ionization cross sections for different elements. The cross sections of such processes can be calculated by using the Bethe formula

$$\sigma_{z-1,z} = \frac{K}{\varepsilon \cdot I(Z)} \cdot \ell n \frac{\varepsilon}{I(Z)} \tag{2.10}$$

where K is a constant, E is the energy of an incident electron, and $I(z)$ is the ionization potential.

The maximum of the ionization cross section is due to the electron energy exceeding the ionization potential by about 2–3 times. The minimal energy for the ionization cannot be lower than the ionization potential.

2) Successive ionization with simultaneous detachment of several electrons

$$A^{(z-1)} + e^- \rightarrow A^{+z} + 2e^- \tag{2.11}$$

Such a process is most probable in collisions of high kinetic energy electrons very heavy atoms and ions. The energy is absorbed simultaneously by several electrons on the shells (or on a shell).

3) Direct ionization up to an ion with a specified charge

$$A + e^- \rightarrow A^{+z} + (z+1)e^- \tag{2.12}$$

The effective cross section of such a process can be calculated by the formula

$$\sigma_{0,z} \approx (E_e - I(z))^z \tag{2.13}$$

This ionization type is intermediate one between cases (1) and (2). Such a process is possible only in the case of high energy of electrons and in absence of energy scattering by other particles.

In reality, the processes of all types occur in the plasmoid but the probability of them differs and depends first of all on such plasma parameters as density, temperature (or electron energy) and lifetime of this plasma at these characteristics. If thermodynamic equilibrium is absent in the plasma, the important ionization parameter will be the time of effective interaction of electrons with atoms and ions:

$$\tau_{z,z+1} = \frac{1}{ne\, V_e\, \sigma_{z,z+1}} \tag{2.14}$$

where V_e is the velocity of ionizing electrons and *ne* is the electron flux density. In this case, the time needed for a specific ion charge to be obtained can be found as follows:

$$\tau_z = \sum_0^z \frac{1}{\sigma_{z,z+1}} \tag{2.15}$$

Let us assume that the thermodynamic equilibrium is absent in the laser plasma. Then, taking into account that the electron velocity can reach the values of 10^7–10^8cm/s at densities 10^{19}–10^{21}cm^{-3} and the cross sections required for the charge Z=20 to be reached are equal approximately to 10^{-18}cm^{-2}, we obtain that the time needed for the ionization of the ions with Z=20 is equal approximately to 10^{-9}–10^{-8}s.

2.1.7. Summary on the Ionization Processes in the Laser Plasma

On the basis of experimental results, we can present a possible scheme of the ionization processes proceeding in the laser plasma.

The energy distribution structure, the dependence of the number of the ions with different charges on the radiation flux density, as well as charge distributions of the ions in the torch indicate the absence of the thermodynamic equilibrium in the laser plasma and an essential non-equilibrium of the ionization process. The main process in the laser plasma leading to the ionization of atoms, molecules and ions is the successive single electron ionization. To this end, the most important parameters affecting the formation of a large enough number of the highly charged ions are the electron energy and the torch lifetime at a high density of electrons and ions. The electrons receive energy directly from the electromagnetic field of the laser, and at high densities of the laser power that average vibrational energy of the electrons can reach very high values:

$$E_0 \gg I(z)\,; \tag{2.16}$$

$$E_0 \sim q\lambda^2 \nu_{ei} \tag{2.17}$$

where E_0 is the laser radiation field, *I(z)* for the ionization potential of atoms and ions, ν_{ei} is the frequency of electron–ion collisions and $\lambda = {}^{c}/_{\omega}$ is the laser wavelength.

It follows from the dependence of the vibrational energy of the electrons on the radiation wavelength that when exposed to 10.6μm radiation, the electrons energy is 100 times higher than in the plasma produced by the neodymium laser. The plasma ions with the equal Z_{max} generated by the action of various wavelengths are observed in case of the same electron energy calculated by the relationship

$$q(1.06)/q(10.6)=100 \tag{2.18}$$

and with other torch parameters assumed equal (in an electric field of high intensity the electron velocities are high), the transport collision frequency can become less than the ionization frequency. Within the time of the distribution function isotropization, several electron generations are produced, which do not have a Maxwellian type of velocity distributions of electrons. As a result, the contribution of the isotropic part into the ionization processes turns out to be small.

An important and necessary confirmation of the controversial point about the absence of the ionization equilibrium in the laser plasma is the experimental confirmation of the influence of the plasmoid lifetime on the expansion product parameters (in particular, on the maximal ion charge).

Chapter 8

2.2. RECOMBINATION

Many theoretical papers [4–6] are devoted to the consideration of recombination processes in an expanding plasmoid produced by laser radiation, whereas the experimental descriptions of such processes is mainly fragmentary, indicating that these processes take place and lead to these or those changes in the torch.

The spectroscopic studies made it possible to obtain a layer-by-layer structure and a temporal structure of the torch glowing during the pulse action, which enables evaluating of some processes proceeding in the torch core and obtaining the values of the constants and probabilities of excitation and ionization. Mass spectroscopy techniques allow one to refine the recombination process and observe the behavior of separate masses, charges and their changes. The most complicated task is the correct observation of the recombination processes against the background of the ionization, thermal and secondary processes, which are due to the interaction of the laser torch with the surface of the sample.

The recombination processes in the laser torch start to manifest themselves after the laser radiation has produced and heated the plasma. The plasmoid is expanded into vacuum with the following density decrease, accompanied by emission and attachment of electrons, atoms and ions with the formation of clusters, neutral atoms and negatively charged ions. To this end, several types of recombination processes well-known in the literature are possible:

a) triple recombination

$$A^{+z} + e + e \rightarrow A^{+z-1} + e \qquad (2.19)$$

b) photo-recombination

$$A^{+z} + e \rightarrow A^{+z-1} + h\nu \qquad (2.20)$$

c) di-electron recombination

$$A^{+z} + e \rightarrow A^{*+(z-1)} \rightarrow A^{+(z-1)} + h\nu \qquad (2.21)$$

d) charge exchange between atoms and ions, proceeding both resonantly

$$A^{+z} + A^{0} \rightarrow A^{0} + A^{+z} \qquad (2.22)$$

and non-resonantly

$$A^{+z} + B^0 \rightarrow B^{+z} + A^0 + dE \tag{2.23}$$

In the region of small energies of particles the probability of the resonance charge exchange is very high and the cross sections are equal to 10^{-14}–$10^{-15}cm^2$.

Domination of each of these recombination processes in the laser plasma depends on the plasmoid characteristics and because the density and local temperature of the laser plasma can change within a short time by several orders of magnitude, then at the different stages of the expansion, all these recombination processes become possible. Different authors describing laser plasma use this fact in solving kinetic equations and consider the conditions of dominance of different types of recombination processes.

A very important aspect of description of physical processes in the laser plasma is a determination whether a thermodynamic equilibrium of any type exists or not in the plasma generated by pulses with a high density of particles and energy. The possibility is considered to establish equilibrium of 3 types between the processes of ionization and recombination in the laser plasma:

1) local thermodynamic equilibrium;
2) corona equilibrium; and
3) collisional-radiative equilibrium.

In the model of local thermodynamic equilibrium, it is assumed that electron velocities are distributed according to Maxwell's equation and the ionization equilibrium is determined by Saha's formula:

$$\frac{N_{z+1}}{Nz} = 2\frac{Wz+1}{Wz}\left(\frac{mTe}{2\pi\hbar^2}\right)^{\frac{3}{2}} exp(\rho_z) \tag{2.24}$$

For the model of local thermodynamic equilibrium, the main process is the recombination in triple collisions, other types of recombination processes being neglected. Such situation is typical for plasmoids with a high density and temperature. In particular, for aluminum ions with charges Z=12 and 13, the local thermodynamic equilibrium applicability lies in the range T_e= 300–1500eV and n_e>$10^{21}cm^{-3}$.

In the corona model the equilibrium occurs between the ionization processes due to the electron collisions, the di-electron recombination and photo-recombination. The atom ionization potential can significantly exceed the electron temperature (Te/I(Z)=0.25-0.2). Unlike the model of local thermodynamic equilibrium, the corona model is assumed to be well working in the region of low densities and high temperatures. The collisional-radiative equilibrium model (calculated in detail in work [14]) suggests using the features of two previous models with some characteristics of the laser plasma taken into account.

In the formulae for velocities, constants and cross sections of different recombination there are two parameters: the torch temperature and the ion charge. The plasma temperature evaluation requires the presence of thermodynamic equilibrium in the plasma, which is rather problematic for such an unstable and inhomogeneous substance. Thus, the temperature measurement in such a torch is a complicated problem, whereas the determination of the ion charge in mass spectrometry is a trivial task. Hence, to determine the type of recombination proceeding in the laser plasma, it is convenient to use the charge composition characteristics of the laser plasma expansion products.

2.2.1. Recombination in the Energy Distributions of Particles

As has been noted above, the recombination processes are first mentions in the studies on the atomic fluxes, when the spectra demonstrate the second maximum and a group of particles emerges in the region of high energies. The appearance of a group of high-energy particles (E>40eV) in a neutral plasmoid indicates the beginning of the recombination process. In this case, a correlation is established between the radiation flux density on the sample, the first ionization potential and the binding energy of atoms in a solid body. The registration of the recombination atoms testifies to the start of ionization process in the material vapors and laser plasma formation.

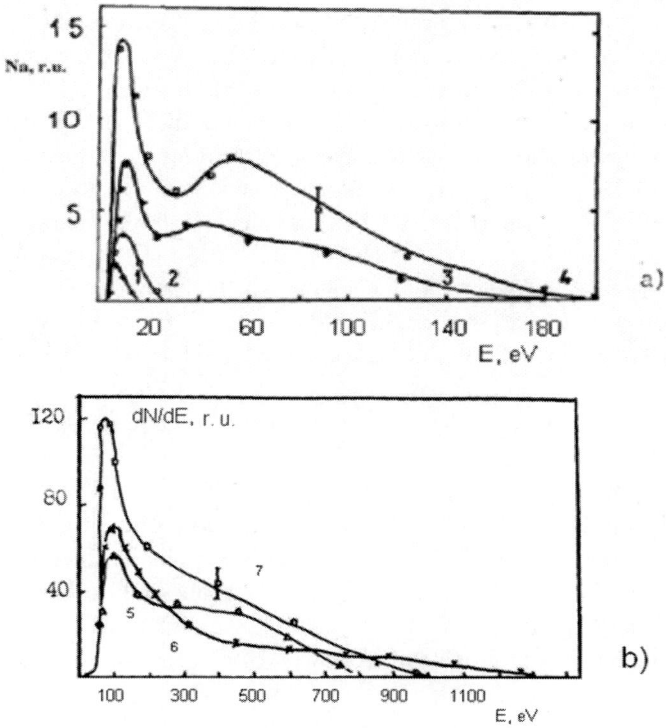

Figure 2.10. Energy distributions of cobalt atoms at different radiation flux densities in W/cm^2 (d=300μm): (a) 1–9×10^7, 2–1×10^8, 3–2×10^8, 4–3×10^8; (b) o -2.5×10^9, x - 3.5×10^9, п -5×10^9.

Let us consider the structure of the energy spectra of cobalt atoms (Figure 2.10). After the laser radiation intensity reaches $3x10^8 W/cm^2$ in a plasmoid, which previously had one intensity maximum, there emerge atoms with energies up to 200eV, and the second maximum of the particle intensity lies in the region of 40eV (Figure 2.10a). Then, the spectrum splits into two parts: with energies in the maxima 10eV and 60eV. In the range $4\times10^8–10^9 W/cm^2$, the number of the cobalt atoms in the plasmoid continues to increase., and one can observe a rise of the upper boundary of the atom energies to 700eV and of the amplitude of the energy spectrum maximum (E=60eV) in comparison with low-energy maximum. Besides, there emerge groups of neutral particles with energies in their maximums at about 200eV and 400eV. With a further increase in the radiation flux density in the range $(1–5)\times10^9 W/cm^2$ (Figure 2.9b), the amplitude growth of the spectra slows down, the main maximum shifts into the region 75–100eV, and the atoms with the energies up to 1keV are registered. Above $(5–7)\times10^9 W/cm^2$, the number of the neutral particles in the plasmoid starts to decrease and the energy spectrum broadens up to 1.5keV. Such a dynamics of development of the energy distributions of cobalt atoms is typical for all materials studied.

Therefore, it is established experimentally that after reaching some threshold energy of the incident radiation, the sample starts to intensively evaporate, and then with increasing radiation flux density up to the 'second threshold' values, the ionization process of the evaporating material starts and the fast recombination fluxes of atoms emerge, i.e. the laser plasma is formed. Recombination processes can be well illustrated if we consider the dynamics of formation of the energy and angular distributions of the particles, starting with threshold flux densities, at which the ions emerge in the plasma torch.

The main characteristics of the energy distribution of singly charged ions at $10^8–10^9 W/cm^2$ are as follows: The intensity maximum of the singly charged ions corresponds to the energies 40±10eV, and the energy changing from element to element; The intensity maximum of the particle distribution has a bell-like shape with the energy spectrum FWHM also equal to 40±10eV.

When the energy of incident radiation increases, the number of ions increases in the plasmoid without an essential change in the energy distributions. With increasing the radiation flux density on the sample, in the expansion products the ions are registered with lower energies than in the main beam and ones with higher energies, i.e., three beam types of singly charged ions are observed; they differ in average energies (10, 40 and 80eV) in the intensity maxima.

The first energy maximum is very narrow: the number of particles in the peak exceeds the value of dN/dE in the region of 40eV, but the integral number of particles of the lowest-energy beam is less than in the main group. With increase radiation flux density, the energies of these intensity maxima increase. The energy spectrum of the third beam is broadened into the region with higher energies (above 150eV). After a certain amount of particles is reached in the third beam, the maximum (at 50–100eV) exhibit the ions emerge the charge Z=2 (Figure 2.11).

Their number in the plasmoid increases with increasing flux density and sometimes (at some energy) even exceeds the number of singly charged ions in amplitude (in some narrow energy zone). In the high-energy region of the spectrum, the curve of singly charged ions and the curve of doubly charged ions overlap. The overlapping of the energy spectra of multiply charged ions in some region of the spectra indicates the establishing of equilibrium between the recombination and ionization processes.

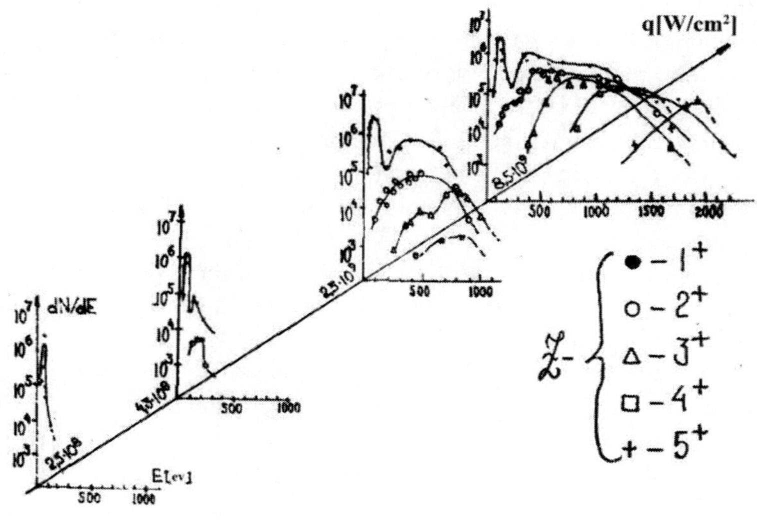

Figure 2.11. Dynamics of the formation of the energy spectra of manganese ions with increasing radiation flux density on the sample.

Initially, as in the case when singly charged ions are generated, the energy distribution of the ions with the charge $Z=2$ has a bell-like shape (close to Maxwell's function) and differs from singly charged ions only in energies of the directed motion. The maximal number of ions (for this period of the spectrum development) with the charge $Z=2$ appears at the energy values 100–200eV and shifts gradually into the region of high energies. The position of the maximum of the doubly charged ions essentially depends on the target material. The further increase in the radiation flux density results in complication of the energy spectrum of doubly charged ions.

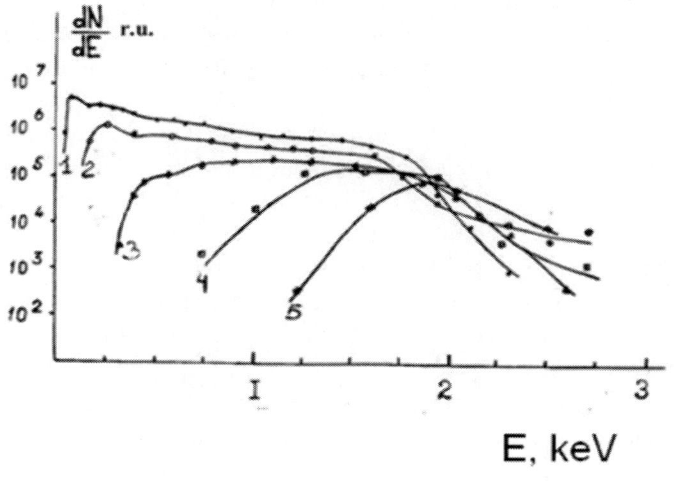

Figure 2.12. Energy distributions of niobium ions at 10^{10}W/cm^2 (the numerals indicate the ion charge).

The energy distribution in the region of higher energies (300–500eV) contains the second intensity maximum, followed by the third intensity maximum (after the intensity increases in both maxima) in the region of the 800eV energies. With a further increase in the radiation flux density, the dips in the particle intensity in the spectra are filled up and the energy distribution with Z=2 becomes smooth, overlapping the energy range from 100 to 1000eV. The ion beams with the charge Z=3 emerge in the energy range corresponding initially to the second intensity maximum of the ions with the charge Z=2. The triply charged ions also initially have only one intensity maximum. The energy spectrum of the triply charged ions develops due to the emergence of the new ion beams in the region of high energies. This process is accompanied by a gradual increase in the number of the particles until they reach equilibrium (Figure 2.12).

Analysis of the energy and velocity spectra of atoms of various elements and of the way they depend on the incident radiation show that recombination strongly influences the formation of the neutral component of a laser plasma. It is therefore most convenient to track the evolution of the plasma by comparing the energy spectra of the atoms with the energy and charge spectra of the ions in a set of experiments on manganese (Figure 2.11). At 10^8W/cm^2 only singly charged ions are produced in the energy interval 25-70eV, with a spectrum maximum at E=55eV; in this case the atom-energy range is 2-60 eV, and the maximum lies in the region of E=20eV. When singly and doubly charged ions appear with energies 100–200eV, the atom spectrum broadens to 150eV (3×10^8W/cm^2) and a second maximum of the neutral particle intensity, due to recombination of singly charged ions, is registered at E=50eV. The intensity maximum in this region exceeds the first one by three times. With a further increase in the radiation intensity, ions with higher charges are produced and contribute to the distribution in energies with Z=2 and 1 and to the energy distribution of the atoms. When the radiation flux density increases to 5×10^8W/cm^2, the total number of the high-energy neutral particles continues to grow, the atom energies increase to 350 and 600eV, respectively, and the ion energies to 400 and 800eV.

A decrease and shift of the first recombination maximum to the region 70-90eV takes place in the distribution of the atoms, and pronounced maxima of the particle intensities in the high energy region (230 and 350eV) appear in the spectra. Consequently, the intensity maxima that appear in the atom spectra near energies typical of the ion component of the laser plasma are produced by recombination of the ions with Z=1, 2, and 3, and the distributions of the atoms duplicate the bell shape typical of ions. A comparative analysis of the distributions of the neutral and charged components of a laser plasma reveals the following main properties:

1) on the whole, the energy ranges of the ions and atoms are the same;
2) the limiting energies in atom spectra are lower than for ions, and the difference between the right-hand boundaries of the ion and atom distributions increases with increasing q;
3) the maxima of the particle intensities in the atom spectra are shifted towards lower energies relative to the maxima of the ion distributions.

Thus, it can be noted that by emergence of the additional intensity maxima in the energy distributions of the ions made it possible to establish the presence of the recombination processes in the laser plasma, which allow one to estimate the effectiveness of such processes in the plasma torch at the early stages of the plasmoid expansion.

Besides, it follows from the dynamics of the formation of the energy spectra that the intensity of the process is not the same for different radiation flux densities (for the ions with a pre-fixed value of *Z*): at the radiation densities corresponding to the appearance of the ions in the torch, a very high recombination velocity is observed, which is found in additional intensity maxima of the ions with the charge *Z*-1 (whereas they themselves are not registered). With increasing flux density, the number of the ions with the charge I grows and reaches the values approximately equal to the recombination values in the spectrum *Z*-1.

2.2.2. Formation of the Charge Composition

To produce the ion beams with a specified charge, it is necessary to increase the radiation flux density (i.e. to create conditions for the appearance of electrons with higher energies) and to increase lifetime of the torch needed for successive stepwise ionization.

To increase the ion charge in the plasma, it is needed to ensure the conditions, under which the lifetime will be large enough. On the other hand, a competitive process (as compared to the ionization process) in the plasma is the recombination, whose time is long enough in comparison with the ionization time. That is why while solving the optimization problems of the processes proceeding in the plasma and of obtaining the maximal number of the ions with the highest charge at the invariable radiation flux density, one should take into account another relationship: between the times of the ionization, the torch lifetime and the recombination processes:

$$\tau_i(z) \leq \tau_0 \leq \tau_r(z) \tag{2.25}$$

It is clear (and more precise calculations show the same) that the recombination times are much longer than the ionization times for the dense high-temperature laser plasma.

For the laser plasma, which exhibits a quick change in its characteristics, it is hard to obtain theoretical solutions coinciding with experimentally obtained results. In work [15], various conditions were explored leading to an increase or decrease in the plasmoid lifetime, which enabled observing the dynamics of development of the ionization process. If we continue to increase the plasmoid lifetime, i.e. decelerate (retain) the plasmoid at the state of higher density, we come to the conditions, when the recombination processes start to proceed intensively in the plasma.

It is convenient to consider the influence of the recombination process on the charge spectrum of the laser plasma by taking into account the dependence of the intensity of the particle yield per unit surface area on the focusing spot diameter. Hence it follows that the plasmoid lifetime increases with increasing the torch diameter. Initially, with increasing torch lifetime, this factor makes an effective impact on the proceeding of the ionization process. But after reaching of some value of the initial plasmoid size (*d*), the recombination processes also start play a noticeable role (Figure 2.13). At small *d*, this becomes evident for the ions with the charge *Z*=1 and 2, and later on with increasing initial diameter of the torch this also becomes evident for the ions with the charge *Z*=3 and 4, such that the recombination processes begin to prevail. The slope of the curve in the dependence showing a decrease in the number of the ions as the plasmoid initial size grows, it is possible to evaluate experimentally the recombination time, which depends on the charge Z for the given element. The higher the ion charge, the higher the recombination rate. The recombination rate in different models of the thermodynamic equilibrium depends in different ways on the ion charge and grows with increasing charge value. When recombination occurs in triple collisions, the recombination rate is proportional to Z^3.

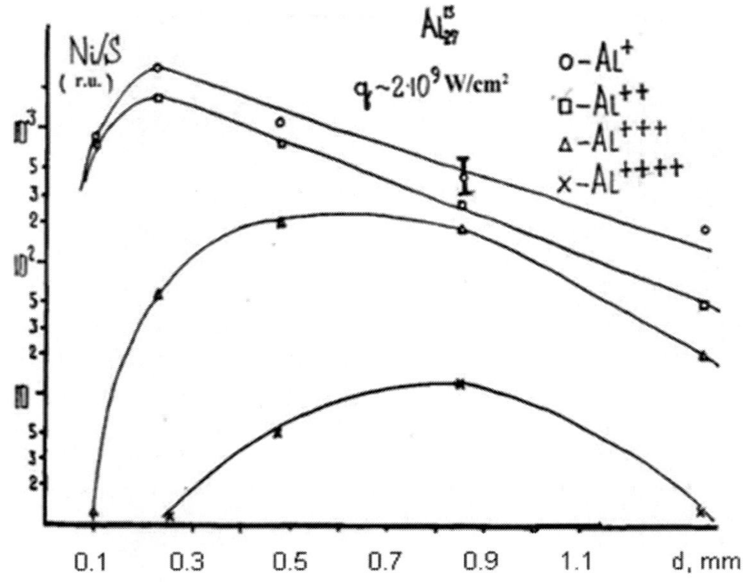

Figure 2.13. Dependence of the number of the aluminum ions per unit surface area exposed to radiation (**S** is the area of the focusing spot) on the plasmoid initial size at 2×10^9W/cm^2.

The change in the number of the ions with time can be written in the from [12]:

$$\mathrm{Ni}/dt = -\zeta_r^2 \; nei \tag{2.26}$$

$\zeta_r(t)$ -for the speed coefficient of the recombination processes,

nei-for the densities of the electrons and the ions.

The solution of the equation for the change in the ion concentration during the expansion has the form:

$$n_i = n_{i0} exp\{-\int_{t1}^{\infty} \Gamma_2\,(t) n_e dt\} \tag{2.27}$$

where ***t1*** is the starting time of the recombination stage of the expansion.

As was established in several theoretical papers, this stage duration makes few typical expansion times, which can be approximately written as:

$$\int_{t1}^{\infty} \Gamma_2(t) n_e dt \approx \tau_0 \langle \Gamma_2 n_e \rangle = \frac{R_0}{u} \langle \Gamma_2 n_e \rangle \quad (2.28)$$

$\langle \Gamma_2 n_e \rangle = \tau_r$ where is equal to the average time of the recombination processes for all the period of the plasmoid expansion. Thus, we obtain

$$\ln \frac{n_i}{n_{io}} = -\frac{R_0}{u} \tau_r^{-1} \quad (2.29)$$

$$\tau_r^{-1} = -\ln \frac{n_i}{n_{io}} \frac{u}{R_0} \quad (2.30)$$

Decreasing of the ion emission with increasing initial size from 300 to 1500μm is related such a regime of the plasma expansion, when the lifetime of the plasmoid turns out to be longer than the time of the recombination process, i.e., the decrease in the ions emission is caused by the recombination losses growing with increasing *d*. The slope of the right-hand side of the curves (Figure 2.13) makes it possible to determine the average times of recombination in the laser plasma for ions with various charges *Z*. Using the corresponding plasmoid size and expansion speed and substituting them into the derived formula, we obtain an estimation of the average recombination time for the ions with the charge *Z*=1, which is equal to 10^{-8}s. The authors of paper [16] present the calculations of the recombination times for a large class of substances, which are in agreement with the results of our experiments. It follows from the experimental results that the recombination time decreases with increasing charge *Z*, i.e. the curve dN/dS =f(d) has a steeper descent for highly charged ions.

From experimental curves like those shown in Figure 2.13, we obtained the estimate of the dependence of the recombination rates on the ion charge, which turned out to be weaker than the quadratic dependence and closer to linear, which, as we can see from the formulae, does not fit with the triple recombination (where such a dependence has Z^3) and photo recombination (Z^2). The charge exchange processes, in particular on the hydrogen atoms, exhibit the linear dependence on Z^n (n=1.0–1.1) [17], in the case if the particle velocities are not too high (10^7cm/s). Hence, the major role in the recombination must be played by the processes of the charge exchange.

The plasma lifetime is determined by its expansion time under action of its own pressure. And the expansion time (as it presented before) depends on the ratio of the plasmoid size to the average velocity of expansion. The average velocity of the laser plasma expansion (at the same absorbed radiation energy) can be controlled by changing the plasmoid average mass M_c(Eq. 2.8).

To understand the influence of the torch average mass on the recombination processes, the experiments were conducted, where the plasmoid average mass was varied by changing the sample composition. To this end, suitable binary compounds were selected, where one of the rank components was kept invariable; for example: NaCl, NaI, NaBr, $CuCl_2$, CuI_2,$CuBr_2$, $PbCl_2$, PbI_2.

The characteristics of the plasma binary compounds were compared with the energy spectra of pure samples of copper and plumb. Such sample selection enables in full force envisaging both the influence of a light component on a main heavy one and the influence of the heavy second component on the light main one.

Figure 2.14 presents the dependence of a change in the number of natrium ions with a charge Z=1, 2 and 3 on the atomic weight of the additive. With increasing atomic weight of the second component, i.e. with increasing torch lifetime, the number of the ions of all charges decreases significantly (down to 10 times). Notably, the higher the charge, the more intensive the recombination process. The increase in the torch lifetime with increasing average mass of the sample M_c results in the fact that in the torch formed by NaI and NaBr, triply charged ions are absent, though they were observed at small values of M_c.

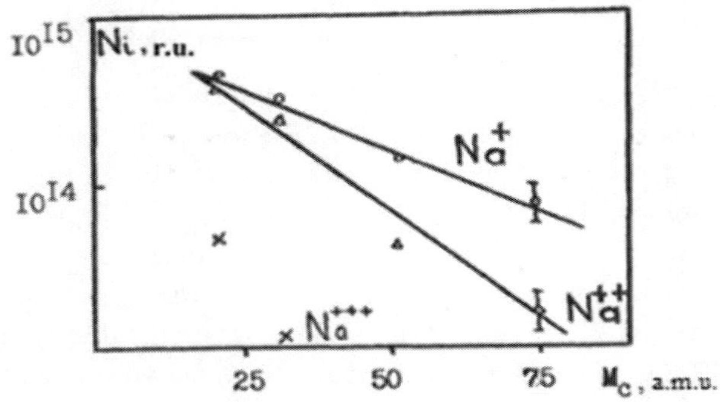

Figure 2.14. Influence of M_{ave} (M_c)of the sample on the emergence of natrium ions in the plasma.

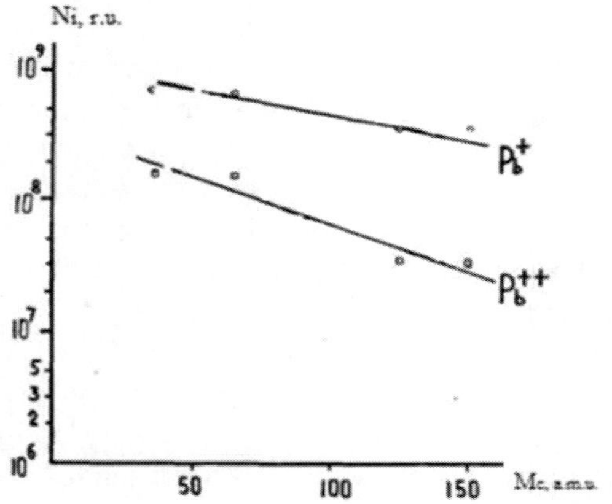

Figure 2.15. Recombination processes in the plasma with different values of Mave(M_c)by the example of plumb ions.

If the natrium ions can be classified as one of light elements, then the influence of a heavy component is convenient to observe on the basis of the plumb compounds. Herefrom follows the same decrease in the ion emission with the charge Z=1 and 2 from the laser torch with increasing atomic weight of the second component or of M_{ave} of the sample. The characteristics of the binary compounds with parameters of a torch produced from mono-samples must be compared with great precaution, taking into account the spectra of the neutral components of the plasma and the model of the ionization process [15], as it is necessary to introduce a coefficient taking into account the ratio of the atoms in the sample.

This is needed in order to align the initial conditions of the experiment on the number of the atoms participating in it. Thus, for instance, in comparing the characteristics of the plasma torch on the basis of the plumb samples with its compounds $PbCl_2$, PbI_2, (Figure 2.15), it is required to decrease the number of ions from a 'pure' sample by one third because initially in the process of the plasma formation from the binary compounds, there will be quantity 3 times less neutral atoms (definitely, if we assume that during such interactions the same number of all particles from various samples is produced).

2.2.3. Control of the Recombination Process

The experiments show that the recombination process in the laser plasma can be controlled by changing of both the initial torch size and the plasmoid average mass. Of interest here is the simultaneous variation of the initial conditions with the aid of both these parameters. Let us consider the characteristics of the expansion of the two-component plasma as a function of the focusing spot diameter in the case, when the radiation flux density is kept at the same level. We will consider this dependence by the example of the energy distributions of the ions of the binary compound $PbCl_2$, which illustrate the evolution of the spectra as a function of the focusing spot diameter. Because an increase in the initial diameter of the torch leads to longer plasma lifetime, this fact is responsible for the increase in the emission of plumb and chlorine ions of all the charges; however, the rate of the increase in the number of the ions with all charges Z is different for these elements. Figure 2.15 presents the integral characteristics of the change in the plumb and chlorine particles in the laser plasma. The maximal efficiency of the formation of different particles corresponds to different values of the initial torch size (d).

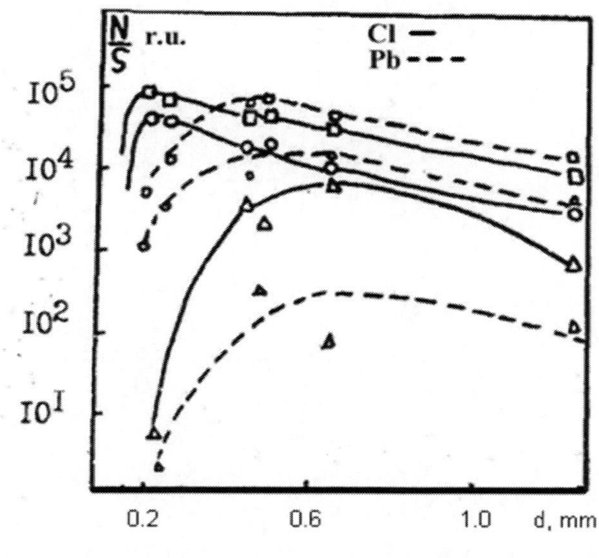

Figure 2.16. Dependence of the emission of plumb and chlorine ions on the initial size of the plasma for the compound $PbCl_2$: solid lines show chlorine ions; dashed lines show plumb ions; **o** – Z=1, **п** –Z =2; **A** –Z =3.

Let us compare the values of *d*, when the emission of the ions of close masses is maximal under conditions of both the presence and absence of the second component. Let us take, for example, the dependence of the number of aluminum (27 amu) and chlorine (35 amu) ions per unit surface area (efficiency) on *d*. The maximum efficiency of the yield of singly charged aluminum ions corresponds to the value *d*=350μm. Hence for the samples with the same atomic weight as for pure chlorine samples, the emission efficiency must be an order of magnitude smaller. In the case of the chlorine and plumb compounds, the expansion of the entire plasmoid slows down, i.e. the plasmoid lifetime increases, which results in the shift of the maximum emission of the particles per unit surface area towards smaller *d*. The recombination processes come to the fore at an earlier stage described by the value *d*, which is be observed in the experiments, i.e. the chlorine ion emission is maximal at *d*=200–280μm.

Due to the fact that the ion yield is maximal at different values of *d*, this fact makes it possible to interfere and control the recombination processes for these or those elements in the plasmoid. Thus, Figure 2.17 presents the dependence of the ratio of the emission of the singly charged chlorine ions to plumb ions on *d* for two values of the radiation flux density on the samples. When changing the focusing spot diameter from 0.2 to 1.5mm, the ratio of the emission of ions of different masses in the laser plasma changes significantly (the ratios change from 25 to 0.5). In this case, at *d*=400μm, the ratio of the singly charged ions in the plasmoid corresponds to the real stoichiometric composition of the atoms in the target (2:1) used for the experiments. This means that in a case of suitable selection of the expansion time of the plasmoid, a favorable situation can occur, when the relative content of the ions in the plasma corresponds to the content of the atoms in the sample.

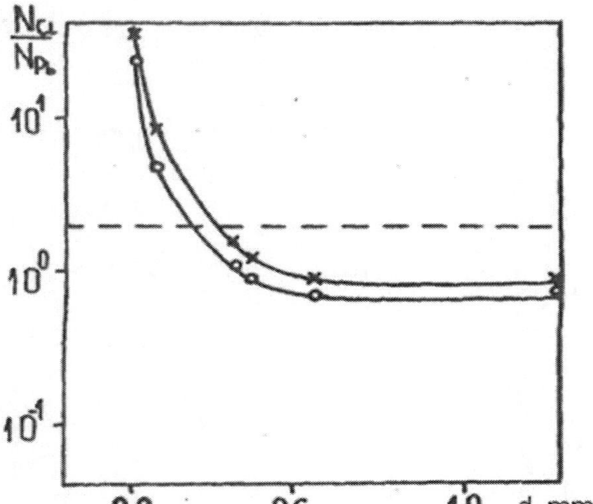

Figure 2.17. Relative content of the plumb and chlorine ions in the torch as a function of the initial torch size (o-2x10^{9}W/cm^2, x–5x10^{8}W/cm^2).

Consider the influence of the recombination process on the charge composition of the torch by example of the dependence (Figure 2.9), which shows the influence of the average mass of the plasmoid on the charge composition of hydrogen ions. Whereas the laser plasma of the pure hydrogen contains the ions with the charge *Z*=5 (at the given radiation density), the compound of light atoms (the $(SH_2)n$ – polyethylene target) leads to a more accelerated expansion and a reduced lifetime of the plasmoid; besides, the ions with the charge Z=5 have no time to form in the laser plasma (it turns out that the ionization time of penta-charged ions is less than the plasmoid lifetime), and therefore the maximal ion charge can be only Z=4. The expansion deceleration and hence the increase in the lifetime leads to an increase in the number of the penta-charged ions (Z=5) and the production of the hydrogen nuclei in the expansion products. The deceleration is achieved by using of teflon samples $(CF_2)n$. The increase in the torch average mass (by making heavier the second component) leads to an even greater deceleration of the expansion and manifestation of the recombination processes. The use of the TiC and CaC carbides as samples makes the plasmoid lifetime longer, which ensures the recombination processes for the hydrogen nuclei that are absent in the plasmoid. An even greater deceleration of the plasmoid development in a target made of ZrS results in the effective recombination of the ions with the charge Z=5, besides in the laser plasma the hydrogen ions with the charge Z=4 are observed (as in the case of a polyethylene sample).

2.2.4. Ion Charge Exchange

In considering the expansion of the laser plasma into vacuum, the processes of the ion charge exchange are usually not taken into account, as in the vacuum 10^{-5} Torr and less such processes are considered as unlikely. This assumption would be correct if not an insignificant circumstance, which is typical for the processes of the laser interaction with a solid body.

The process of the charge exchange of multi-charged ions of the laser plasma into the ions with a smaller charge Z is possible in presence of a cold layer of a neutral or weakly ionized gas formed in front of the 'hot nucleus' of the torch due to (most frequently) slowly growing part of the laser pulse, which takes place under the experimental conditions. Thus, despite very high contrast of the radiation pulse (10^5), such a layer was observed by Boiko et al. in the dependence of the density distribution of the particles on their distance from the target surface. If the laser pulse contrast is rather low (in the case of LTI lasers), the presence of such a layer can be considered obligatory. The cross section of the charge exchange depends on the mass composition of this layer (of the near-surface layer). For example for a 'cold' gas consisting of the sample atoms, the resonance charge exchange is possible. In this case, the cross section is approximately equal to 10^{-14}cm^2 and this size increases with decreasing energy of arriving particles. As a rule, in the course of interaction of dissimilar atoms, the cross sections are equal to 10^{-17}cm^2 and have threshold values in the region of low energies of interacting particles. In studying the neutral atom fluxes of the laser plasma, it was found that the energy spectrum consists of atoms evaporated from the sample and recombined from ions with different charges Z. The number of neutral atoms makes not less than 95% of the total number of the particles at 10^{10}W/cm^2. Nonetheless, it is understandable that neutral atoms formed in recombination cannot significantly affect the charge exchange processes. As for thermal fluxes of the atoms, they will participate in such processes only in those cases when their formation occurs in front of the main torch, or, in other words, when they would impede the multiply charged ions expansion (the multiply charged ions should penetrate through them because the multiply charged ions have higher velocities of expansion).

Good illustrations of the charge exchange processes in the laser plasma are the results of the experiments on the transversal expansion of the ions of the laser torch. It was noted in these experiments that except the ions of the main element the mass-spectra exhibit a significant number of hydrogen, carbon, oxygen and nitrogen ions. These ions in such amounts are not present in the sample matrix but are contained in adsorbed layers on the sample surface at 10^{-5}–10^{-6}mm hg (preliminary pumping by an oil pump and high vacuum pumping by NORDs). To remove absorbed gases from the surfaces, a laser purification procedure is applied with the use of a second *Q*-switched laser (the radiation spot is 1mm in diameter) synchronized with the first one (with the lead time of 100μs). With switching on the second laser, the number of the absorbed ions drastically decreased in the mass-spectrograms.

Note that in the case of a polluted surface, the observed ions of carbon, azotes and oxygen have a charge up to Z=3, the energy distribution of these ions differing significantly from the ion distribution of the main element. The maxima of the energy distribution of these ions are shifted towards low energies. In velocity distributions, the intensity maxima of the ions of these elements coincide with the maxima of the velocity distributions of multiply charged ions of the main element of the sample. Hence, the ion energies of similar elements being present in the sample are higher than the energies of the ions produced from adsorbed gases of the sample surface. Thus, we can speak about different mechanisms of formation and acceleration of ions of adsorbed gases and ions of the main torch. Using the experimental results of transverse expansion and impact of initial conditions on the laser torch parameters, it is possible to assume the following model of spatial expansion of the laser plasma torch.

The main 'hot' part of the torch plasmoid is a source of X-rays and ultraviolet radiation as well as the source of the multiply charged ions, which irradiate the sample surface around the place of their formation. As a result, the atoms of the adsorbed gases and some part of the sample material start to evaporate intensely.

After the laser plasma is produced and heated and in front of the target (supposing that the expansion has not started yet), we have a hot nucleus (consisting of multiply charged ions) and a 'cold' layer surrounding it (consisting of the target atoms and adsorbed on the surface of the gases formed by both the pre-pulse and the fundamental radiation. In addition, the main 'hot' part of the torch plasmoid is a source of X-ray and ultraviolet radiation and is also a source of multiply charged ions, which irradiate the sample surface around place of its formation. As a result, the atoms of the adsorbed gases and some part of the sample material start to evaporate intensely. The electron density distribution as a function of the distance from the sample surface, measured by spectrometric methods, confirms the presence of low density electrons in front of the surface. This fact also testifies to the emergence of neutral and low-ionized gas in the initial phase of irradiation. During the expansion the particles of the 'hot' part of the torch have a higher directed expansion velocity and penetrate through the 'cold' gas layer. As a result of interaction of the fluxes of the multiply charged ions, target atoms and desorbed gases, the following processes are possible: the charge exchange (both resonant and non-resonant) of the multiply charged ions of the hot nucleus, ionization of the 'cold' layer of the gas, the increase in this layer ion energy due to their interaction with the main ions fluxes, and the scattering of the arriving particles on a cold cloud. The analysis of the experimental data on the transverse and reverse expansion of the laser plasma particles shows that the recombination efficiency increases linearly with increasing ion charge. The effects of the charge exchange in the gas layer formed by the atoms desorbed from the target surface are especially conveniently observed during the torch expansion along the sample surface as the thickness of this layer grows significantly due to the above-mentioned reasons.

2.2.5. Mechanisms of the Formation of Negatively Charged Ions in the Laser Plasma

A drastic decrease in the density and temperature of the laser plasmoid realize the conditions typical of the formation of negatively charged ions: a relatively low average energy of the electrons at a significant density value.

Taking into account the fact that one can observe a significant increase in the number of neutral atoms, the process of the electron attachment must play an essential role at this stage of the plasma expansion. After the laser pulse terminates, the process of the ion recombination, whose efficiency determines the final plasma composition at large distances from the place of its formation, dominates. As the density changes accompany this expansion, there exists a probability of the 'continuation' of the recombination process into the region of negatively charged ions (recombination + attachment), which leads to the formation of a significant number of negatively charged ions in the laser plasma.

To elucidate the effect of the recombination and acceleration mechanisms on the relative yield of negatively charged ions from the laser plasma, special experiments were performed. As the targets we selected one-element samples consisting of elements with a close affinity energy to an electron, equal to 1.2–1.4: carbon, silicon, germanium and copper. Figure2.17 presents the results of these experiments. The value of the relative yield of the negatively charged ions decreases with increasing M_{ave} of the plasma torch. In this case, the maximum energy (Figure 2.19) falls from 1.2keV to 400eV and the average energy from 500eV to 150eV, respectively.

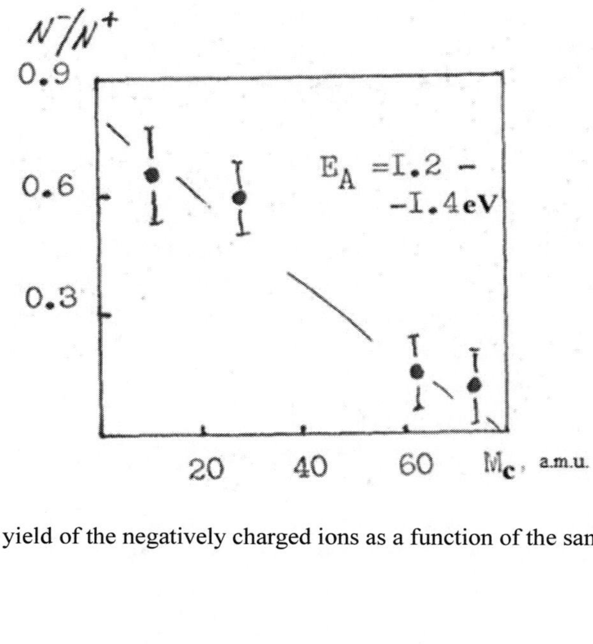

Figure 2.18. Relative yield of the negatively charged ions as a function of the sample mass.

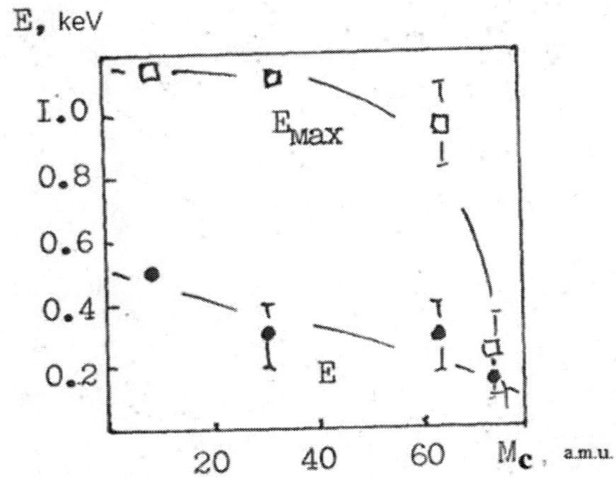

Figure 2.19. Dependence of the maximum and average energy of the negatively charged ions on the average sample mass.

These results exist are consistent with the results of another experiment conducted with the same purpose. As the targets, we used the binary compounds containing the element under study in crystallized form. The energy distributions of the negatively charged ions are shown in Figure 2.20. Figure2.21 presents the total yield of the negatively charged ions as a function of the average mass of the ionic component of the plasma; in this case, the maximum energy of the ions decreases approximately linearly.

One can see that the decrease in the total number of the ions is due to the decrease in the number of the high-energy states. These experimental results well fit into the entire picture of the recombination processes: the negatively charged ions of the laser plasma are produced as a result of attachment to other particles, mainly to the recombination atoms, and their hardening is related to the decrease in the time of their presence in the high density zone.

Thus, one can observe a natural connection of the recombination of positively charged ions into the neutral atoms, which in turn are responsible for forming the negatively charged component of the laser plasma in the processes of attachment:

$$A^{+} \Rightarrow A^{0} \Rightarrow A^{-} \quad (2.31)$$

The observed shift of the registered atoms towards smaller energies range (in comparison with the distributions of the negatively charged ions) is an additional reason in favor of such a mechanism. It should be noted that the attachment of the electrons to an atom (as recombination) is also exothermic process. That is why to some extent attachment can be considered a next step of the recombination and it is fair to speak about presence of a joined 'recombination + attachment' process.

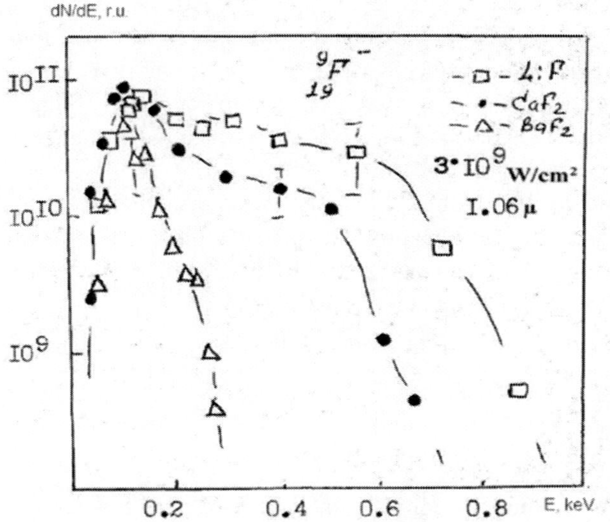

Figure 2.20. Energy spectra of negatively charged fluorine ions.

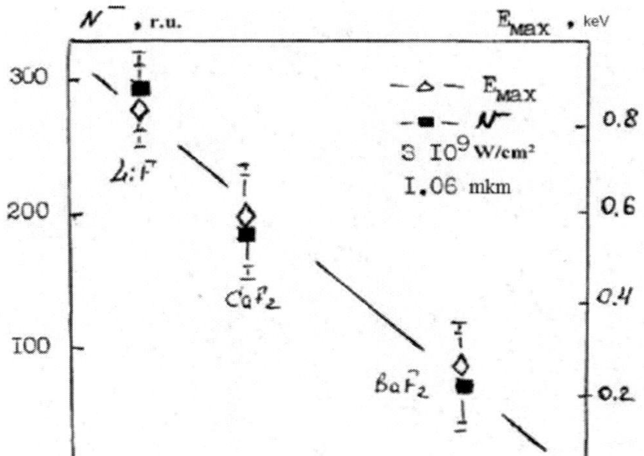

Figure 2.21. Quantitative and energy dependences of the negatively charged fluorine ions on the average mass of the ionic component.

Thus, we can present the following order of the processes dominating during the laser plasma expansion:

$$e + A^0 \rightarrow A^+ + e + e \rightarrow A^{+z} + e + e + \cdots \rightarrow A^+ + e + e \rightarrow A^0 + e \rightarrow A^- \qquad (2.32)$$

The electrons participating in the acceleration of positively charged ions lose their kinetic energy because of their attachment to the recombination neutrals, and the thus produced negatively charged ions (due to the velocity they have already gained) have time to leave the region of the high density without any destruction in the collisions. The same fact is indicated also by the dependence of the relative yields of the negatively charged ions on the average mass growth.

As is known the neutral atoms are mainly produced in the laser plasma due to two mechanisms: thermal and recombination; hence, the observed transition will deform the spectra of the negatively charged ions in accordance with the presence of the two-channel scheme of the resonant charge exchange of the ions:

$$A^+ + A^0 \rightarrow A^0 + A^+ \tag{2.33}$$

$$A^- + A^0 \rightarrow A^0 + A^- \tag{2.34}$$

upon achieving the optimal conditions in laser plasma for transitions like this, the distributions of the negatively charged ions must have a well-pronounced ‘double-humped’ structure, which indeed was observed in some cases.

Chapter 9

2.3. PROCESSES OF THE ACCELERATION OF THE PARTICLES IN THE LASER PLASMA

Earlier in the experiments on the interaction of laser radiation action with a solid body surface, during the explosion of thin wires in vacuum and in various types of electrical discharges, the ions were observed with a very high energy of directed motion exceeding the thermal energy, which indicates the presence of different acceleration mechanisms in the plasmoids with high temperature and density. The experimental study of the acceleration of the particles in the laser torch upon their free expansion into the vacuum is one of the most interesting problems, which are important both from fundamental and purely application-oriented points of view, i.e., the problems of the inertial thermonuclear fusion, film sputtering techniques and fabrication of particle sources.

In theoretical and experimental works devoted to the expansion of plasmoids produced, in particular, under laser irradiation, two main mechanisms of acceleration of atoms and ions are considered: due to the existence of a large pressure gradient in the plasma–vacuum system and as a result of collective processes of interaction between electrons and ions.

Paper [19] devoted to the expansion of plasmoids indicates three mechanisms and regimes of electromagnetic plasma acceleration: due to the existence of a large pressure gradient, as a result of electron–ions friction, and as a result of the action of an electric field.

In paper [20], the energy spectra of the ions are constructed on the recombination heating of the plasma torch. In the other works, the acceleration of the laser plasma particles was explained within the framework of gas-hydrodynamic effects [21].

The most probable explanation is that the laser plasma particle is accelerated due to the electric forces acting in the torch at early stages of the plasmoid formation. Consider some models of electromagnetic nature [22–24]. Thus, Silin [22] presented a formula to calculate the ion velocity by using the effect of an increase in the electric field in the region of the plasma critical density due to the existence of an electric field vector component, which is normal to the target surface:

$$v_{max} \sim \frac{E_0}{\sqrt{4\pi n_i M}} \frac{1}{v} \sqrt{\frac{c\omega_0}{2\pi a}} . \qquad (2.35)$$

where E_0 is the amplitude of electric field in the light flux, ω_0 is the effective frequency of electron collisions, *v* is the radiation frequency, *ni* is the density of the ions in the region of the critical point, *M* is the ion mass, and *a* is the typical change in the plasma density.

It follows from paper [25] that the plasma is accelerated by an extremely heavy wall having a velocity V; that is why the plasma energy is proportional to the ion mass. As was pointed out by Silin [22], the velocity of the accelerated ions depends on the coordinate axis. Thus, at some instant V(r) will be equal to the speed of sound. On this base Silin made a conclusion about the existence of the acceleration range: from the speed of sound to V_{max}. At the same time, he estimated the total number of the accelerated ions by the formula:

$$\delta n_i \sim n_i \frac{4\pi r_0^2 \alpha}{\sqrt{2\pi\rho}} \left[\frac{E_0^2}{4\pi/(n_e k T_e + n_i k T_i)}\right]^{\frac{1}{2}} \tag{2.36}$$

The acceleration time is extremely short and is evaluated, depending on the experimental conditions, to be 10^{-9}-10^{-8}s. Numerical calculations [26] based on this model indicate an increase in the ion energy with increasing radiation flux density and ion charge.

Gurevich and Pitaevskii [27–29] considered the ion acceleration in the case of the expansion of rarefied plasma into vacuum; such acceleration is due to spatial splitting of the charges. The splitting appears because of the presence of the electrons with the velocities higher than the velocities of the ions. Due to this reason the electrons tend to exit from plasma.

As a result, a self-consistent electrostatic field appears, which transfers high kinetic energy to ions. Such effects were studied also in [30], which indicates some features forthcoming in the plasma in the presence of such electrostatic acceleration: ions with a large charge pass the accelerating region at high velocities; low-charged ions (especially singly charged ions) must have a more sophisticated energy spectrum due to a change in the electrostatic field potential during the expansion process; and ions with different atomic weights have different acceleration velocities. Degtyarenko [3] used the calculations by the model of electrostatic self-consistent ion acceleration in the laser plasma to obtain the dependence of the ion energy on the charge:

$$Ei = Z^2 \tag{2.37}$$

whence it follows that the expansion velocity is proportional to the square of the ion charge.

The first mass-spectrometric studies of the laser plasma showed that in the expansion products contain particles with velocities much exceeding thermal velocities. Even at threshold densities of the radiation flux, the torch consists of atoms having a velocity of (3–8)×10^5cm/s.

2.3.1. High-Energy Multiply Charged Positive Ions

To illustrate the influence of acceleration processes on multiply charged ions of the laser plasma, Figure 2.22 presents the energy distribution of the cobalt ions at 10^{13}W/cm^2 [31]. The plasmoid of such a plasma contains particles with a maximal energy up to 40keV. Let us describe the most important features of the energy spectra of multiply charged ions:

1. Energy distributions of the ions of each charge have their left low-energy boundary, which is not equal to zero.
2. Most often the right high-energy boundary of the distribution is common for all the ions (if to consider it at the same level of the sensor sensitivity) and differs if to consider it at the level 0.1 or 0.5 in relation to the intensity maximum of the particles.
3. The spectrum contains such an intensity maximum of these particles intensity whose energy position depends on many factors.
4. The entire plasmoid consists of a set of ions whose energy spectra with increasing charge Z are successively located after each other and overlap the region of high energies.

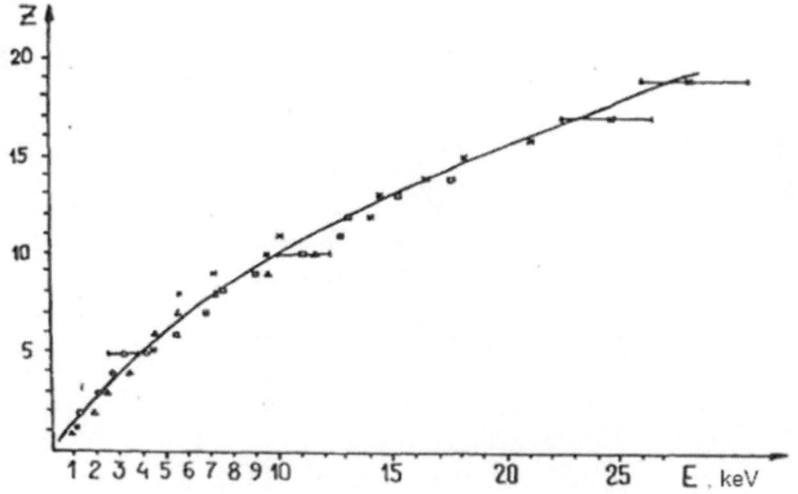

Figure 2.22. Ions energies as a function of the charge.

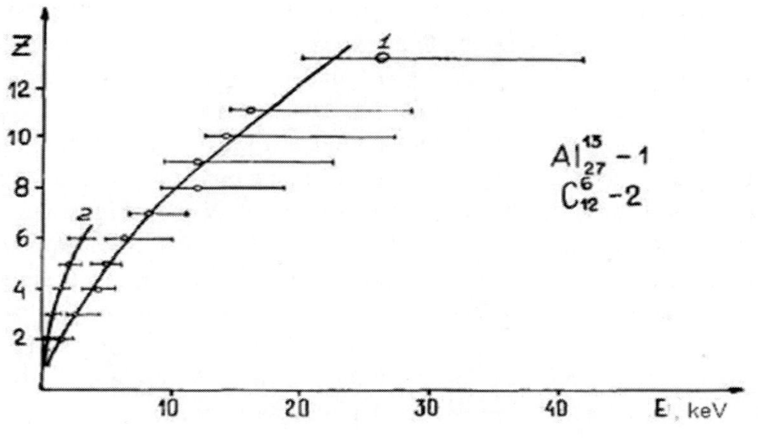

Figure 2.23. Ions energies as a function of the charge for samples made of (1) aluminum and (2) graphite.

The values of multiply charged ion energies leave no doubt about the existence of acceleration processes in the laser plasma. If we consider the plasmoid as a whole, there is a common energy boundary. This boundary depends essentially on the radiation flux density and ion charge: the higher the latter, the higher the ion energy (Figure 2.23). Besides, this value is a function of a sample element: the higher the sample atomic weight, the larger the value of the energy boundary of the spectrum. In this case, an important characteristic of the value of the energy boundary of the distributions is the appearance of the ions with maximal charge in the plasma.

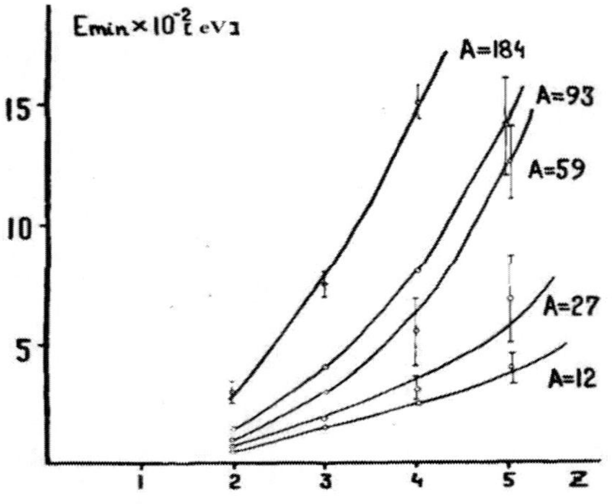

Figure 2.24. Value of the low-energy boundary of the ion distribution versus ion charge for carbon, aluminum, cobalt, columbium and wolfram at 10^9–10^{10}W/cm^2.

To single out the parameters affecting the acceleration processes, we consider the influence of the initial conditions on the maximum energy of each charge. In this case, the boundary of each distribution was considered at a sensitivity level of one tenth and a half of the particle intensity in the distribution maximum. Figure 2.24 presents the dependence of maximum energies of carbon and aluminum ions on the value of the charge. It is seen that with increasing atomic weight, the energy of the ions of the same charge grows. This dependence was obtained at a fixed radiation flux density equal to 5×10^{13}W/cm^2. From above it follows that the formation of the intensity maximum of the particle distribution and its right high-energy boundary occurs mainly as a result of recombination processes from ions with higher values of the charge *Z*. Hence the characteristics of right high-energy distribution boundaries reflect not only the acceleration of the ions of the considered charge but also make the contribution of many ion distributions into the spectrum. A more exact and unambiguous characteristic of the acceleration processes of the laser plasma ions of this or that charge is the value of the low-energy boundary of the distribution. The ions emerging in the region of the left distribution boundary are due only to ionization and acceleration processes. The ionization process takes place before the acceleration starts; that is why, apparently it is not important so much that before the acceleration starts these ions have had the charge less by unity because these particles have not had the thermal velocity. Thus, the value of the low-energy boundary must not depend on the recombination processes. It is found and confirmed that the value of the left energy boundary of the spectrum is independent of the angle at which the particles leave the torch and is weakly dependent on the radiation flux density. The latter statement is correct at least for the radiation flux densities 10^8–10^{13}W/cm^2 because when the radiation flux density changes by three or four orders of magnitude, the value of the low-energy boundary could change by no more than 2 times. With increasing ion charge, the low-energy boundary of the distribution shifts to the region of high energies. Figure2.24 presents the change in the low-energy boundaries of the ion distribution as a function of the charge for several samples. When the atomic weight is increased, the value of low-energy distribution boundary grows as well. Figure2.25 shows the dependence of the change in the low-energy distribution boundary on the ion charge and the element atomic weight at a large range of the radiation flux densities. Within spread of the experimental points we may assert that the velocity of the laser plasma expansion products (judging by the low-energy boundary) does not depend on the sample element and grows proportionally to the ion charge.

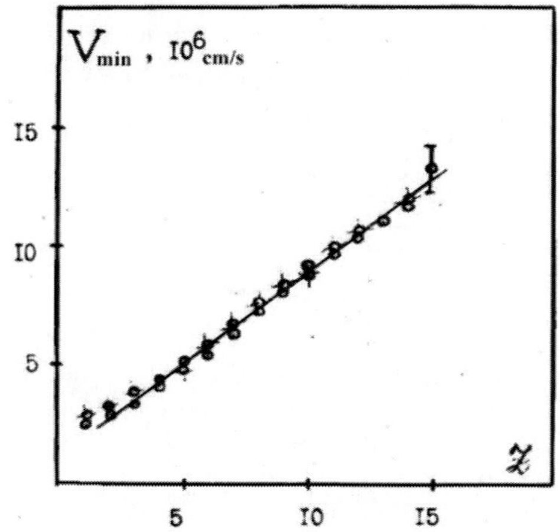

Figure 2.25. Change in the minimal velocity of the ions as a function of the charge for the 10.6-μm radiation (o–cobalt, ф –copper).

2.3.2. Influence of the Electrons of the Laser Torch on the Acceleration Processes

The electrons of the laser plasma play an essential role in the formation of spectral and charge distributions of the ionic component: they determine the charge composition and the number of the ions of a particular charge. The presence of a large number of high-energy electrons affects also the formation of the energy spectra of the ions despite of a great difference in the particle masses. Even if we suppose that the ion and electron energies are equal (at the stage of the plasmoid heating), the electron velocities will be higher than that of the ions by several orders of magnitude. The electrons are located in the plasma cloud and cannot freely leave the cloud without the violation of the plasmoid neutrality, because separation of the ions and electrons in a dense plasma (10^{19}–10^{21}cm^{-3}) is prevented by electric forces, which leads to mutual influence and the change in the velocities of the directed motion of the electrons and the ions.

For comparison of the electron pulses with the ion characteristics it is convenient to consider the temporal structure of the expansion products obtained at a distance of 450cm (Figure 2.26).

The plasmoid fore-body is formed of the ions of all charges from Z=1 up to Z_{max}, and their number drastically increases as the plasmoid moves inside into the torch, just as in the case of the electron pulse.

The higher the ions charge, the less the duration of the particle flux. The temporal distribution of ions of different masses is given by the example of the binary compounds of lead dichloride ($PbCl_2$).

The temporal distributions of the particle fluxes reflect not only the dynamics of the torch in time but also in the drift space. Despite great differences in atomic weights of all these elements (M_{Pb}/M_{Cl}=208/35=6), there exists a common front boundary passing through the plasma–vacuum interface.

This comparison allows one to come to a conclusion that either a great value of friction exists between ions in the plasmoid or the ions experience collective acceleration up to electron velocities, the values of which (as will be shown below) depend on the radiation flux density, the sample material M_{ave} and other parameters.

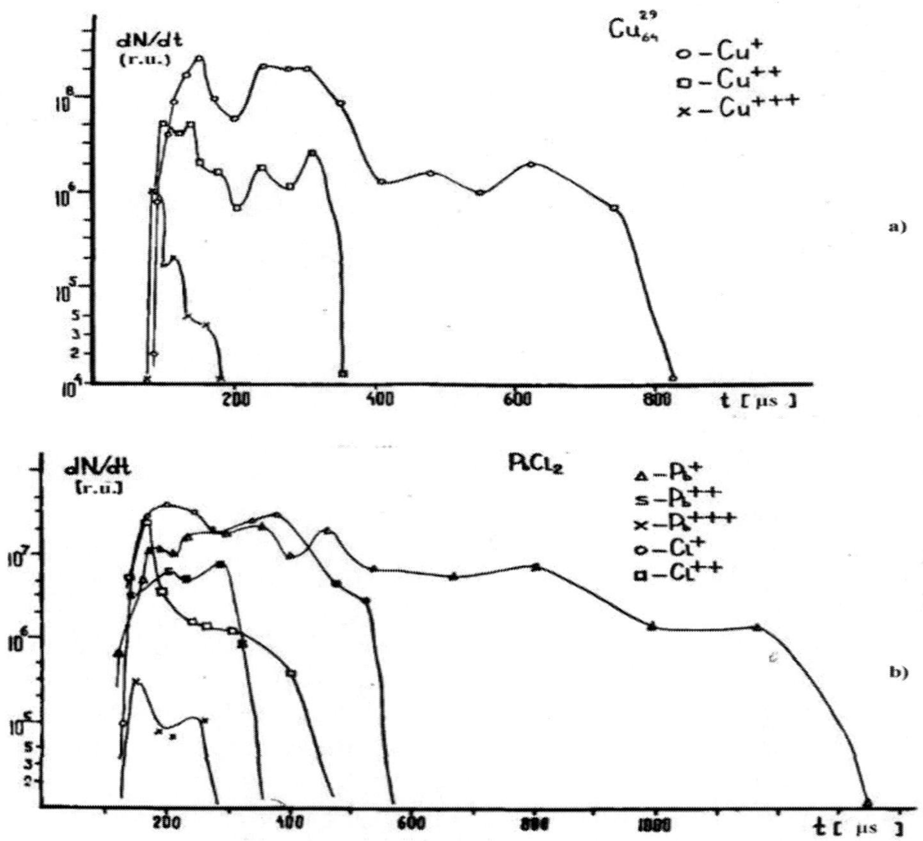

Figure 2.26. Temporal distribution of ions in the plasmoid for (a) single-components samples and (b) binary samples.

By comparing the temporal electron distributions (Figure 2.26) with the temporal ion distributions, the process of the leading edge steepening of the electron pulses becomes clearer. The increase in the radiation flux density is accompanied by an increase in the number of the higher-energy electrons, which leads to the production of multiply charged ions and their motion to the fore-body of the plasmoid. The heavy ions in turn decelerate the most energetic electrons by creating of a steeper leading edge of the pulse.

The value of the high-energy boundary of the electron component spectrum is a very convenient characteristic of the velocity of the plasmoid expansion and formation. This correspondence is indicated in Figure 2.27, where shows the equality of maximal velocities of the electron and ion expansion measured in the range of almost three orders of magnitude with respect to the radiation flux density.

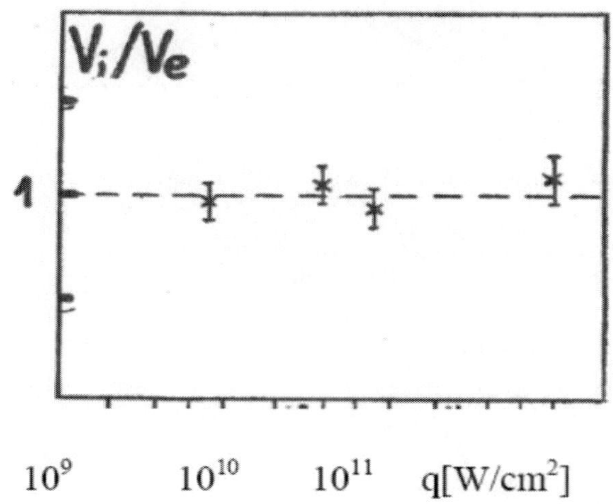

Figure 2.27. Comparison of the velocities of the directed motion of ions and electrons in the torch.

Consider the influence of the radiation flux density on the maximum velocity of the electrons in the laser plasma expansion (Figure 2.28) for several materials. For lighter elements, the particle velocity increases significantly. With increasing atomic weight, the electron component expansion slows down and the velocity of the electrons increases slower with increasing radiation flux density on the targets. The experimental curves in Figure 2.28 make it possible to find the relation between the three parameter: velocity (energy) of the particles, atomic weight of the sample ions and radiation flux density:

$$\mathcal{U}_e \approx \sqrt{(E_i/M_i)} \approx \ln q \tag{2.38}$$

Having obtained this relation and substituted the relevant coefficients into it, one can determine in advance the velocity of the moving plasma front and the maximum energy of the multiply charged ions of a particular element. It follows from this relation that only a small part of the radiation energy is spent to increase the velocity of the directed motion of the particles.

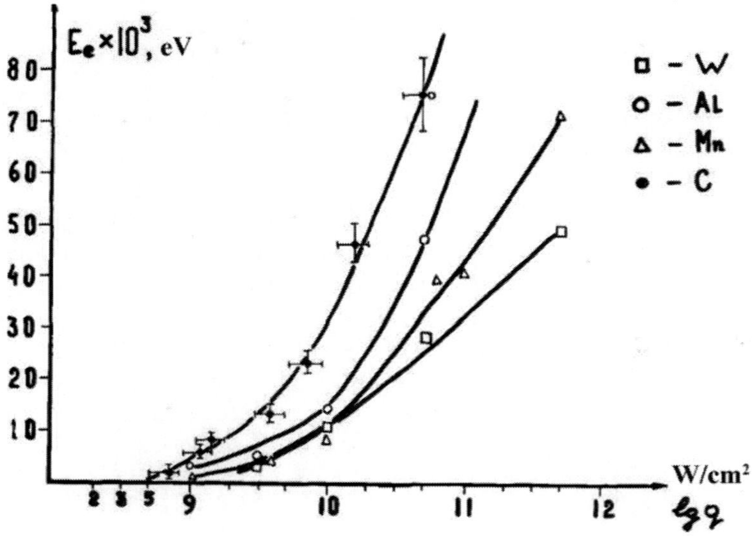

Figure 2.28. Radiation flux density versus electrons energy for different samples.

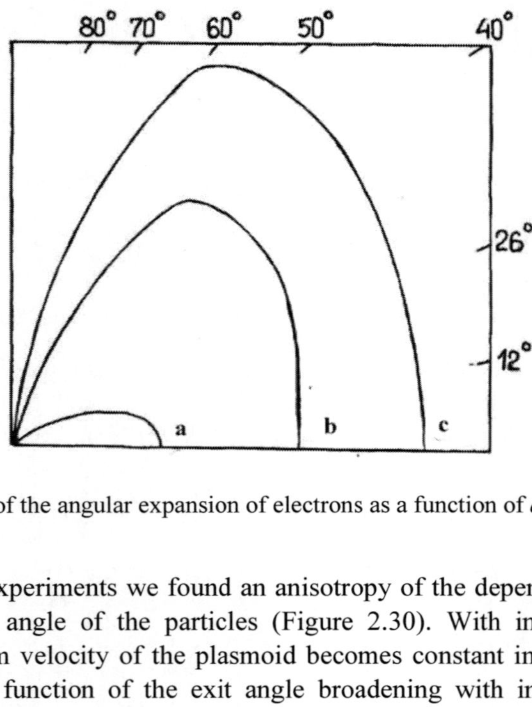

Figure 2.29. Dynamics of the angular expansion of electrons as a function of q at W/cm^2 (a–4x10^9; b–1,6x10^{10}; c–5x10^{10}).

Besides, in our experiments we found an anisotropy of the dependence of the maximum velocity on the exit angle of the particles (Figure 2.30). With increasing radiation flux density, the maximum velocity of the plasmoid becomes constant in the entire range of the exit angles, and the function of the exit angle broadening with increasing radiation flux density can be written as:

$$\varphi^2 \sim \ln q \tag{2.39}$$

Using the data obtained we can conclude that faster electrons fly apart in a comparatively narrow angular interval along the normal to the target.

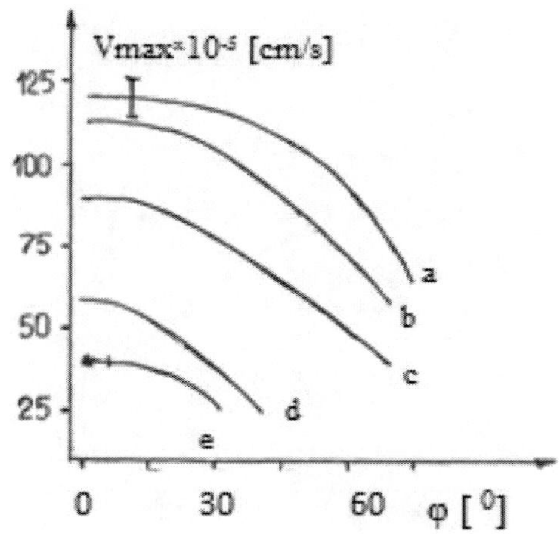

Figure 2.30. Maximum velocities of the ions and electrons as a function of the exit angle analysis: (a) electrons at $5x10^{10}W/cm^2$, (b) ions at $3x10^{10}$ W/cm^2, (c) electrons at $1.6x10^{10}$ W/cm^2, (d) electrons at $4x10^9$ W/cm^2, (e) ions at $3x10^9W/cm^2$).

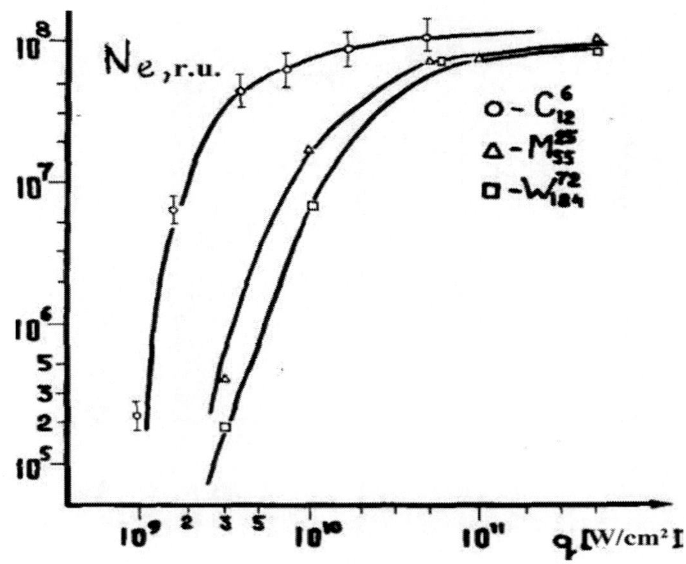

Figure 2.31. Number of the ions and electrons in the plasma as a function of the radiation flux density for different samples.

Let us continue the comparison of the electron and ion characteristics by using velocity distributions of the ions of all the charges and the electrons (Figure 2.32). Within the limits of the error, the coincidence of both distributions is quite satisfactory, which indicates the joined expansion of these two components and the electron neutralization in the laser plasma. Therefore, this makes it possible to conclude that the electrons and the ions experience collective interactions, typical of this type of plasma torch (Figure 2.31).

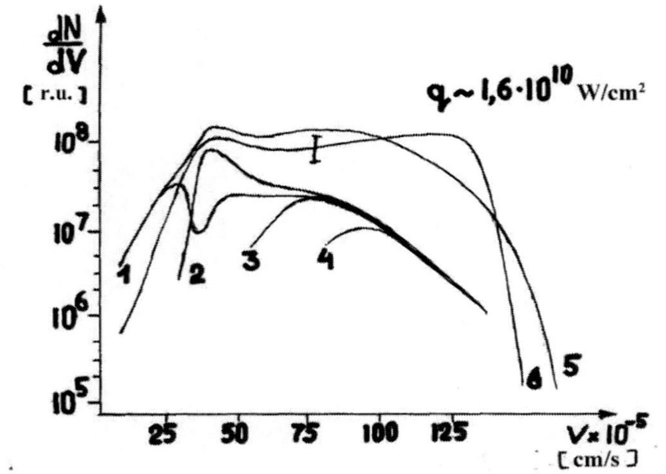

Figure 2.32. Velocity distributions of the ions and electrons (*1, 2, 3, 4* – ion charges; *5* – total charge; *6* – electron distribution).

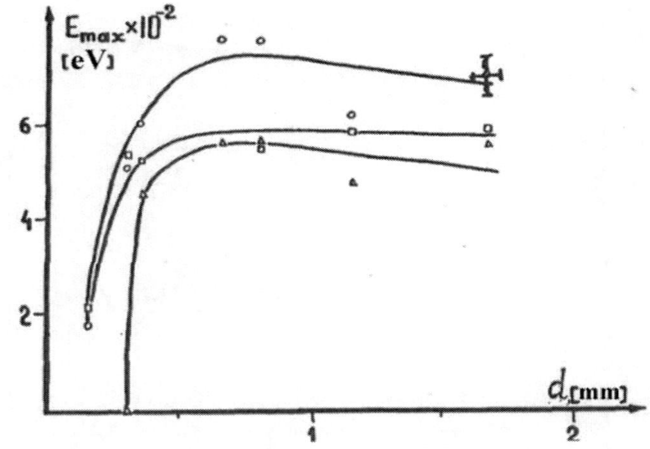

Figure 2.33. Change in the maximum energy of aluminum ions as a function of the initial size of the torch (o–Z=1, п–Z=2; ^–Z=3).

2.3.3. Influence of Initial Conditions on the Particle Acceleration in the Torch

Let us return to the experiments, in which we changed the initial conditions of the plasmoid formation, namely to the studies, where the influence of the focusing spot diameter on the laser torch characteristics was examined. Changing the initial size of the torch leads not only to a change in the plasma lifetime but also can significantly affect the plasmoid expansion geometry, i.e., with increasing focusing spot diameter, the initial phase of the expansion must approach to a plane case, which is reflected in the change of the geometrical parameters of the torch expansion, namely, the angular and energy expansion spectra. Figure

2.33 presents dependence of the change in the maximum energies of the ions of different charges on the focusing spot diameter (taken at maximum possible level of the sensitivity). Initially, when the ratio of the diameter to heated layer depth was increased, we observed a sharp increase (by several times) of the high-energy boundary of the ion spectra. Besides, the shift is observed also for the low-energy boundary of the ion spectra, i.e., the energy distributions move into the regions with higher energies. Table 2.4 lists the values of the low-energy boundary of chlorine ions obtained on the samples made of lead dichloride.

Table 2.4. Values of the low-energy boundary of chlorine ions obtained on the samples made of lead dichloride

Initial size of the torch (*d*)	0.2	0.25	0.45	0.55	0.65	1.4
Value of the low-energy boundary (eV)						
Z=1	25	30	35	70	70	70
Z=2	60	60	90	150	150	-

If changes in the high-energy boundary can be explained by ionization and recombination processes (by varying the focusing spot) as a result of emergence of ions with higher charges *Z*, the shift of the spectrum as a whole with the displacement of the low-energy distribution boundary indicates the changes in the acceleration processes at different value of the relation *d/L* (where *L* is the depth of the heated layer or the plasma torch thickness at the instant of its heating by radiation). It follows from these experimental results that under flatter initial expansion, the acceleration processes are more efficient. To confirm this statement, we performed additional experiments. We measured the angular expansion of the laser plasma for two different cases: tight focusing and large *d/L*. As a result of this experiment (providing that the radiation flux density coincided in both cases) we found that the angular expansion is limited at large *d/L* (Figure 2.34)

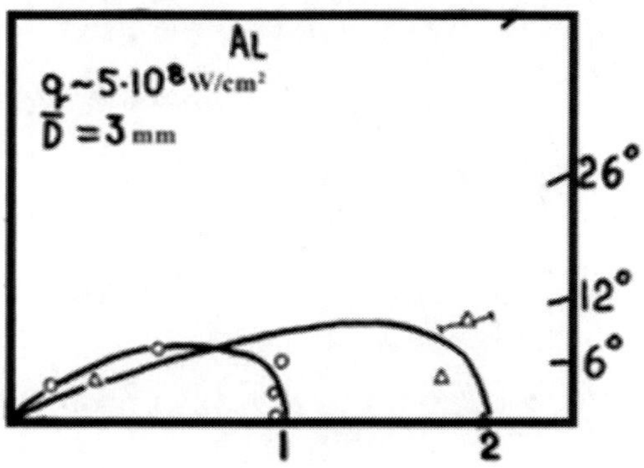

Figure 2.34. Angular distributions of aluminum ions at q=5x10^{8}W/cm^2 for a defocused beam.

The change in both boundaries of the energy spectrum testifies to the change in the acceleration processes when the initial experiment conditions are varied. Such spectrum changes cannot be explained due to the strengthening of recombination with increasing radiation density. One of possible mechanisms of the particle acceleration in the torch is the collective acceleration of ions by the laser plasma electrons. The energy, which can be transferred to the ions, depends on the number of the electrons and their velocity. Note that the ions can be accelerated up to electron velocities only in the presence of the field constantly acting on electrons that compensate for the energy loss spent on the ion acceleration (or else the electrons will decelerate in the ion field). During the action of the laser pulse on the plasma, the electrons can absorb the energy lost as a result of collisions or deceleration in the ion field (Figure 2.35). As was shown above (Figure 2.24), when the atomic weight of the plasmoid changes, one can observe a significant change in the energy spectrum of the plasma ions. Figure 2.36 presents the change in the maximum and minimum energies of natrium ions as a function of the atomic weight of the second component of the binary compound. With the increase in the atomic weights of the entire plasmoid, the velocity of the entire plasmoid decreases, i.e., the energies of natrium ions fall. The electrons, which have received the energy from laser radiation, transfer it to the ions due to electrostatic acceleration only in certain limits (part of this energy is spent on the plasmoid ionization). Thus, we can note that the higher the average atomic weight of the plasmoid, the less the plasmoid velocity, up to which it can be accelerated (other experimental conditions being equal).

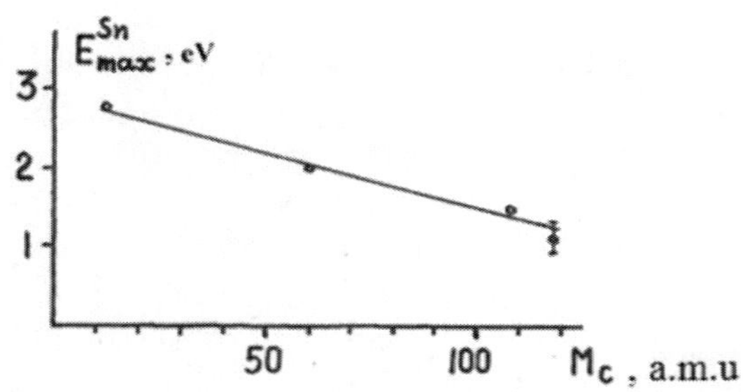

Figure 2.35. Dependence of the maximum energy of Sn ions on the plasmoid average mass.

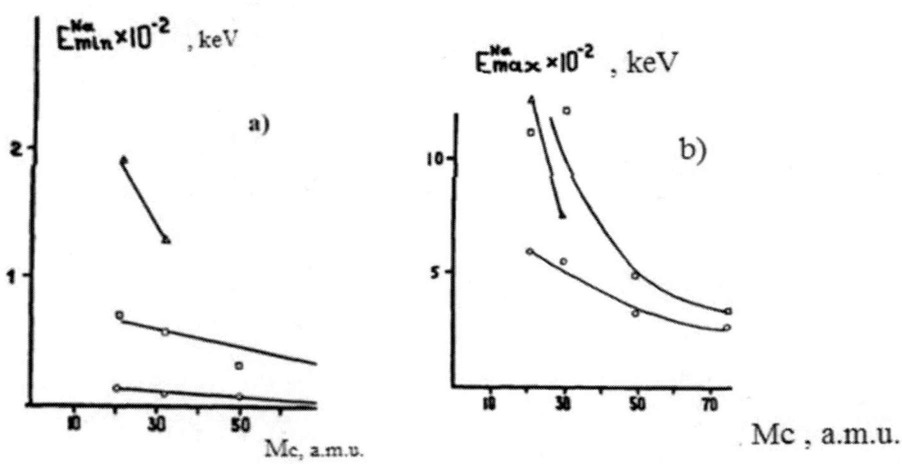

Figure 2.36. Influence of the average atomic weight of sample on the value of the (a) minimal and (b) maximal ion energies (o –particles with the charge Z=1; □ –Z=2; Δ– Z=3).

Thus, when the average atomic weight of the plasmoid is increased, one can observe a proportional decrease in the ion energy (part of the energy is spent on the acceleration of the second component; see Figure 2.35). One can observe not only a change in the high-energy boundary of the spectrum (which can be due to different recombination processes), but also a shift of the low-energy part of the spectrum into the region of low energies (Figure 2.36), i.e. the shift of the entire ion beam is observed on the energy scale.

2.3.4. Effect of 'Focusing' of Ions with a Maximal Charge

In studying the angular spectrum of the torch expansion products we can refine and determine the main mechanisms of the laser plasma formation. In the case of thermal heating of a material, i.e., under conditions of thermodynamic equilibrium, the angular expansion must be isotropic. If in the plasma torch the acceleration processes of the ions or electrons are essential, the angular distribution will be anisotropic with the direction motion determined by the torch expansion geometry.

Figure 2.37 presents the energy distributions of the aluminum and carbon ions at fixed values of q for different angles of the analysis. The energy distributions for each angle and a charge Z can be used to calculate the number of the particles exiting at the given angle. The main features of the angular expansion of the laser plasma in the plane Lo are (Figures 2.38-2.41):

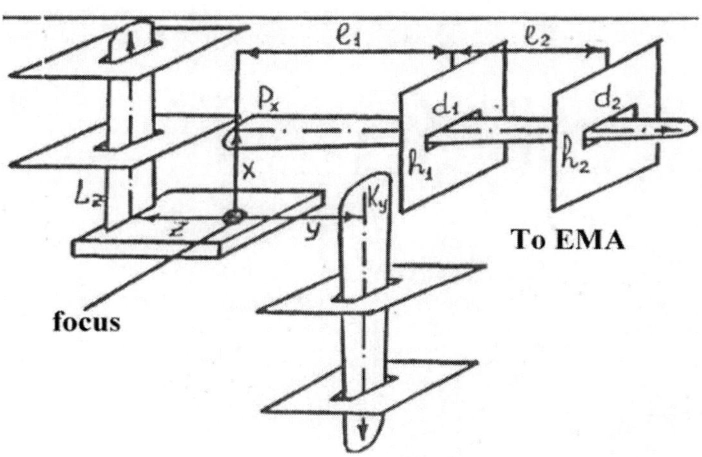

Figure 2.37. System of apertures to select regions of the laser plasma torch.

The main features of the angular expansion of the laser plasma in the plane Lo are (Figures 2.38-2.41):

a) The particle expansion is symmetric with respect to the normal of the target. Either the sample plane or the initially formed plasma layer, which can be represented as a flat disc, plays the major role for the expansion structure; and the normal to this plane is the pivot, around which we obtain the symmetric angular distribution.

b) The angular expansion of the particles narrows with increasing ion charge, which is an indicative of the acceleration effect on the laser plasma ions.

Thus, the laser plasma at any fixed radiation flux density on the sample contains the ions with the charge from Z=1 up to Z_{max} determined by initial experimental conditions (Figure 2.39). The ions having a maximum charge fly apart in a narrow interval, i.e. narrow beams of highly charged ions are formed. The ions having a maximum charge form narrow beams and focus themselves along the normal to the target. This phenomenon first discovered by us in the laser plasma was called the 'focusing' effect of the highly charged ions beams (Figure 2.40). This effect can be used in various fields of science, engineering and technology and characterizes the acceleration processes in the laser plasma.

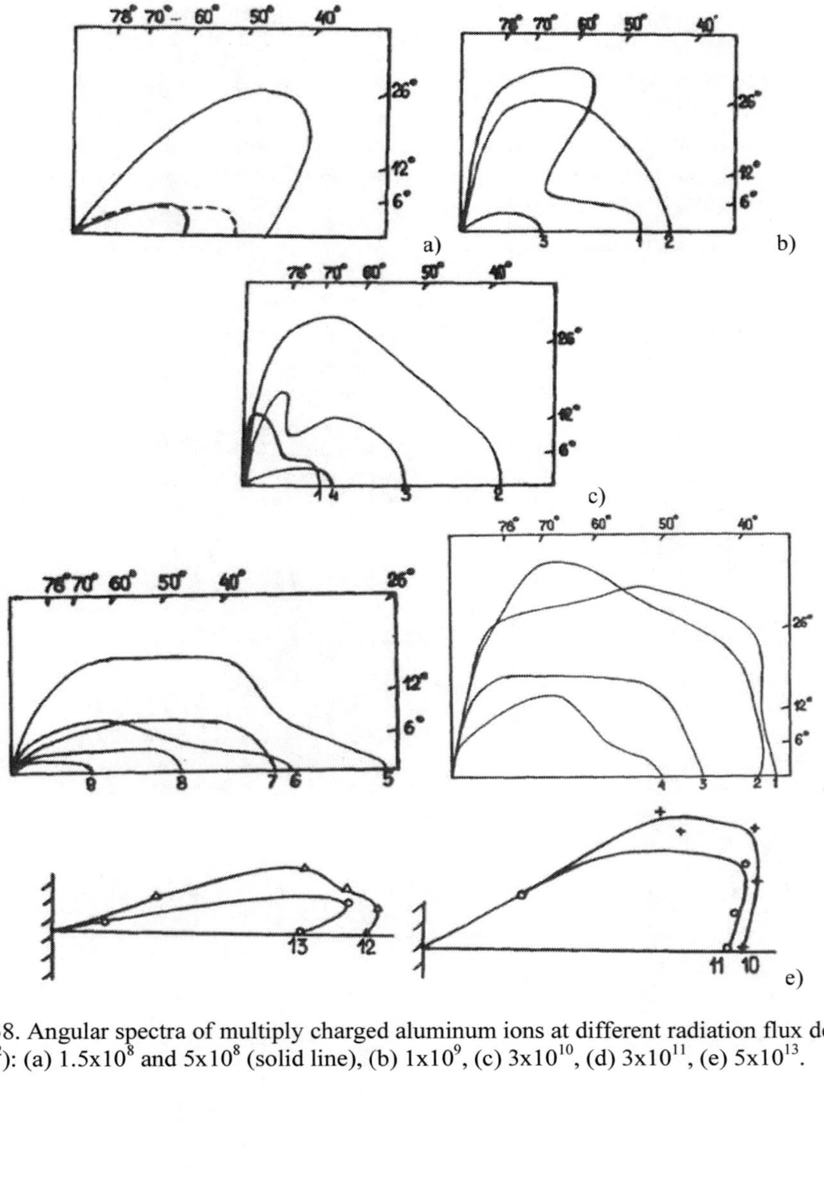

Figure 2.38. Angular spectra of multiply charged aluminum ions at different radiation flux densities (in W/cm^2): (a) 1.5x10^8 and 5x10^8 (solid line), (b) 1x10^9, (c) 3x10^{10}, (d) 3x10^{11}, (e) 5x10^{13}.

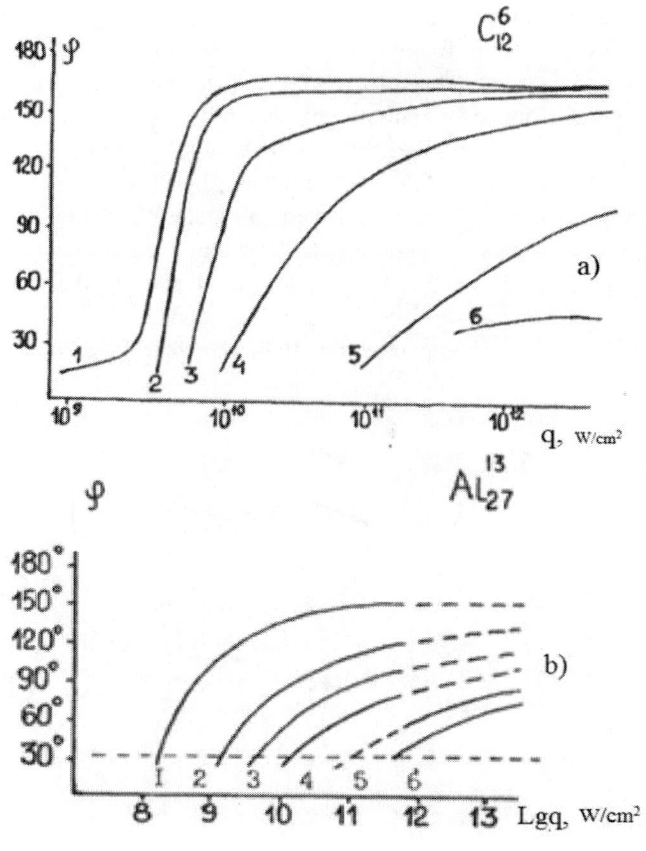

Figure 2.39. Dependence of the exit angle of (a) carbon and (b) aluminum ions from a laser plasma on the radiation flux density.

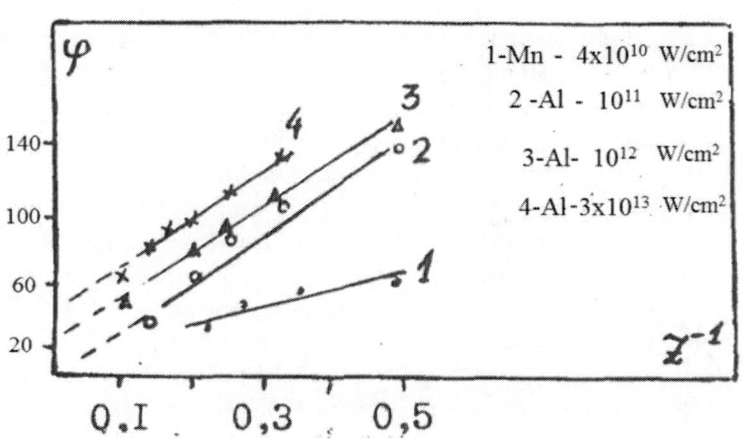

Figure 2.40. Angular scattering of ion beams of different elements as a function of the charge for fixed *q*.

The verification of the influence of the crater (formed under irradiation of a solid body surface) on the ion focusing along the normal to the target surface was conducted as follows. The laser radiation was defocused in such a way that the interaction area of light with the surface increased by two orders of magnitude in comparison with the common case (100–200 μm). To keep the flux density at the previous level, the radiation energy was increased correspondingly. The crater diameter exceeded its depth by several orders of magnitude. The angular distributions resembled the lobes and were elongated along the normal to the sample surface. These experimental data indicate that the 'focusing' effect is not a result of the crater formation process but depends on the acceleration conditions in the laser plasma at early stages of the plasmoid expansion. There exists the dependence of the angular emission of the particles with a given charge on the position of these particles in the energy distribution (Figure 2.41).

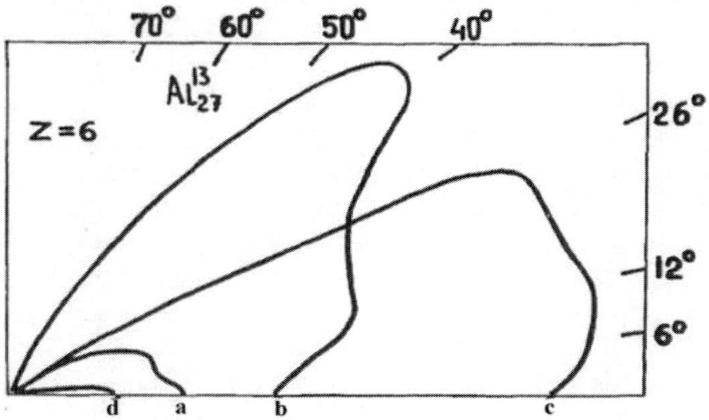

Figure 2.41. Angular distributions of the hexa-charged aluminum ions at various fixed values of energy: (a) 1300eV, (b) 1950eV, (c) 2880eV, (d) 5200eV.

These experiments indicate that the effect of 'focusing' of ions having a maximum charge is a consequence of the fact that the ions situated close to the low-energy boundary are intrinsic ions in contrast to the recombination ones, and so they have a very narrow angular distribution. These ions are formed due to non-elastic collisions of atoms and ions having the charge Z–1 with electrons at the initial sizes of the torch (in the form of a disc) with a subsequent acceleration. The second part of the energy spectrum (namely, its middle part including the intensity maximum of the particles) is a result of recombination of ions with higher charges Z. This process occurs later when the torch becomes spherical. Thus, the angular exit of the ions from the middle part of the energy spectrum is wider. The last high-energy part of the distribution is composed of the ions formed also as a result of the recombination process but from ions with the maximal charge Zmax, which have had initially a very narrow angular distribution. The ions with the considered charge in the region of high energies have again a very narrow angular distribution.

Therefore, the location of the ions in different energy intervals reflects the acceleration processes at different stages and with different particles, which can be effectively split and separated in the plasmoid.

2.3.5. ENERGY OF MULTIPLY CHARGED IONS AS A FUNCTION OF THE HEATING RADIATION WAVELENGTH

It was shown experimentally that the processes of laser plasma heating and formation depend significantly on the wavelength of incident radiation. Therefore, to understand the acceleration processes in the laser plasma, it is important and necessary to establish dependences of the changes in ions energies on the incident radiation wavelength. While comparing the energy spectra of ions of the laser plasma formed under irradiation at different wavelengths, the following questions arise:

- On what part or on what boundary of the spectra to make comparisons?
- Which flux density range to select for comparison?
- How to compare the energies values arising due to differences in the initial experimental conditions?

The intensity maxima of the particles and the right high-energy boundaries of the spectra are not suitable for such a comparison because plasmoids formed by radiation with a higher wavelength exhibit the ions with a higher charge, which would essentially change the location of these values due to the difference in the course of the recombination processes. Therefore, the most suitable parameter for the comparison is the value of the low-energy boundary of the spectrum. In selecting the low-energy boundary of the spectrum for comparing the results, one can also take into account a weak dependence of the left boundary of the ion distributions on the flux density; hence, one can compare the values of the low-energy boundaries of the spectra in a wide range of radiation densities. Now the third question: How to take into account the dependence of the parameters compared on the selection of the sample material and the focusing spot diameter? As was shown above, the best parameter for the comparison is the characteristic of the minimum velocity of the ion expansion, which (see Figure 2.25) is independent of the atomic weight of the element. Hence, the comparison is possible for various elements. The both conditions can be easily met because information about the minimal expansion velocities is available and measurements a performed for many substances.

The most difficult task to accomplish is the selection of the conditions, which would provide the same focusing spot diameter because in the lasers emitting at 10.6μm, the focusing spot in our experiments could not be less than 1 mm. However, most ion energy spectra of the plasma formed by the radiation at 1.06 and 0.69μm were obtained at the initial torch sizes equal to 0.1–0.2mm. To match these two values we shall use the dependences of the low-energy boundary on the focusing spot diameter, described in this Chapter (Table 2.3).

Figure 2.42 presents the change in the minimum energy of the ions as a function the charge for plasmoids consisting of the aluminum ions and formed by radiation at 1.06 and 10.6μm. It follows from the comparison of these dependences that the ion acceleration in the plasmoid formed by the radiation at 1.06μm is more pronounced. If we take into account that the initial size of the plasmoid formed by radiation at 1.06μm is less by an order of magnitude, then in accordance with Table 2.3 the ion energy increases with increasing focusing spot diameter. Thus, the ion energy in the plasma produced by radiation at 1.06μm is twice that in the plasma produced by radiation at 10.6μm.

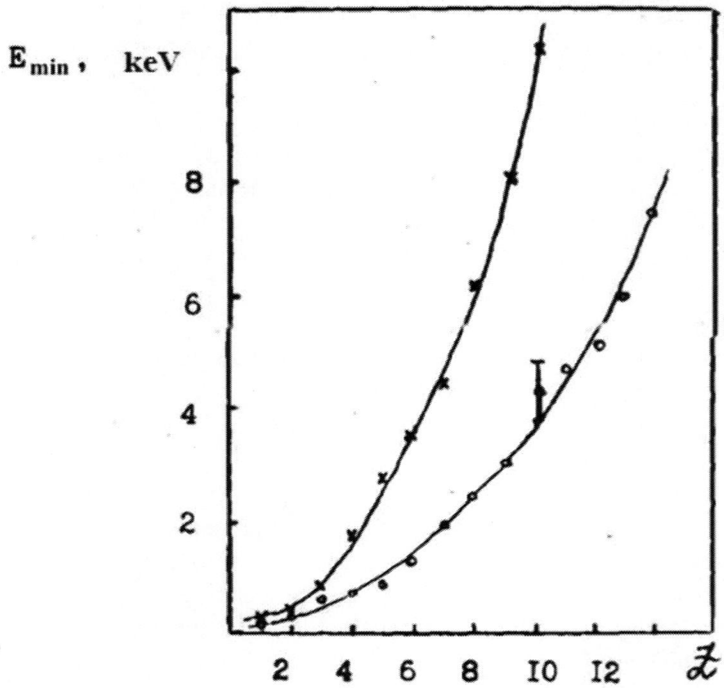

Figure 2.42. Dependence of the minimum energy of the ions on the charge for plasmoids formed by radiation at 1.06 (x) and 10.6 (o)μm (the sample is aluminum).

It follows from the formulae of work [22] that

$$V_{max} \sim \lambda^{-1/6} \tag{2.40}$$

Using this formula, we obtain that the ratio of the ion energies at 1.06 and 10.6μm with is equal to 2.15, which indicates good agreement with the experimental results.

2.3.6. Conclusion

In the literature devoted to the acceleration process of the particles in the laser plasma, it is often assumed that this phenomenon has a threshold nature with respect to the radiation flux density. Using the results of mass-spectrometric studies, we draw the conclusion that acceleration of the particles is a continuous process and starts at the moment of the plasmoid formation. Gas-dynamic (neutral fluxes) and electric (the effect of the charge on the particle energy) fields are involved in the acceleration. Thereby, the laser plasma ion energies reach 50keV for the given radiation flux densities. Distinctive characteristics of these processes are the independence of the low-energy velocity of the ion expansion on the atomic weight of the element as well as the direct proportionality of the low boundary of the particle velocity and the ion charge. The electrons and ions participate in these processes and influence the characteristics of each other.

The characteristics of the angular expansion of the laser plasma ions fit to the common scheme of description of the acceleration processes in the torch, where it is necessary to take into account intrinsic processes in the electrostatic field of electrons emerging at early stages of the plasmoid expansion during the times compared with the radiation pulse duration. All the recombination processes of the ions with a higher charge into the ions with a less charge occur after the acceleration terminates. This fact is experimentally confirmed by the structure of the energy distributions (the energy displacement is absent of the recombination intensity maxima) and by the 'focusing' effect as well as with by the presence of many accelerated particles (and not only in the thin layer of the torch, i.e. in a 'crust'), having very high energies of directed motion. In addition, the presence of the electrostatic field effectively pulling the ions out, which has a positive effect on the process of hardening of the multiply charged ions, allows (among other products of the laser plasma) obtaining the directed beams of ions with a high charge (Z=25) and nuclei of various elements (aluminum, carbon, etc.).

The experimental data make it possible to represent the acceleration process of the laser plasma particles as follows. First, the acceleration process occurs mainly at an early stage of the torch development, following directly the ionization process. Second, initially the laser plasma electrons accelerate in the field of the electromagnetic wave, roughly speaking according to the mechanism pointed out by Silin [22]. Under certain circumstances these superfast electrons have been apparently observed in the experiments on the laser fusion. And third, during the pulse action, there occurs a transfer of the kinetic energy of the electrons in a self-consistent electrostatic field [32].

2.4. References

[1] Afanas'ev Yu. V., Belenov E.M., Krokhin O.N., Poluektov I.A. *Zh. Ekps. Teor. Fiz.*, 63, 121 (1972).

[2] Beigman I.L., Vainshtein L.A. *Preprint FIAN* (Moscow, 1967).

[3] Degtyarenko N.N. Study of the Ion Emission from a Solid Target Irradiated by *Nanosecond Laser Radiation* (Cand. Diss.) (Moscow, 1974).

[4] Afanas'ev Yu.V. *Theory of the Formation of High-temperature Laser Plasma* (Doct. Diss.) (Moscow, 1977).

[5] Derzhiev V.I., Zakharov A. Yu. Ionization and Recombination Kinetics. – *Preprint IPM Akad. Nauk SSSR* No. 60 (Moscow, 1978).

[6] Boiko V.A., Krokhin O.N. Study of Laser Plasma Parameters and Dynamics. – *Preprint FIAN* No. 121 (Moscow, 1972).

[7] Anan'in O.B., Afanas'ev Yu.V., Bykovskii Yu.A., Krokhin O.N. *Laser Plasma* (Moscow, 2003) [in Russian].

[8] Bykovskii Yu.A., Elesin V.F., Kozyrev Yu.P., Degtyarenko N.N., *Sil'nov S.M. Zh. Eksp. Teor. Fiz.*, 60 (4), (1971).

[9] Anan'in O.B., Bykovskii Yu.A., Sil'nov S.M., et al. *Pis'ma Zh. Eksp. Teor. Fiz.* 16 (10), (1972).

[10] Sarantsev V.P., Perel'shtein E.A. Collective Acceleration of Ions by Electron Rings (Moscow, Atomizdat, 1979, pp. 79-80). [in Russian]

[11] Silnov S.M., Suslov A.I. Possibility of Developing a Multiply Charged Ion Source on the Basis of a Pulsed CO_2 Laser.- Deposit manuscript VINITI, 1978 No.2010 – 78 D.N. – 22 p.
[12] Bykovskii Yu.A., Degtyarenko N. N. Fetisov V.K., et al. Zh. Tekh. Fiz., 44, 73 (1974).
[13] Donets E.D., Il'ishenko V.I., Al'pert V.A. Preprint JINR, No. P7-4469 (Dubna, 1969).
[14] Bykov V.N., Pergament A. Kh. *Preprint IAE*, No. 56 (1978).
[15] Bykovskii Yu.A., Sil'nov S.I. Ionization of Atoms and Ions in Laser Plasma. – *Preprint MEPhI,* No. 004-86 (1986).
[16] Derzhiev V.I., Zakharov A. Yu., Ramendik G.N. On Influence of Chemical Composition on Ionization and Recombination Kinetics in Spark and Laser Plasma. – *Preprint IPM Akad. Nauk SSSR*, No. 23 (Moscow, 1980).
[17] Myuller A., Shevel'ko V.P. *Zh. Tekh., Fiz.,* 50 (5), 985 (1980).
[18] MacDaniel E.W. *Collision Phenomena in Ionized Gases* (New York: Wiley, 1964).
[19] Luk'yanov G.A. *Supersonic Jets of Plasma* (Moscow:, 1985). [in Russian]
[20] Bykovskii Yu.A., Buyanova E.A., Sil'nov S.M., Fetisov V.S. Ion Acceleration and Recombination in a Plasmoid of Multiply Charged Plasma. – *Preprint MEPhI*, No. 005-87 (1987).
[21] Linlor W. *Appl. Phys. Lett*, 3, 210 (1963).
[22] Silin V.P. *Pis'ma Zh. Eksp. Teor. Fiz*., 21, 333 (1975).
[23] Denisov N.G. *Zh. Eksp. Teor. Fiz*., 31, 609 (1956).
[24] Guinzburg V.L. *Propagation of Electromagnetic Waves in Plasma* (New York: Gordon and Breach, 1962).
[25] Gurevich A.V., Silin V.P. *Yadern. Fiz.,* 2, 280 (1965).
[26] Boloneu, Dilona et al. *Phys. Lett*., 57a, 247 (1976)
[27] Gurevich A.V., Mechsherkin A.P. *Zh. Eksp. Teor. Fiz.,* 80 (5), 1810 (1981).
[28] Gurevich A.V., Pitaevskii L.P. *Zh. Eksp. Teor. Fiz.,* 49, 647 (1965).
[29] Gurevich A.V., Pariiskaya L.V., Pitaevskii L.P. *Zh. Eksp. Teor. Fiz*., 63, 516 (1972).
[30] Horse, Nielson, *Phys. Fluids.,* 16, 909 (1973).
[31] Apollonov V.V., Sil'nov S.M., et al. *Zh. Eksp. Teor. Fiz*., 377 ().
[32] Bykovskii Yu.A., Sil'nov S.M. Acceleration of Ions in Laser Plasma. – *Preprint MEPhI,* No. 001-84 (Moscow, 1988).

Part 3. High-Power Laser Radiation as a Tool in Science and Engineering

Chapter 10

3.1. LASER WEAPONS – MYTHS AND PROSPECTS

The Wall Street Journal has recently published an article in which it was suggested that the anti-missile airborne laser (ABL) project suffers cost overruns and delays and may fall victim to budget cuts. Russian mass media have immediately happily reported that the Pentagon refuses to develop combat lasers. However, this euphoria seems very misleading, if not tougher.

Today, chemical lasers are being widely replaced by solid-state diode-pumped lasers. It is these lasers that the Pentagon counts on, because they are more compact, simpler and cheaper to use than chemical lasers. Besides, they are reliable, easily compatible (without any transformation of the output voltage) with nuclear and solar energy, allow for further scaling of output parameters and the efficiency of their operations is significantly higher. Northrop Grumman Corporation has already presented a 10^5-kW solid-state laser and intends to significantly increase its output power. According to data from U.S. laboratories work is on the way to develop a prototype of a 500-kW laser. Subsequently, 'hyperboloids' will be truck-mounted [high energy laser technology demonstrator (HEL TD) program], ship-mounted [maritime laser demonstrator (MLD) project], and airborne [high energy liquid laser air defense system (HELLADS); a laser for F-35, B-1, CH-47 aircrafts). Another direction is largely supported by Raytheon, which stakes on fiber systems. A 50-kW laser weapon system (LaWS) will be integrated with the Phalanx Close-In Weapons System (CIWS) against anti-ship missiles and its land based version of Centurion C-RAM. In addition, the officials in the U.S. have recently reported on the progress of works on a combat free electron laser. At the same time, we should not forget about "Alpha" system (a laser with an output power of 4.5 MW), lying on the ground and waiting for the decision to be launched. Brining solid-state lasers to megawatt powers takes time and considerable resources. However, the experience of designing strategic laser systems in the previous years and a strong belief in the attainability of the goal – the development of high-power laser weapons – help to significantly accelerate the pace of work in the field of new technologies. It should be noted, however, that tactical laser systems with lower power outputs have already been tested in the U.S. and now find applications in the army. All this testifies that the Pentagon experts clearly do not think about cancelling the promising laser programs. We are talking here about an effective system of disinformation. Last year's report 'Technology Horizons' by the Pentagon refers to the global changes in the 'rules of the game' after the proliferation of 'high energy weapons', which will turn the traditional symbols of military power into outdated stuff like cannonballs and cavalry... And while the U.S. develops laser programs, Russia shows 'laser apathy.'

Incompetent bloggers and pseudoscientific workers, who had something to do with the laser program thirty years ago, speculate that 'lasers are a bluff.' As a result, there has appeared an epic set of myths about combat lasers. Consider the most ridiculous of them.

3.1.1. Ten Myths

Myth 1. 'Combat lasers have been developed for four decades with no progress in sight.'

Let us quote one of the Russian papers: "In the 1970s the Americans took a 150-ton Boeing-707 and 'stuck' a laser to it, which successfully burned small rockets. In the 2000s, they took a 350-ton Boeing-747 and 'stuck' a heavier and more powerful laser, which successfully burned larger size rockets. In 20 years they will buy in Ukraine a decommissioned "Antonov AN-225 Mriya" (640 tons), and here it is, the Death Star. Yes, probably it will be able to shoot down a "Scud", and even a "Taepodong". However, only at the site, and once, no more."

Under the '150-ton Boeing-707,' to which a laser was 'stuck,' is apparently meant a 137-ton Boeing KC-135 stratotanker based on the" 707 "of the first), modified in 1973 into NKC-135ALL [airborne laser laboratory (ALL)]. In 1983, a laser mounted on the aircraft shot down several Sidewinder air-to-air missiles at a distance of 5 km and some other 'little' things. What has changed since then? According to the above quotation – only the size of the aircraft.

But what about the reality? Even the so-called 'megawatt' cw lasers of the 1980s did not emit light at megawatt power levels, they consumed more power. The 2.2-megawatt Miracle laser system, which later in the combat version was called a tactical high energy laser (THEL) ('MIRACLE' with the 'SEALITE' guidance system) did not possess a supernatural power. What to say about the earlier and five times weaker ALL. Has there been any progress since then? An ABL has a power of 1.1 MW and it is not the power to be consumed; this is the power in the beam. Thus, a 'more powerful' (50 times) laser was 'stuck' to a 350-ton Boeing... However, it should be understood that the actual capabilities of the laser rely not on the power as such, but on the concentration of the radiation, i.e., the ability of the 'guns' to emit not only a powerful, but also a narrow beam. The ALL had the level of the radiation concentration equal to 10^{13} J/(sr·s). An ABL has the level of about 10^{18} J/(sr·s) – that is, 10 thousand times higher. These achievements are made up not only of the straight-line growth in the power. The last 30 years have seen a period of extremely rapid development of adaptive optics to compensate for the effects of atmospheric turbulence and laser path on the transmitted beam. In addition, the lasers of the same class are radically reduced in size. The first version of a THEL weighed 180 tons and could be hardly installed into six trailers. The laser used a hydrogen–fluoride mixture which is very unfriendly to the environment. The second generation of advanced tactical lasers (ATLs) relies on oxygen–iodine mixtures [so called chemical oxygen–iodine lasers (COILs)] and is more compact. Finally, a new solid-state laser of Northrop Grumman weighs 1.5 tons, including the cooling system. In the future its weight is expected to be reduced to 750 kg. As a result, the land based version of the system consists of a heavy expanded mobility tactical truck (HEMTT A3), command post on a high mobility multipurpose wheeled vehicle and a towed single-axle trailer with the AN/MPQ-64 radar. At the same time, the U.S. works hard to convert the CW regime of the lasers in the pulse-periodic one, which will dramatically increase their range of actions.

Talks about the fact that "combat lasers have been developed for forty years and so they are hopeless," only show ignorance in technical matters. Breakthrough technologies are first tested for a few decades before the entry into a phase of maturity. Thus, aircrafts at the time of the first flight had almost 60 years of history – the first flying models were built in 1840 and the attempts to build full-size airplanes date back to 1868. It is, in fact, a classical scheme of development of any technology using new physical principles: First, a long 'incubation period' with no apparent practical results, and only then a 'great leap forward.'

Myth 2. "Lasers cannot be used for a long time, usually they work several seconds."

It is not so! In fact, chemical and solid-state combat lasers ensure continuous wave operation for minutes and tens of minutes. The next step in the development of high-power laser systems, of course, will be the implementation of a variable temporal structure of radiation in order to raise the peak radiation power for the ablation mechanism and to eliminate the effect of the screening of the target by plasma.

Myth 3. "Energy of laser weapons is negligible compared to the fire arms."

For comparison, the power of a 76.2-mm F-22 divisional gun (1936) is at 150 MW. This power is 150 times higher than that of the ABL!... Besides, we ignore here the energy of the explosive in the shell, which is the same. Ponder upon this simple fact: A small ancient gun of the Second World for the price of scrap metal is hundreds of times more powerful than the ultramodern 'combat' laser weighing tens of tons and valued at over five billion USD. Only one shot from the ABL costs millions of dollars. And this shot comparable in energy with the burst of a machine-gun fire.

Such a comparison of the power achieved in 0.01 s with the power of CW light, and the 'proof' of inferiority of 'long-play' weapons (by using this comparison) contradicts even the school physics course. Let us try to compare everything in the correct way, i.e., by counting the energy sent to the target.

The muzzle energy of a 12.7-mm heavy machine gun is 15 – 17.5 kJ at a rate of 80 – 100 shots per minute. In other words, a 100-kW laser can be 'replaced' by three and a half heavy machine guns (6000 kJ/min vs. 1750). But let us return to the gun. The muzzle energy of F-22 is 1.35 MJ, while the power of the ABL is 1.1 MW, i.e., 1.1 MJ every second. Thus, the laser shoots 48 'shells' per minute. By converting the MW power to the TNT equivalent, we will obtain 240 grams of explosives per second and 14.4 kg per minute, which corresponds to the content of 18 high-explosive shells from the same gun. However, the actual 'value' of the laser is higher. The matter is that even with accurate firing of firearms, most of the 'energy' goes not to the enemy, but dissipates. The reason for that includes a dozen factors (wind, fluctuations in humidity, air pressure and temperature, the Coriolis force, etc.), making the bullet or shell spread inevitable. A steady flux of photons flies to where it was sent, excluding the huge amount of unnecessary loss.

Myth 4. "Efficiency of lasers is a few percent."

In fact, the efficiency of combat lasers is about 20.6%, and this is not the limit. According to the RELI program, the efficiency is planned to be raised up to 25%. Fiber lasers, which are adapted by Raytheon for military applications, now have an efficiency of about 30%. The efficiency of firearms is 20%–40%. At the same time, the efficiency of (smaller but steadily growing in power) solid-state diode-pumped systems is more than 50%, and soon it will approach its physical limit of 85%.

Myth 5. "The laser beam has a huge diffraction divergence."

Here we deal a physically insuperable law of diffraction, which states that the laser beam always diverges with an angle proportional to the ratio of the wavelength to the diameter of the beam. If we take a specific combat IR laser with a wavelength of 2 μm (this wavelength is typical of combat THELs, etc.) and a beam diameter of 1 cm, we obtain a divergence angle of 0.2 mrad (this is a small angular divergence; for example, standard laser pointers/rangefinders have a divergence angle of 5 mrad and higher). However, because of 0.2-mrad divergence the beam spot diameter will increase from 1 cm to 3 cm at a distance of 100 meters. That is, only 100 meters away the beam density will decrease (proportionally to the area of interaction) by 7 times. At a distance of a kilometer the density of the beam falls by 300 times.

Actually, a combat laser emitting a beam with the initial diameter of 1 cm is a fruit of an unhealthy imagination, not burdened with some knowledge in this area. In fact, when using the focusing optics the diffraction divergence is about λ/D, where λ is the wavelength, and D is the diameter of the mirror and also the initial beam diameter, tapering due to focusing as the beam approaches the target; a large beam diameter (meters in this case) provides a low diffraction divergence. In the case of the ABL the wavelength is 1.315 μm, and the diameter of the mirror is 1.5 m; by dividing one by the other, we obtain the divergence of about 10^{-6} rad. In other words, at distance of only one kilometer the laser beam from the Boeing diverges by (oops!) 1 mm. At a distance of 200 km, the diffraction divergence will be 20 cm. The actual ABL beam divergence exceeds the diffraction limit by only 1.2 times.

Myth 6. "One can easily be protected from laser weapons using, for example, an aluminum mirror."

Yet another malapropism. Indeed, some metals may have near-100% reflection coefficients. First, however, these coefficients are not equal 100%. Thus, at a wavelength of 1 μm the reflection coefficient drops to 75% for most structural metals. A real missile after its launch will be also significantly contaminated with combustion products. Meanwhile, modern 'hyperboloids' emit light in the vicinity of 1 μm (the ABL has a wavelength of 1.315 μm). In this case, 25% of hundreds of kilowatts will be sufficient even in the cw regime to heat up and melt down the thin skin layer of the missile. Thus, reflectivity will decrease because the absorption of laser radiation increases rapidly with increasing temperature, and abruptly jumps after the start of melting. In the pulse-periodic regime the situation is even more favorable.

Besides, there arises also a 'childish' question: If the laser beam can be focused and directed by a mirror, why cannot the same mirror protect us from the laser beam? In lasers use is usually made of multilayer dielectric mirrors that can reflect very much – but in a very narrow range and only under strictly defined angles. In addition, they are cooled, which is impossible to do with the entire surface of the target. In other words, simple, effective and affordable protection from high-power lasers does not exist.

Myth 7. "The problem of overheating of lasers cannot be solved."

Four megawatts of heat, which can heat red-hot an aircraft and burn to the ground, are generated per each megawatt of laser power. The cooling system with the gas flow rate of 1800 m/s (de Laval nozzle) is unable to release the generated heat from the fuselage.

In reality, the 'disposal' of the amount of heat in the megawatt units is quite trivial. Has anyone seen a red-hot diesel locomotive? Meanwhile, a decent diesel engine with a capacity of 2 MW releases more than 1 MW of heat into in the oil and cooling system. Far less easy is the problem of heat release from the limited volume of the laser weapon. In the case of a chemical ABL, the heated reaction products are simply blown out of the cavity by the well-known de Laval nozzle, and then liquid ammonia is used for cooling. Fairly cumbersome is the system with cryogenic components; however, it really is able to 'recycle' a very impressive amount of heat. Tactical solid-state lasers, which need more than 400 kW of heat to be released, do well without cryogenic 'refrigerators.' Thus, the HELLADS makes use of a unique cooling technique; the circulation of a liquid releases excess heat outside the 'laser cannon.' Remarkable also is General Atomics' advanced thermal energy storage device capable of cooling directed energy weapons. Heat is stored in the 35 kilogram module by melting a wax-type phase change material. As a result, the HELLADS provides for the interception of missiles within two minutes at the specified cw power followed a thirty-second break.

Myth 8. "High-power and compact energy sources for combat lasers do not exist."

This is partially true: It is not yet possible to mount a 100-kW solid-state laser on anything less than a truck because of the need to have at hand a 500-kW generator and suitable capacitors. In fact, this is a real problem, which has nothing to do with fantasy. In practice, the hybrid version of the HEMTT (HEMTT A3), even in the basic version has a 350-kW generator, which can provide up to 200 kW. When the engine power is increased up to 505 hp, HEMTT A3 can produce 400 kW of continuous exportable power. A nice addition is the 1.5-MJ capacitor bank. In other words, where the bloggers fancy an electric power plant, in fact, there is a high-tech truck. However, the issue of energy in the space can be solved by other, more efficient ways.

For example, well developed are nuclear power sources, solar energy, with its unlimited possibilities.

Myth 9. "Every shot a laser is worth millions of dollars."

In fact, one shot of the ABL costs 10 thousand USD, whereas the domestic 16 million rubles are the propaganda and exaggeration. This is comparable with the cost of a Fagot anti-tank guided-missile system. More complex anti-tank guided-missiles cost tens of thousands of dollars; for example, a Maverick air-to-ground missile for hitting targets ranging from a distance of a few thousand feet to 13 nautical miles at medium altitude costs 154 thousand USD, a Patriot missile – 3.8 million USD. The cost of a shot from tactical lasers is less than from the ABL. Even in a hydrogen fluoride THEL it was 2-3 thousand USD; in this case, the laser used not hydrogen but deuterium which is quite expensive.

Myth 10. "All the problems that can be solved with laser weapons, are easier and cheaper to solve by traditional means."

This conclusion has already been proven to be ineffective; see, for example, at Israel's attempts to defend against missile attacks by Hamas missiles (the Iron Dome system). One interception costs 30-40 thousand USD. The cost of a rocket for a Grad multiple rocket launcher system is about 1 thousand USD, the cost of a Qassams rocket does not exceed 200 USD.

Thus, the interception will be 40-200 times more expensive than the rocket launched by the enemy. In this connection Tariq abu Nazar, the Hamas spokesman, once said "if every rocket launch of our missilemen costs tens of thousands of dollars for Israelis, we will assume that the goal is achieved." As a result, some newspaper people accuse not developers of laser systems but those who are responsible for termination of the Israeli-American program. The Centurion missile defense system has not found wide application because of the short-range and great ammunition consumption.

3.1.2. Legends

Of course, this is not a complete list of the legends about lasers. Most of them are built on the same principle: either outrageous lie or painstaking making of a mountain out of a molehill. In fact, lasers on the battlefield are real and the army, which possesses such weapons, will receive an impressive advantage. Thus, the air force, able to actively defend against anti-aircraft and air-to-air missiles, will be much less vulnerable to air defenses. Thus, the development of laser technology is critically important not only for Americans. Combat lasers are an obvious asymmetrical response to the superiority of the West in precision weapons. The 'ideology' of the latter in a very rude manner is to ensure that instead of a dozen of shells, a guided munition (though much more expensive) is used to precisely hit a specific target, and to minimize collateral damage. However, such a scheme is particularly vulnerable to laser defense systems, which make no difference between an archaic shell for two hundred dollars or an expensive high-tech device or military equipment. The number of targets is not so great, and their cost is ten times higher than that of the most expensive laser 'shot.'

However, instead of trying to catch up with the U.S., Russia continues to develop ideologically outdated lasers, which are inefficient, cumbersome and do not allow the achievement of the posed goals in the near future. Besides, anti-propaganda flourishes. The reasons for this development of the situation are quite clear:

First, the highly successful Soviet laser program of 1970s-1980s was literally in ashes in the early 1990s and the officials, for obvious reasons, are not too eager to take responsibility for their opportunistic decisions, and are now involved in much more profitable and safer businesses.

Second, if the production of traditional weapons is largely supported in our country by well-defined business interests, the laser lobby does not exist at all, because some people are far distance and some are dead.

Third, a significant part of the Russian political elite is always ready to turn a blind eye to strengthening of the emerging 'asymmetry' in the field of strategic weapons in order not to annoy 'partners' and always have a guaranteed access to their accounts in Western banks.

Fourth, to continue to fight for the interests of the defense of the country today is not safe for personal career and health. One must possess an enviable courage, great scientific outlook, intuition and expertise in the field of high technology, as well as a good vision of further development of the strategic situation in the world to defend his position in the modern world.

3.1.3. Afterword

It is already evident that the world has entered a new round of technological race. Unfortunately, it does not depend on us, and we do not start it. Most developed countries, based on their technological advantages, are spending billions of dollars to develop high-tech next-generation lasers. Their investments in new technology of laser weapons are not comparable to what we do. They are ten times greater. Thus, according to Japanese media (2008) U.S. invested more than 85 billion USD in the development of semiconductor laser pumping of solid-state lasers.

The need to accelerate the development of high technology was the major topic of President Dmitry Medvedev's speech at an enlarged meeting of the State Council (2010). It is important to note the opinions of American experts, who say that today laser technology is one of the most effective ways to gain technological superiority in the world. Due to the efforts of Nobel laureates A.M. Prokhorov and N.G. Basov Russia has always been one of the world leaders in this field. And today, laser technology makes it possible:

- to provide a new level of development of industry, science and technology, the revival of scientific and technological strength of Russia on the basis of modern high-techs;
- to make Russia a leader in the field of technical, scientific and technological progress;
- to revive a large number of enterprises in many sectors of Russian industry, which are well known for their developments in previous years;
- to strengthen Russia's leading position in space and to ensure strategic and geopolitical priorities in the modern world; and
- to derive benefit from the sale of a wide range of laser programs, which is today comparable to the benefit from the sale of natural resources.

3.2. Photos and Figures

Air-based military laser.

Mobile laser system.

Sketch space based laser system.

Chapter 11

3.2. TO THE SPACE BY LASER LIGHT

3.2.1. INTRODUCTION

The LJE is the most promising rocket engine of a new class. In our article we discuss the creation of an automatic spacecraft engine, which will pass its trajectory under the action of a train of high repetition rate short laser pulses directed toward it from the Earth's surface.

The importance of this problem is caused by the fact that the LJE is substantially more cost-effective than conventional chemical-propellant engines. In the initial stage of a flight as the working medium adapts to atmospheric air, and beyond the atmosphere an insignificant onboard reserve of gas or easily sublimated substance is required. In this case, specific expenditures for the launch of loads into space can be reduced to 5000–10,000 rub/kg, which is approximately 100 times less in the comparison with the contemporary scale of world prices. Especially, the possibility of maintaining the orbital parameters at the assigned level with the aid of the same laser system, intended for the starting, can be evaluated very highly.

In the present time, works on the study of LJE designing possibility is under way. Within the framework of the 'Lightcraft' project such systems are being intensively developed in the USA, Germany and Japan. Thus, in November 2000, 'Lightcraft Technologies' (USA) successfully conducted model tests of a rocket, which rose to a height of 70 m during 12.7 s under the action of an exhaust jet, which appears due to high-power laser emission. In the experiments a low-frequency 20-Hz CO_2 laser with an output power of 10 kW was used, which in the present time is being upgraded to the level of the 100 kW output power. Mechanical impulse appeared due to ablation of a special polymeric material from the concave surface of the reflector, located in the tail end of the rocket, where the laser beam pulses were accepted.

In the beginning of 1973, the USSR under the scientific guidance of Acad. A.M. Prokhorov performed works on the study of possibility of LJE designing. The reflector, located in the tail of the rocket prototype, concentrated the obtained radiation in air and ensured a micro-burst, thereby producing a reactive thrust. Different models of the reflectors, which served at the same time as the laser light receivers, were successfully. One should note that all experiments pointed out above were conducted with the use of electric discharge CO_2 lasers with a power up to 10kW, while for launching different highly and technologically efficient equipment (satellites ensuring global network communications, Internet, photo-monitoring of the Earth surface, debris cleaning) into orbit, a substantially higher radiation power is required; thus, for example, to launch a 1000-kg spacecraft the laser with a power of no less than 10–15 MW is needed. Currently, such a laser can be only a gas-dynamic laser (GDL), which stems from the fact that the laser technology, to a significant degree, intersects the rocket technology, which for the last 50 years has noticeably advanced forward in the creation of super-power gas generators and now makes it possible to pose similar problems. Furthermore, the laser must operate in the pulse-periodic regime at a high repetition rate of short pulses. It helps to except the process of laser emission screening entering by generated plasma in vicinity of reflector surface and due to that to increase the engine work effectiveness [1]. In the opinion of specialists—classical missilemen—the LJE can find an effective use in creation of cheap launching single-stages for 5–50-kg nano- and micro-satellites, which according to the opinion of the experts will constitute the major part of commercial launch program in the nearest future. In the first stage of the spacecraft flight at the heights up to 30km, as a working medium in the engine use can be mad of atmospheric air, and then to launch a satellite into orbit the onboard reserve of the special fuel—ablating substance—in the quantities not exceeding 15% -20% of the spacecraft started weight is needed.

The experience of high-power laser development is accumulated at the A.M. Prokhorov General Physics Institute of RAS, at the OJSC 'NPO Energomash' and other organizations of Russia. During the last few years the Private Joint Stock Company 'Energomashtekhnika' performed very successful experimental studies of the pulse-periodic regime in the high-power and well-developed CW laser systems. This makes it possible to approach the experimental realization of super-power pulse-periodic laser source on the basis of the gas-dynamic principle and LJE, which forms part of a light carrier with control system. The forthcoming works must become a major step towards the future of super-lightweight spacecrafts being launched into the low near Earth orbit. The realization of the project will make it possible to create highly cost-effective reusable LJEs for orbital injection of different payloads.

The key advantage of the new approach stems from the fact that the source of the kinetic energy and payload are united in the space and the launching weight of the rocket can be lowered to the payload weight only. K.E. Tsiolkovsky, a Russian and Soviet rocket scientists, foretold that in the future spacecrafts will e launched with the aid of the electromagnetic waves directed from the external energy source (the laser at that time was not known to the world at all). The great interest of scientists and specialists, as was confirmed by a recent symposium in this field of research, is mainly due to the successful solution of the problem of construction of high-power lasers with high repetition rates (50–100 kHz) of short pulses (150–250 ns) in our country.

High-frequency pulse-periodic regime of laser operation is developed and tested on the basis of a high-power CO_2 GDL and a lamp-pumped Nd:YAG laser and can be successfully used for other types of the high-power lasers, such as HF/DF lasers, diode-pumped Nd:YAG lasers, COILs and CO lasers. The results of computer modeling for high repetition rate pulse-periodic lasers are presented below. The modeling data depend on the characteristics of media, way of pumping, resonator geometry, and many other parameters of the laser under consideration [2].

1. COILs, pulse-periodic regime starts at frequencies >20 kHz, the modulation depth of 100% starts at frequencies >100 kHz and pulse duration <250 ns. Ratio: P_{peak}/P_{aver}=100–1000.

2. HF/DF lasers, pulse-periodic regime starts at frequencies >100 kHz, the modulation depth of 100% starts at frequencies >250 kHz and pulse duration <150 ns. Ratio: P_{peak}/P_{aver}=1000–10000.
3. Nd:YAG lasers, pulse-periodic regime starts at frequencies >4 kHz, the modulation depth of 100% starts at frequencies >40 kHz and pulse duration <250 ns. Ratio: P_{peak}/P_{aver}=100–1000.
4. CO lasers, pulse-periodic regime starts at frequencies >10 kHz, the modulation depth of 100% starts at frequencies >100 kHz and pulse duration <250 ns. Ratio: P_{peak}/P_{aver}=100–1000.

To date, the project of the GDL with an output power of 20MW with the varied high-frequency structure of emission is located in the stage of active study in our country. We have obtained significant results on a notable increase in the effectiveness of laser energy use in the developed regime of emission, which made it possible, according to our estimates, to pass from tens of kg (for payloads) to a hundred, or even a thousand kg. Thus, one should note that the works in this direction, on the basis of their large prospect for different applications, have already begun in Germany, Japan, England, France, China, South Korea, Brazil and a number of others countries. In this case, during the last Symposium in Ludvigsburg practically all participants of these works in the above-mentioned countries noted that a GDL is the most promising system from the point of view of its scalability on the basis of the rocket technology to the level of several ten MW and other parameters, important for this kind of applications.

To the present time two directions, in which the possibility of applying the laser emission in the aerospace tasks is investigated: launching lightweight spacecrafts to the space orbit, reduction of aerodynamic drag of the flying bodies, which move in the atmosphere with the high speed. In the problem of developing a LJE, the emission generated by a pulse-periodic laser is focused by a reflector near the rear end of the spacecraft and the periodically repetitive laser sparks are created. Sparks generate shock waves, which transfer the part of their mechanical pulse to the reflector, located near the tail of the rocket. As the characteristic of the previous years the gas-discharge laser systems demonstrated the pulse repetition rate in the range of 100–300Hz, which was limited by the time of the gas exchange in the discharge zone. To achieve a high average power of 10 MW for this rate of operation it should be necessary to use laser pulses with the energy about 100 kJ. With the reduced air pressure, for example at the height more than 15 km, long living plasma sphere, created by each pulse, occupies almost the entire volume of the reflector, which leads to the screening of the subsequent pulses during 10 ms. A technical difficulty of that method of energy extraction for the laser is also a very strong impact loads with such a high energy of pulses. The use of high energy pulses with the small repetition rate and, consequently, with the very high peak power is limited also by optical breakdown as on the route, so on the surface of reflector. The method proposed by us is the way of overcoming all that difficulties on the indicated basis: usage of laser emission with short pulse and high pulse repetition rate and mechanism of generated by OPD shock waves resonant association. Under these conditions of the LJE, the laser pulse energy (100 ns) can be effectively absorbed (95%) and converted (~30%) into shock waves. In addition to this, it is shown that the specific thrust can be increased several times due to the artificial transformation of radial component of shock waves into the longitudinal one. There are many other advantages of high repetition rate pulse-periodic lasers, which should be mentioned here as well [3].

They are:

1. Much higher efficiency of energy extraction from a media and conversion into laser light, which is important for scaling up high aperture lasers;
2. Plasma screening effect decreasing due to the decreasing in the laser pulse duration and increasing in the number of pulses up to a few tens of kHz;
3. Decreasing in the laser beam thermal defocusing due to the optimal temporal structure of laser radiation;
4. Localization of energy deposition in space and time and elimination of stressed state of solids under laser action due to replacement of melting by ablation;
5. Possibility of unrolled 3D geometry of breakdowns in space due to a very high repetition rate of laser pulses, creation of optical, acoustical and electro-magnetic fields far away of the laser source;
6. Much longer distances of optical breakdown conditions for the same optical systems.

3.2.2. Spark Parameters in the LJE

Laser emission is focused by the reflector, which can have a form of hemisphere or parabolic. Figure 2.1(a) shows the dimensions of the reflector, focusing one spark, and also plasma zone, created by spark.

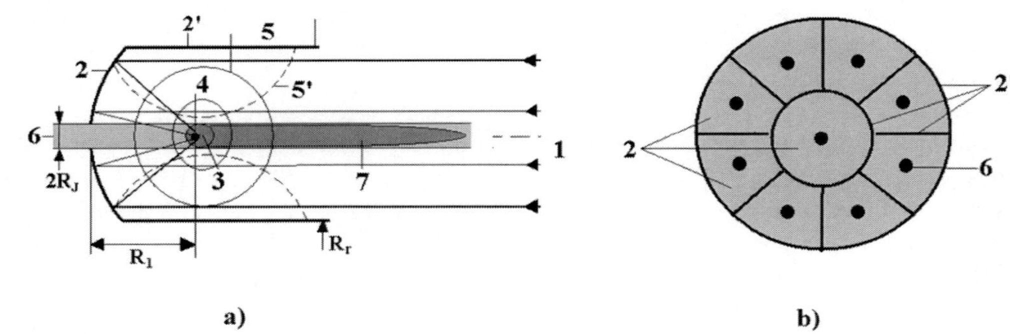

Figure 2.1. (a) Scheme of the reflector and (b) possible structure of a matrix reflector for the LJE. 1: high repetition rate pulse-periodic laser radiation; 2: back side of the reflector interaction zone of laser pulses; 2': side wall of the reflector; 3: cavern; 4: OPD; 5: shock wave; 5': reflected shock wave; 6: gas stream; 7: plasma stream.

The distance Ff must be small between the focal point and the reflector Ff/Rd <0.2, which follows from the condition of achieving the high value of the specific impulse of the thrust. If the emission intensity exceeds the optical breakdown threshold, then the plasma front is extended towards the ray, air is heated and ionized as a result of the laser emission absorption. Since the intensity in the tightly focused ray rapidly decreases (shape factor), then already at the small distance from the focus there occurs the disruption of the light-detonation regime of the plasma front propagation. Then, the emission is absorbed for a while in the decaying plasma. It follows from the calculations carried out by us that for the LJE of greatest interest are the pulses with the duration of ~0.2 s. The maximum energy of laser pulses is limited by the condition of the high efficiency achievement when using laser emission for the thrust application. In the free gas space, the pulse transferred by the shock wave is different from zero only at the small distances from the burst center. The same data were observed also in the early experiments. For the parabolic reflector and short duration of the pulse of ~100-200ns the specific impulse is maximum J_1~550 H·s/J at the distance R1, equal, approximately, the tenth of a dynamic radius, or in another words, the distance, where the value of pulse diminishes significantly (factor-3). With increasing R_1/R_d from ~0.1 to 0.33 the value of J_1 rapidly decreases ~ by 550-200 H·s/J. In the case of the LJE the possibility of using the low values of R_1/R_d is limited by the formation of the long-living plasma, whose characteristic radius is compared with the dimensions of the reflector. At the last stages of the thermal expansion of a spark the plasma zone is formed with the low density and the high temperature ~8000 ^{0}K of the ionized gas. The contact boundary of the hot region stops with the pressure balance in the plasma and the surrounding gas. In the approximation of the spherical form of spark it is possible to estimate the radius of plasma formation for the entire operating range of the pressures 0.1–1 atm.

$$\frac{R_k}{R_d} = \frac{0.49 \cdot r_0^{0.29}}{(Q_r/P_0)^{0.097}}$$

$$\frac{t_k}{t_d} \approx 0.374$$

The ratio of the plasma zone radius to the dynamic radius in this case is

$R_k/R_d \approx 0.25\ (F_f/d_f)\ 0.19 \approx 0.15–0.25$.

This relation does not depend on the energy and gas pressure. Moreover its value is located in the same range, where the maximum value of the specific recoil momentum is attained, which is produced by a shock wave. In the reflector, which has the form of hemisphere, the maximum specific impulse of the thrust is reached at $R_1/R_d \approx 1$. Let us present the values of the plasma formation radius and time of its formation for the energy 10^5 J and the atmospheric pressures of 1 and 0.1 atm. The parameters are as follows: the peak pulse power—2×10^7 W and the pulse repetition rate—100 Hz, and besides the start and the end of the spacecraft acceleration stage in the LJE regime should be taken into account. At a pressure of 1 atm, the size of the plasma sphere will be 25 cm with the time of formation on the order of 1 ms. At a pressure of 0.1 atm, the size of the plasma sphere will be already 50 cm with the time of formation on the order of 2 ms. Laser plasma is cooled due to the turbulent mixing with the cold surrounding gas. The characteristic time of this process exceeds by more than an order of magnitude the time of its formation. Here the spherical sparks have been examined. The spark in LJE has the form of a cone with a large apex angle. This worsens the situation even more. With the specified conditions at the last stage of the expansion of the spark a cumulative jet can be formed, in which the gas moves in the direction to the reflector. Plasma ceases to be singly connected, taking the form, similar to the torus. Thus, the use of a regime with a maximum pulse, presented in the early works of many authors, will be accompanied by the contact of the plasma with the surface of reflector, which, undoubtedly, will lead to its destruction.

3.2.3. Mechanism of SW Combining in LJE

The effect of resonance merging of separate shock waves into the low-frequency quasi-stationary wave (QSW) in the general case consists in the following. In a continuous medium, periodic disturbances—shock waves are consecutively are produced, whose initial velocity is more than the speed of sound. If the speed of the displacement of the field of pulsations is lower than the speed of sound in the medium, then shock waves are merged and are produced QSW. Depending on the time-space structure of pulsations, the mechanism manifests itself in the form the effects, basic property of which is the large length of the high-pressure area. The mechanism of shock wave merging does not contain limitations on the type of the medium and source of pulsations. Depending on the structure of pulsations, QSWs can have various forms.

The realization of the mechanism QSW makes it possible to remove the problems examined above: the screening of emission and the thermal action of the laser plasma in the LJE. Furthermore, a QSW makes it possible to substantially increase the effectiveness of laser emission in the calculation of an increase in the specific thrusts per unit power. Let us determine the requirements on energy and pulse repetition rate of the pulse-periodic laser emission, which satisfy the solution of this problem. In this case, it is reasonable to examine two methods, based on the use of 'spherical' and 'flat' OPDs. In both cases flat QSW is formed due to the use of both the geometry OPD and the geometry of reflector, and also due to the organization process of the laser emission energy introduction into the engine.

3.2.4. LJE on the Base of SW Resonance Combining for the Matrix Reflector

The appearance of the matrix reflector (MR) is presented in Figure 2.1(b). The problem of developing a LJE on the basis of shock wave merging and pulse-periodic laser emission with a high (to 100 kHz) pulse repetition rate can be solved by using a MR. The engine consists of several monoreflectors. The pulse-periodic regime ensures the emission with the pulse energy q at an average power CW, incident on the MR. In the general case, the elements of the MR are identical. The MR produces matrix OPDs, each of which is stabilized by the air jet of the incident flow. The number of matrix elements in the general case determines the high-speed regime of the gas jet. In our case we used 8 elements in the matrix.

In our calculation the power of laser emission was taken equal to 20 MW, the pulse repetition rate – to 100 kHz. The energy of each pulse in this case was equal to 200 J. Calculations showed that for the accepted parameters in the case of normal atmospheric pressure the complete engine thrust will be equal to 100 thousand N, and at a pressure of the order of the tenth atmosphere its value will fall to the value of 36 thousand N.

Big number of the matrix elements is not reasonable. The expressions given below yield the values, which correspond to the following conditions: WC=20 MW (Wm=2.5 MW), $f=10^5$

Hz, q=200 J (qm=40 J), arm=0.3. Index 1 corresponds to P0=1 atm (launching of the spacecraft), whereas index 2 corresponds to P0=0.1 atm (end of the LJE operation).

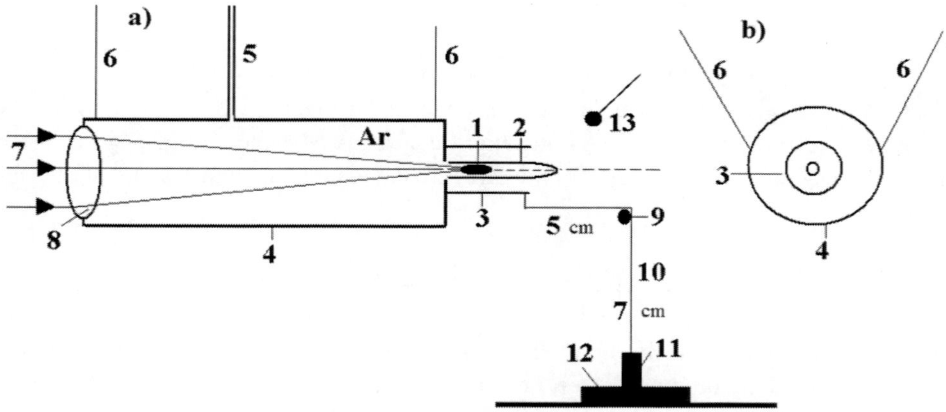

Figure 2.2. Scheme of the experiment: (a) side view and (b) front view: 1 – OPD, 2 – argon jet, 3 – replaceable cylindrical attachment (reflector), 4 – high-pressure chamber (rocket model), 5 – elastic hose for argon supply, 6 – four thin wires suspending the model, 7 – laser radiation, 8 – focusing lens, 9 – block, 10 – wire connecting cylinder 3 with weight 11, 12 – balance, 13 –shock-wave pressure gauge.

The radius of the cylinder of the elementary reflector is equal to Rrm

$$R_{rm} \approx \frac{0.2 \cdot R_{dJm}}{\delta_m^{\frac{1}{2}}} = \frac{0.43 \cdot (q_m/P_{J2m})^{1/3}}{\delta_m^{1/2}} = \frac{R_r}{N^{1/3}} = 15.5$$ cm

The distance from the reflector to the focus is equal to ~5 cm

The characteristic size of the MR is equal to ~90 cm. The excess pressure is

$$\delta P_m \approx 14 P_{0m}^{0.45} \cdot P_{J2m}^{0.55} \frac{\delta^{0.82}}{a_m^{1.64}} = 1.56 \cdot P_0^{0.45}$$ atm

δPm1=1.56 atm and δPm2=0.55 atm.

The first element of the MR is subjected to the average thrust force

$$F_{am}[H] \approx \frac{81.4 q_m^{0.67} P_{0m}^{0.45}}{a_m^{1.64} P_{J2m}^{0.12} \delta_m^{0.18}} = \frac{F_a}{N^{0.67}} = \frac{F_a}{4}$$

Fam1=12.5·10^3 Nuton, Fam2=4.45·10^3 Nuton. The MP is subjected to the average force Fam=N·Fam=$N^{1/3}$ ·Fa. Fam1=100·10^3Nuton, Fam2=35.6·10^3 Nuton

The specific force of the MR is the same as for a separate element: JM=Fam/WC=N $^{1/3}$· J, Jm1=4170 Nuton/MW, Jm2=1460 Nuton/MW. (N-Number of reflectors).

$$V_{J2m} = 86 \cdot f^{2/3} \cdot \left(\frac{W_{Cm}}{P_{J2m}} \right)^{1/3} = \frac{V_{J2}}{N^{1/3}}.$$

The jet velocities in the MRs are equal to V_{J1}=2520 m/s, V_{J2}=5440 m/s.

The optimum gas flow for both cases proved to be equal to 2.5 km/s and 5.4 km/s, respectively. One should also note that the flight control is possible with the aid of the laser engine for the calculation of a change in the thrust in the equivalent components of the MR.

3.2.5. Experiments

However, an increase in the number of MR elements is accompanied by a significant increase in aerodynamic drag on the air ducts and the selected number of elements is close to the optimum In the experiments the model of a rocket with the laser engine was a ~26-cm, 1.1-kg cylinder ~8 cm suspended by four thin wires with total length of 1.1 m (Figure 2.2). The construction allowed the cylinder displacement only in the axial direction. The chamber end had a fixed reflector— a cylindrical cap. Laser emission was introduced into the camera through the lens with a focal length of 17 cm. The jet of argon was formed by the high-pressure chamber through the opening with a diameter of ~3 or 4 mm. The speed was regulated by the pressure of argon, which was given into the camera through a flexible hose. The force, created by the jet and shock waves, was transferred with the aid of a thin (~0.2 mm) molybdenum wire to the load, which stands on the scale weights.

The OPD was formed by the gas jet from the camera through the opening in the reflector. The OPD was produced by pulse-periodic laser emission: f=50 kHz and 100 kHz; in this case, the average power was P≈1200 W. Thrusts at f=50 kHz and V=300 m/s proved equal to 40 g, and at V=400 m/s – to 69 g. The value of the thrust coupling coefficient proved equal to 1.06N/kW. The experiment confirmed that pulse-periodic laser emission creates stationary thrust with a very high coupling coefficient—1000N/MW. A difference between the experimental results and theoretical estimates, whose value is >4500 H/MW, is explained by the need for further optimization of the process of the introduction of laser energy into the plasma OPD.

3.2.6. Resonance Properties of the 'Laser – Space Vehicle' System

An important parameter in the estimation of the resonance properties of 'Laser-space vehicle' system is the ratio of the laser pulse duration to the transit time of sound throughout the entire length of the rocket. In the practical sense there is the greatest interest in the case, when uniform load under the action of the laser pulse is created throughout the entire length of the rocket. If the pulse is very short, then acceleration is non-stationary, the length of that excited in space vehicle waves are much less than its length. The case, when the pulse duration and the transit time of the acoustic wave are compared, corresponds to the resonance conditions.

Let us assess the resonance properties of the system in the case of a 20-kg spacecraft whose length is equal to 200 cm and the acceleration – to 100m/s^2. The average power of pulse-periodic radiation is 4 MW, the energy of pulses generated at a repetition rate of 200 Hz is equal to 20 kJ. Calculations show that at small pulse repetition rates of pulse-periodic laser radiation the acceleration regime is accompanied by extremely strong dynamic structural loads. They have resonance nature, since the wavelength of compression is compared with the length of the spacecraft. Furthermore, with an increase in the spacecraft length to about 4 m and the repetition rate of the laser pulses up to 1 kHz, the eigenfrequency of oscillations of the rocket is equal to the pulse repetition rate of the laser and the resonance conditions are fulfilled. This is very dangerous, since resonance loads can exceed the safety factor of the rocket, which can be destroyed in the flight.

Thus, the estimates show that the thermal contact of the plasma with the reflector and strong dynamic loads are inevitable at a low pulse repetition rate. The situation is aggravated by the excitation of resonance oscillations in the spacecraft. These difficulties are surmounted only with the use of a method, based on the merging of shock waves [4]. Calculations and the experiment confirmed the possibility of significant stationary thrust designing with the use pulse-periodic emission at a high pulse repetition rate.

3.2.7. Conclusion

In conclusion, one should say that as a result of the conducted investigations of the LJE it becomes possible to solve the following interesting problems:

1. Creation of the interceptor of manmade space debris and other dangerous space objects, such as asteroids, comets, meteorites and so on;
2. Launch of micro-objects with the super-high acceleration to the space;
3. Realization of orbital scale conducting channels for energy delivery from space to the ground and vice-versa.

3.2.8. References

[1] Apollonov V.V., Tishchenko V.N. Kvantovaya Elektron., 34, 1143 (2004) [Quantum Electron., 34, 1143 (2004)]; *SPIE Proceedings of GCL/HPL*, vol. 5777, Prague, (2004).

[2] Tishchenko V.N., Apollonov V.V., Grachev G.N., Zapryagaev V.I., Gulidov A. I., Menshikov Ya. I., Smirnov A. L., Sobolev A. V. *Kvantovaya Elektron.*, 34, 941 (2004) [Quantum Electron., 34, 941 (2004)]..

[3] Apollonov V.V. *SPIE Proceedings of ISBEP*-3, vol. 5779, Troy, (2004).

[4] Ageev V.P., Barchukov A.I., Bunkin F.V., Konov V.I., Prokhorov A.M., Silenok A.S., Chapliev N.I. Kvantovaya Elektron., 4 (12), 2501 (1977) [*Sov. J. Qauntum Electron.*, 7 (12), 1430 (1977)].

[5] Apollonov V., Baturin Yu., Bashilov A., Katorgin B., Mizin P., Shurov A., *ISBEP*-4,Nara,Japan,(2005)

[6] Apollonov V. *SPIE Proceedings of ISBEP*-3, vol. 5779, Troy, USA,(2004)

[7] Schall W.O., Bohn W.L., *SPIE Proceedings of ISBEP-3* V.4065, Troy, (2000).
[8] Apollonov V., Tishchenko G., *SPIE Proceedings of ISBEP-4*, Nara, Japan, (2005)
[9] Apollonov V.V., Kiiko V.V., Kislov V.I., Suzdal'tsev A.G., Egorov A.B. *Kvantovaya Elektron.,* 33, 753 (2003) [Quantum Electron., 33, 752 (2003)].
[10] Borzov V. Yu., Mikhailov V.M., Rybka I.V., Yur'ev A.S. et al. *Inxh. Fiz. Zh.,* Vol. 66, No. 5 (1994).
[11] Tret'yakov P.K., Garanin A.F., Grachev G.N., Krainev V.L., Ponomarenko A.G., Tishchenko V.N. *Dokl. Akad. Nauk*, Vol. 351, No. 3 (1994).
[12] Apollonov V., TishchenkoV. *Kvantovaya Elektron*., 36 (7), 673 (2007) [Quantum Electron., 36 (7), 673 (2006)].
[13] Zemlyanov A.A. Effect of laser radiation and RF radiation on air (In Russian) (Novosibirsk: Nauka, 1992).
[14] Apollonov V.V., TishchenkoV.N., Proceedings of GCL/HPL, Gmunden, Austria, (2006)
[15] Korobeinikov S.R. Theory of point explosion (In Russian) (Moscow: Nauka, 1995).
[16] Bufetov I.A., Prokhorov A.M., Fedorov V.B., et al. *Dokl. Akad. Nauk SSSR*, Vol. 261, No. 3 (1981).
[17] Kondrashov V.N., Rodionov N.B., et al. *Zh. Tekh. Fiz*., 56, 89 (1986).
[18] Apollonov V.V., *SPIE Proceedings of ISBEP-5*, Hawaii, (2007)
[19] Grachev G. N., Tishchenko V. N., Apollonov V. V., QE 37(7), 669, (2007)
[20] Schall W.O., Bohn W.L., *SPIE Proceedings of ISBEP-3* V.4065, Troy, (2000).
[21] Tishchenko V.N., Apollonov V.V., Grachev G.N., Zapryagaev V.I., Gulidov A. I., Menshikov Ya. I., Smirnov A. L., Sobolev A. V. "*Kvantovaya electronika*" 34, 10, (2004).

Chapter 12

3.3. Laser Plasma in Accelerators

3.3.1. Effect of the Electrode's Electric Field on the Laser Plasma Focus

In previous chapters, the results were presented of experimental studies of the characteristics of the products resulting from the expansion of plasmoids formed under irradiation at wavelengths of 0.69, 1.06 and 10.6μm. The radiation flux density was varied in a very wide range, which made it possible to obtain (different in intensity and energy) fluxes of charged and neutral particles. In this case, the positively charged ions have a charge from Z=1 to Z=25, which allows the use of laser plasma products in various nuclear physics experiments.

In the experiments on the use of laser plasma as a source of charged particles and radiations for a variety of accelerators or facilities for plasmoid confinement and heating, many problems arise due to the impact of electric and magnetic fields of various configurations on the plasma torch. To solve these problems, experiments, elucidating the problem of interaction of a plasmoid with such fields, have been conducted.

Of particular interest is the effect of an electric field on the laser plasma in experiments in which the electrode is placed at the back of the sample [1]. For a detailed picture of the influence of the electrode potential on the characteristics of the products resulting from the expansion of laser plasma, use is made of the time-of-flight mass spectrometric technique.

The experimental setup is shown schematically in Figure 3.1a. The main elements are an LTI PCh-8 laser, a chamber of interaction of radiation with matter and a time of-flight mass spectrometer with a magnetic analyzer. Laser light in the interaction camber was focused on a target made of copper, molybdenum, and carbon. Electrodes, to which a voltage from the power supply was applied (Figure 3.1b), were located at different distances from the plasma torch formation and at different angles to the plane of the sample. The targets were in the center of the chamber and rotated so that the point of the plasma formation did not change its position with respect to measuring devices and the electrodes. Figure 3.2 shows the dependence of the number of singly charges copper ions on the voltage applied to the electrode [electrode (7) in Figure 3.1a] at an ion energy of 180eV. Upon reaching the voltage of 600V the number of ions in the beam is significantly reduced, and at a voltage of 2.5kV the number of ions emitted in the direction of the analyzer is five times less than initially. When considering the impact of the field of the electrode with the 3-kV voltage applied (Figure 3.3), it is found that the overall structure of the energy distribution remains the same, but the number of particles in the left low-energy part of the spectrum decreases.

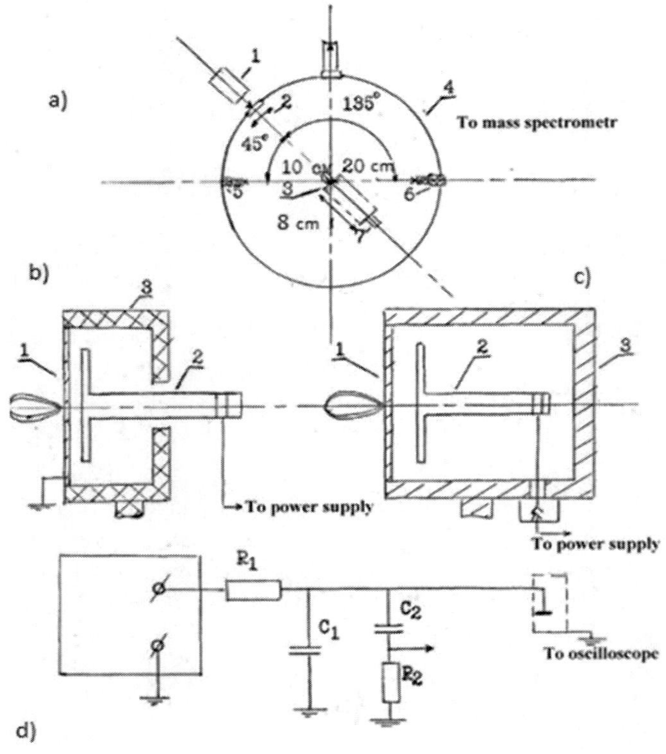

Figure 3.1. (a) Schematic of the experimental setup: (*1*) LTI PCh-8 laser, (*2*) focusing lens, (*3*) target, (*4*) interaction chamber, (*5*, *6*, *7*) electrodes. (b) Target unit design with an open electrode: (*1*) target, (*2*) electrode, (*3*) housing. (c) Target unit design with an isolated electrode: (*1*) target, (*2*) electrode, (*3*) housing. (d) Scheme of electrode connection to the power supply.

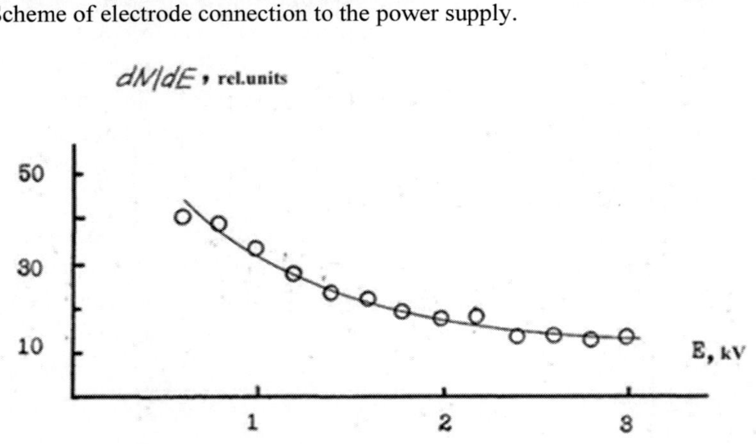

Figure 3.2. Dependence of the number of singly charged copper ions on the voltage applied to the electrode (for the ion energy *E*=180eV).

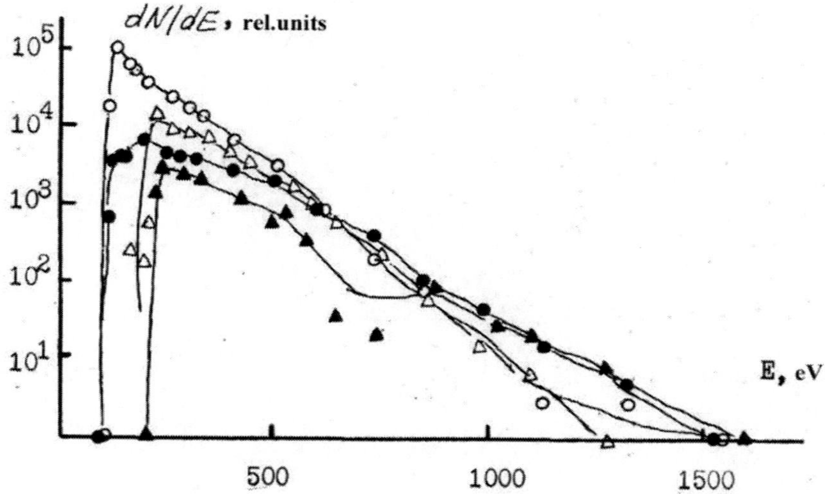

Figure 3.3. Change in the energy distributions of copper ions under the action of the electrode's electric field [(○) Z=1, (Δ) Z=2 at U_{el}=0 (open circles), and (●) Z=1, (▲) Z=2 at U_{el}=3kV (filled circles)].

In addition, the effect of the electrode's electric field on the laser plasma is accompanied by the flow of the current in the electrode shown in Figure 3.1b. Typical oscillograms of such pulses are shown in Figure 3.4. An increase in the voltage applied to the electrode was accompanied by an increase in the current; at the same, the effect on the plasma also increased.

The effect on the plasma was observed only in the case of a positively charged electrode. When the electrode potential was negative, there was no current in the circuit and the effect of the field on the characteristics of the laser plasma was absent.

An important factor that significantly affects the products resulting from the laser plasma expansion is the value of the current of the storage capacitor in the circuit of the electrode. Thus, the inclusion of an additional storage-capacitor electrode into the power supply circuit leads to a significant increase in the field effect, while reducing the voltage on the electrode (Figure 3.5).

Thus, it should be noted that the appearance of the effect of the electrode's electric field on the plasmoids is due to the contact of the electrode with the plasma and the value of current flowing in the circuit (in the case of the electrode screening its effect on the plasma was absent).

3.3.1.1. Influence of the Electrode Position in a Plasma

An important parameter that significantly affects the degree of the effect of the electrode field on the plasma is the position of the electrode in relation to the point of the plasma torch formation. Figure 3.1a shows the general arrangement of the electrodes. Two electrodes are located behind the plane of the sample. The distance from the point of the plasma torch formation to the electrodes is 8, 10, and 20cm. The tilt angles relative to the normal to the sample surface are 45, 155, and 180°.

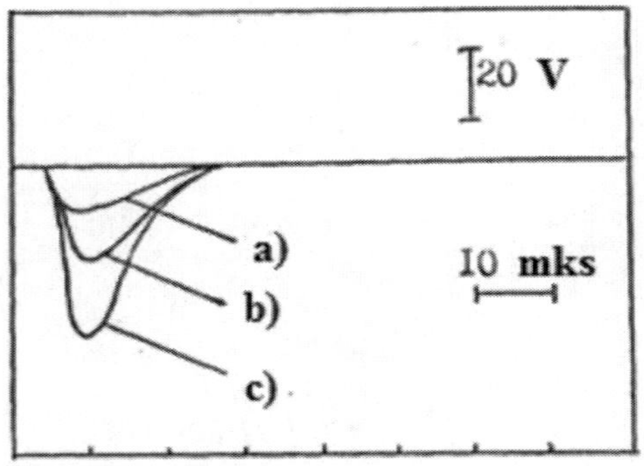

Figure 3.4. Typical oscillograms of the current pulses in the electrode circuit [(a) U=0 V, (b) U=100V, and (c) U=150V].

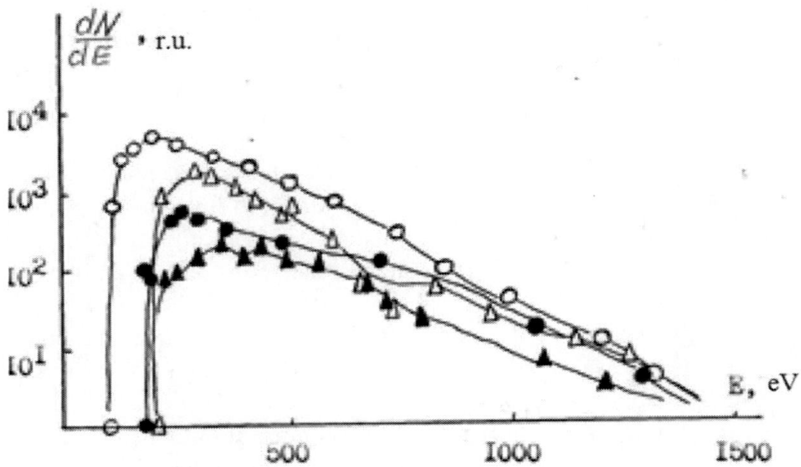

Figure 3.5. Energy distributions of copper ions [(○) Z=1, (Δ) Z=2 at U=3kV (open circles), and (●) Z=1, (▲) Z=2 at U=1kV and C=4μF (filled circles)].

As follows from the spectra in Figure 3.6, the greatest effect on the plasma was exerted by the electrode that is closer to the point of the plasma torch formation. In this case, the greatest effect was achieved at the electrode (*5*) for a voltage that is 10 times smaller than at the electrode (*7*). Besides, the shape of the electrodes also changed. The influence of the electrode position on the plasma expansion products was studied using a steel electrode in the form of a ring (*2*) with an outer diameter of 30mm and an internal diameter of 26mm, and thickness of 6mm (Figure 3.7). The electrode was moved along the 1-1 axis, and the distance between the electrode and the sample varied from 140 to 55mm. In the vicinity of the point of the plasma formation, it was impossible to mount a ring electrode, and so use was made of the electrode (*3*), which represents a segment of an 8-mm-long RK-50 cable with a diameter of 0.5mm. The electrode was placed at a distance of 8mm from the point of the plasma formation torch along the 1-1 axis and 10mm lower. Figure 3.8 shows the oscillograms of the ion signal obtained from a collector installed at a distance of 14cm from the sample on the 1-1 axis. The first grid before the collector is grounded, and a voltage of 100V is applied to the second grid. The oscillograms of the ion signals were recorded for two positions of the ring electrode [(a) 55 mm, and (b) 85mm]. The radiation flux density in these experiments was 10^9W/cm2, and the sample was made of molybdenum. The potential on the ring electrode was +50V. The comparison of the oscillograms indicates that the number of ions in a beam with an energy of 200eV decreases when the electrode is moved to the point of plasma formation.

3.3.1.2. Effect of the Potential Value on the Ion Characteristics

First, it is convenient to consider the effect of the potential on the plasma expansion products at the ring electrode, registered in a distance of 362cm (Figure 3.9). For singly charged ions at a voltage of 15 the energy distribution narrows down both on the left and the right (wings) with a simultaneous decrease in the number of particles in the distribution maximum. A subsequent increase in the potential affects only the reduction in the intensity maximum without changing the 'wings' of the distribution.

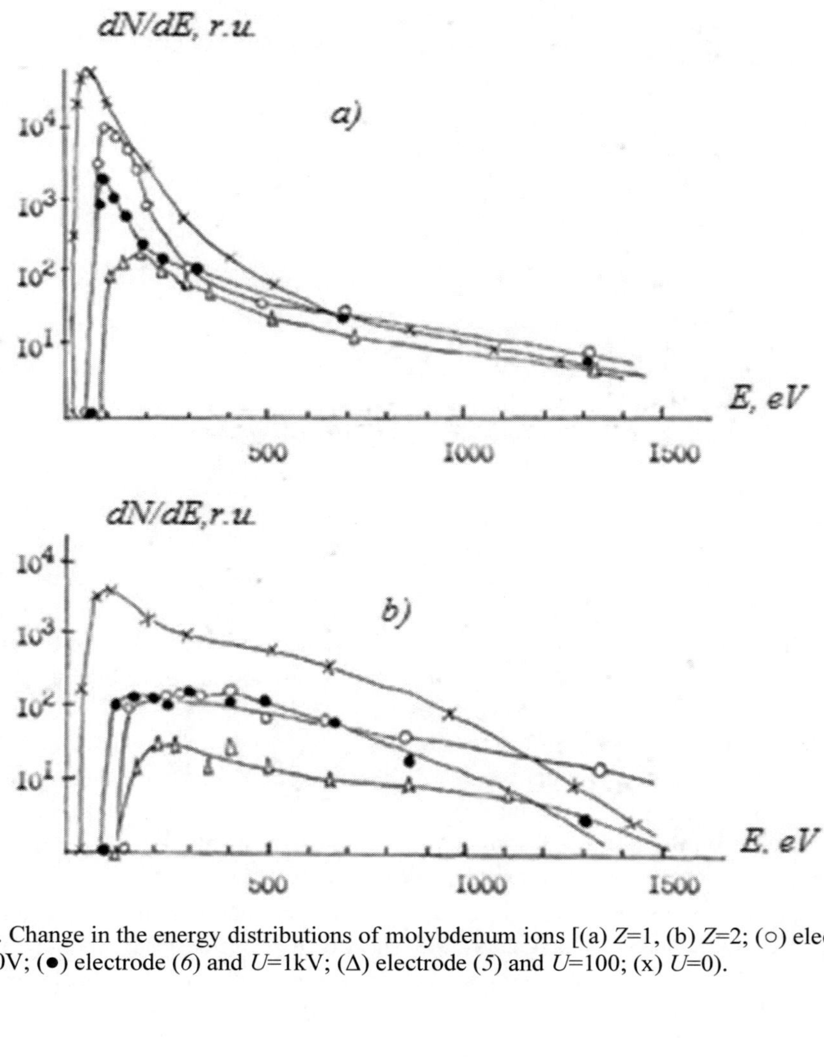

Figure 3.6. Change in the energy distributions of molybdenum ions [(a) Z=1, (b) Z=2; (○) electrode (*7*) and U=500V; (●) electrode (*6*) and U=1kV; (Δ) electrode (*5*) and U=100; (x) U=0).

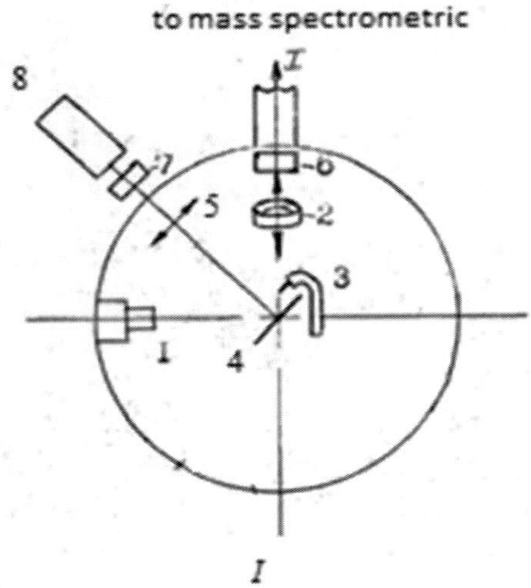

Figure 3.7. Scheme of the electrode arrangement: (*1*, *2*, *3*) electrodes, (*4*) target, (*5*) focusing lens, (*6*) ion collector, (*7*) filter, (*8*) LTI-5 laser.

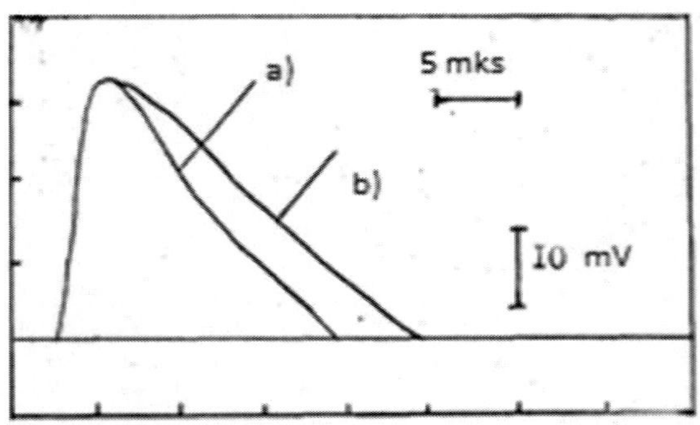

Figure 3.8. Oscillograms of the ion current under the action of the electric field at different distances between the electrode and the target [(a) 55mm, (b) 85mm].

For two- and triply charged ions at a +15V voltage applied to the electrode, there is a shift to higher energies of the left low-energy boundary of the spectrum and the number of particles in the intensity maximum decreases. It should be noted that within the measurement accuracy the electrode position (10 and 14cm from the interaction point) had no effect on the dynamics of the spectra in the range of voltages up to 45V.

Consider the oscillograms of the ion and electron current, obtained from the collector (Figure 3.11), which was at a distance of 140mm from the point of the plasma torch formation. The distance from the point of the plasma formation to the electrodes was 85mm and the distance from the electrode to the collector was 55mm. The radiation flux density on the sample was equal to 2×10^9W/cm^2. As the voltage at the electrode was increased, the collector measured a decrease in the number of ions in the region of low energies, in agreement with the data on changes in the energy distributions obtained by the mass-spectrometric technique. When applying a positive potential of +100V to the second grid, an electron signal was observed at the collector at *U*=0. With increasing potential on the electrode, the electrons gradually disappear from the plasma and the ions, which overcome the energy barrier of 100V, are registered. The electron signal (Figure 3.11b) in this case is identical to the ion signals observed (Figure 3.11a) at a potential of −100V on the second grid of the collector.

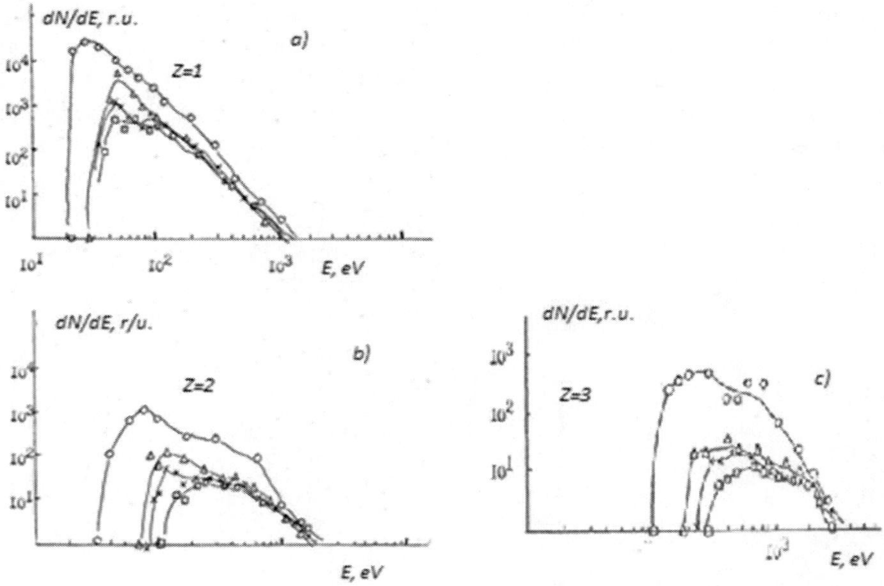

Figure 3.9. Energy distribution of molybdenum ions [(a) Z=1, (b) Z=2, and (c) Z=3) as functions of the electrode's electric field [(○) U=0, (Δ) U=15V, (x) U=30V, and (□) U=45V].

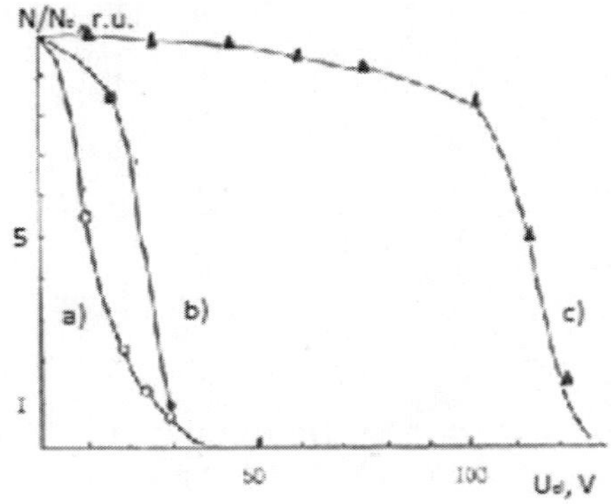

Figure 3.10. Relative number of electrons as a function of different distances between the electrode and the target [(a) 55, (b) 85, and (c) 8mm].

Figure 3.10 shows the changes in the quantitative characteristics of the electrons as a function of voltage applied to the different shape electrodes located at different places of the chamber. Curve (a) in Figure 3.10 shows the dependence of the number of electrons in the plasmoid on the potential applied to the electrode (*3*). On approaching the point of the plasma torch formation, it is needed to raise the voltage on the electrode, in order to obtain a significant change in the number of electrons in the plasma.

Figure 3.12 shows the changes in the quantitative characteristics of the electrons in the plasma for different target materials, obtained by the collector. The electrode (*2*) is located at a distance of 85mm from the target, and the radiation flux density on the target is $2\times10^9 W/cm^2$. One can see from the figure that the dependence of the number of electrons in the plasma on the potential at the electrode does not change with the atomic mass of the ions. When it comes to the mass spectrometric analysis of the laser plasma, at large distances from the point of the plasma torch formation and the electrode it can be noted that the deformation of the energy distribution functions of molybdenum ions is more significant compared to the deformations of the distribution functions of copper ions (comparison is conducted for the case when the electrode is behind the surface of the sample).

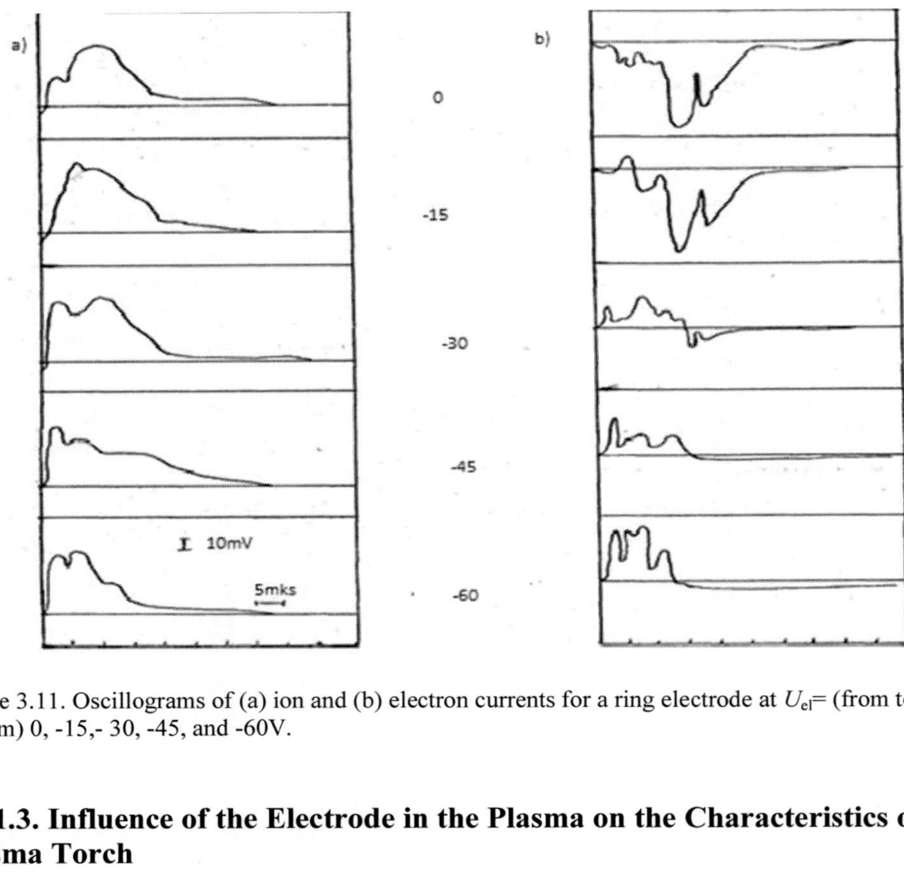

Figure 3.11. Oscillograms of (a) ion and (b) electron currents for a ring electrode at U_{el}= (from top to bottom) 0, -15,- 30, -45, and -60V.

3.3.1.3. Influence of the Electrode in the Plasma on the Characteristics of The Plasma Torch

It was shown above that when a positive potential is applied to the electrode, which can be located in different parts of the vacuum chamber, we can observe the effect of some potential on the characteristics of products of laser plasma expansion. It was found that the effect is observed only when the electrode is immersed in the plasma (the electrode can be located behind the bulk sample). Only when the electrodes are completely isolated from the plasma, no effects are observed. Let us analyze the oscillograms of the currents that flow in the circuit of the electrode. The electric circuit of connection of the electrode to the power source is shown in Figure 3.1d. The current amplitude increases with increasing voltage on the electrode. The pulses appear with a delay with respect to the time of exposure to laser radiation and plasma torch formation. The value of the time delay is 7μs for the case when the electrode is at a distance of 10cm for the copper sample. The current maximum corresponds to 27±3μs and drops to the level of 0.1 during 150±10μs. The delay time decreases and is 5μs for the electrode located at a smaller distance (5 cm) from the sample (molybdenum). The time of the current maximum corresponds to 11±3 μs and falls during 33±5μs.

It follows from the current characteristics that the temporal structure of the current pulse and the amplitude depend on the rate of the plasmoid expansion, the number of particles in the plasmoid, and the velocity distribution. These laser plasma parameters are determined by the radiation flux density on the sample, the atomic composition of the sample and the geometrical arrangement of the electrode.

It was found in the experiments that the electrode placed in the laser plasma focus not only takes measurements, but also affects the laser plasma expansion products: a sharp decrease in the number of ions of all charges is observed, there is a shift to higher energies of the low-energy boundary of the spectra of all the ions present in the laser plasma, and collector measurements reduce the number of electrons in laser plasma. The effect of the electric field of the electrode placed in the plasma, apparently, can be attributed to the case when the processes occurring near the Langmuir probe placed in the plasma are considered. It is only necessary to note that such an electrode in plasma perturbs the plasmoid.

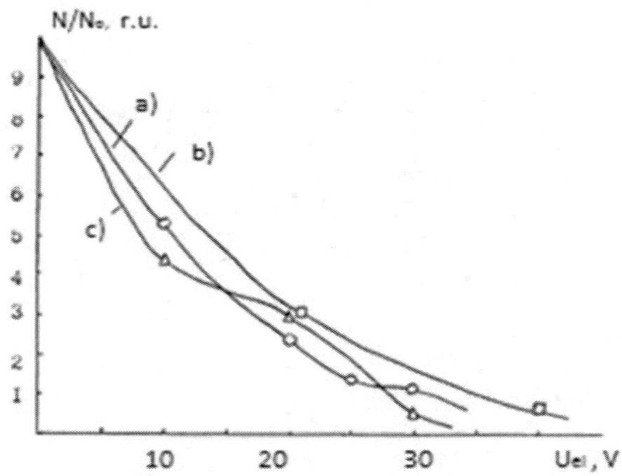

Figure 3.12. Relative number of electrons as a function of the electric field of a)-(○) molybdenum, b)-(□) copper, and c)-(Δ) carbon.

To enhance the effect of the electrode's field on the plasmoid, a storage capacitor with C=4μF is installed in the electrode circuit. Evaluation of the electric charge stored in the capacitor (when the capacitor voltage is 100V) is equal to $Q = 4\times10^{-4}$C, which corresponds to the number of charges, 6×10^{15}. In the laser plasma 10^{15}–10^{16} electrons are formed at a flux density of 10^{10}W/cm^2. These estimates allow us to understand why, in order to obtain a comparable effect on the plasmoid, it is needed to increase (by a on order of magnitude) the voltage on the electrode with decreasing capacitance by an order of magnitude. Based on the experimental results it can be assumed that as soon as the electrode is immersed in the plasmoid, there will be an outflow of electrons from the laser plasma; this process will be terminated by the emerging field of positively charged ions. In this case, the quasi-neutrality of the plasmoid may be violated. The resulting uncompensated ion charge leads to a decrease in the ion beam intensity.

Another possible mechanism involves the processes in the plasmoid affected by the electrode's field. At a field strength in the plasma, 10–100V/cm, there appears the regime of 'runaway' of electrons [2], which leads to the formation of an unstable electron beam with a large increment. The conditions are created for the growth of small initial perturbations and the development of turbulence. In the case of a plasma with a current, the determining factor is the development of ion-acoustic oscillations. Thus, the experimental results may be due to the growing turbulence in a laser plasmoid with a current.

3.3.2. Effect of a Transverse Magnetic Field on the Formation of the Laser Plasma Focus

The behavior of the laser plasma from the moment of its formation and in the initial period of formation and expansion in the transverse magnetic field [3] is both of physical interest, which makes it possible to understand more fully the processes occurring in the plasmoid in the early stages of formation, and of technical interest for application of beam particles in accelerators.

3.3.2.1. Region of the Field – Plasma Interaction

The setup used in the experiments (Figure 3.13) consisted of a laser with a focusing system; a vacuum chamber; an electromagnet producing a magnetic field, which is transverse to the direction of the laser plasma expansion; and a time-of-flight mass spectrometer with a magnetic energy and mass analyzer and detection of ions by the secondary electron multiplier (SEM).

The Nd: glass laser produced a power density in the range from 2×10^9–2×10^{11}W/cm^2. The vacuum chamber with a target inside (as a target material use was made of tungsten and molybdenum) was placed between the poles of the electromagnet and made of a copper waveguide segment (cross section of 3×1cm^2), which guaranteed the penetration of the magnetic field in the vacuum volume. Besides, the target could be moved along the axis of the chamber, thereby making it possible to study the effect of the magnetic field to form a laser plasma inside or outside the region occupied by the magnetic field with the value in the range from 0 to 10kE.

The transverse (to the direction of the plasma expansion and to the magnetic field lines), size of the region occupied by the field e3xceeded the size of the chamber; in this case, the change in the field could in this direction be virtually neglected. The longitudinal size was about 3cm, i.e., it was approximately equal to that of the chamber, and, consequently, to the transverse size of the zone of the magnetic field – plasma interaction.

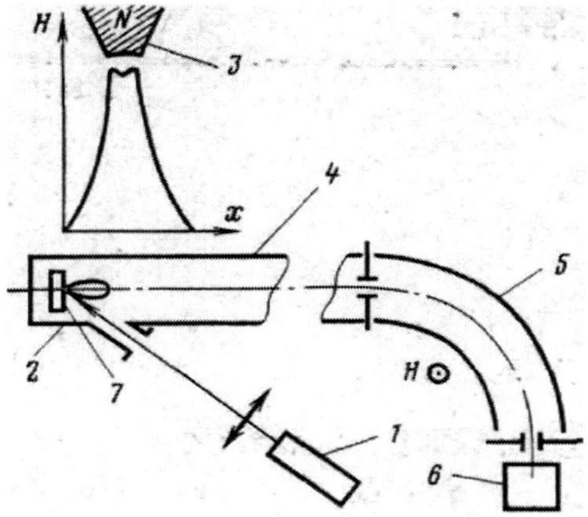

Figure 3.13. Scheme of the experimental setup: (*1*) laser, (*2*) vacuum chamber, (*3*) electromagnet pole, (*4*) time-of-flight tube, (*5*) magnetic energy analyzer; (*6*) SEM, (*7*) target.

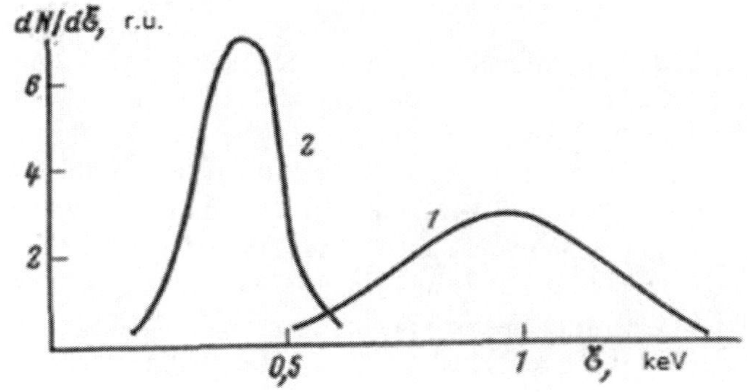

Figure 3.14. Energy distribution of tungsten ions.

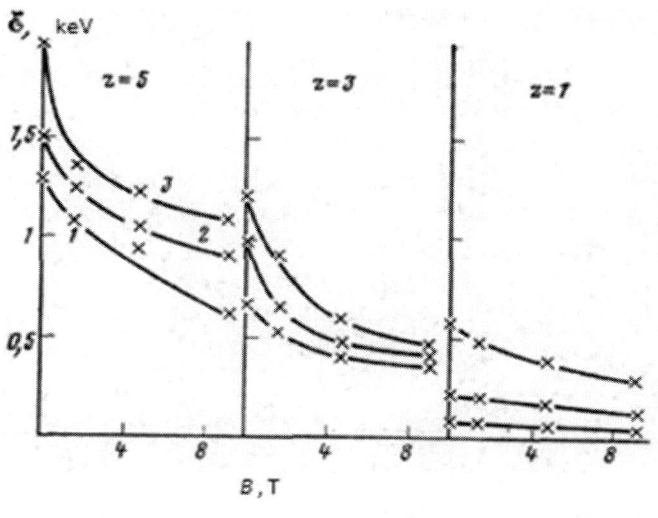

Figure 3.15. Dependence of the ion energy max(3), min(1) on the magnetic field.

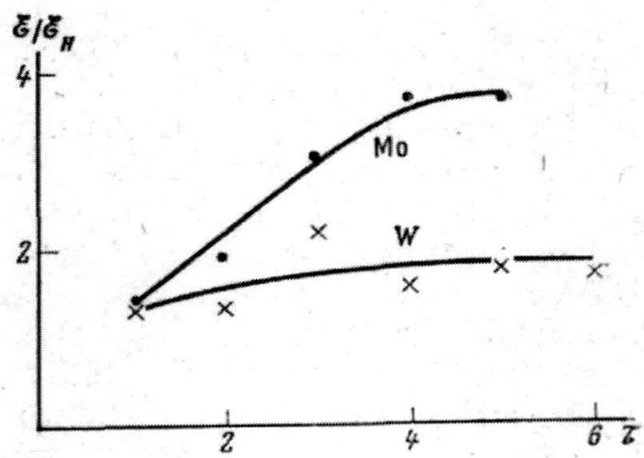

Figure 3.16. Effectiveness of ion slowdown as a function of the distance from the axis of the chamber.

3.3.2.2. Effect of a Transverse Field on the Ionic Component

With the laser plasma flying through the region occupied by a transverse magnetic field, the ion energy decreases. Thus, if the freely expanding laser plasma contains tungsten ions with a charge from 1 to 6 (10^{10}W/cm^2), the energies of which reach 2.2keV, then at a field of 10kE, the charge and the number of ions remain the same within the limits of experimental error, and the ion energies decreased by about half to 1.2keV (Figure 3.14); lighter ions (molybdenum) are slowed down more effectively. Thus, the ratio of the average energy in the palsmoid at 0 and 10kE reached 2.3. The most significant effect is exerted by the magnetic field on the expansion of ions with a maximum charge.

The size of the plasma region exposed to the field was changed by shifting the electromagnet along the vacuum chamber. These experiments showed that the energy shift increases with increasing interaction region of the field and the plasma. Changes in the direction of the magnetic induction vector in the transverse plane are not affected by these effects.

Apparently, the effect of the laser plasma slowdown in a transverse magnetic field can be qualitatively described as follows. When the plasma flies into a magnetic field, the electrons are subjected to the Lorentz force, which leads to the bending of their trajectories; this results in a transverse velocity component and in a decrease in a perpendicular velocity component of the electron motion to the sample surface, which will automatically reduce the ion velocity. But this mechanism will only work in a small, dense region of interaction where there occur the processes of acceleration of ions with electrons.

It is possible that the action of the Lorentz force on the electrons leads to the emergence of the electron current in a direction perpendicular to the motion of the plasmoid and the magnetic field lines. The current flows in the plasma and is short-circuited by the conducting walls of the chamber.

The model problem of the plasma incoming into the magnetic field is presented in papers [4,5]. Here we neglect the angular expansion of the plasma, and we assume that it moves in the direction of the X axis, and the magnetic field is directed along the axis Y.

Using simple calculations presented in work [3], it is easy to obtain the estimate:

$$\mathcal{E}_0/\mathcal{E}_H \approx \exp(\frac{2ZH^2\sigma L}{n_0 v_0 c^2 M}) \sim 2 \tag{3.1}$$

The estimate agrees well with the experimental data, and the formula qualitatively explains the fundamental features. For the quantities in (3.1), we used Z=2, 10^4E, V_0=3x10^6cm/s, M~4×10^{-22}g, G=10^{13}Ohm^{-1}, n_e=10^{16}cm^{-3} (q=10^{10}W/cm^2, the diameter of the focusing spot is 300μm). From (3.1) it follows that the efficiency of the plasmoid slowdown should increase sharply with increasing magnetic field strength and, in addition, it should be enhanced by reducing the electrical resistance during the heating of the electrons due to the Joule heating.

$$\left(v_y \frac{dT}{dy} \sim v_y \frac{T}{L}\right) \tag{3.2}$$

The transverse magnetic field can slow expansion of the laser plasma, the most effective slowdown being observed for the high-energy part of the plasmoid containing multiply charged ions, which is a good feature when using the laser plasma as a source of multiply charged ions for accelerators.

3.3.3. Laser Plasma in a Plasma Focus

In plasma physics, use is widely made of pulsed plasma accelerators, where the plasma is formed by the breakdown of the interelectrode gap which is filled by a gas in a stationary or pulsed manner and accelerated by electrodynamic forces. At certain parameters the plasma sheath can collapse on the axis in front of the injector, forming a so-called 'plasma focus' (PF). The PF is a source of high-energy particle fluxes (up to several MeV), X-rays and neutron radiation.

Since the laser plasma is a source of particles of any elements, in particular containing hydrogen and deuterium, this (using the PF design) makes it possible to design a setup for additional heating and ionization of the laser plasma [6].

Such a method of producing the plasma in the PF injector, compared to the conventional technique, has some advantages:

1. Control of the parameters of the plasma sheath: length, density, target material. You can change the shape, composition and number of particles in the sheath at the stage of the plasma formation.
2. Symmetry in space with a controlled degree of ionization.
3. Reduction of the current losses beyond the accelerated sheath.
4. Increase in the speed of the sheath motion before the collapse.

Figure 3.17 shows schematically a setup for studying the processes in the laser plasma focus. The beam of a pulsed CO_2 laser (*10*) focused by the lens system is directed into the ring target (*1*) and causes the release of a plasmoid consisting of the target materials (in the experiments use is made of the materials containing isotopes of hydrogen, which are impregnated in a variety of metals) into the interelectrode gap in the axial direction.

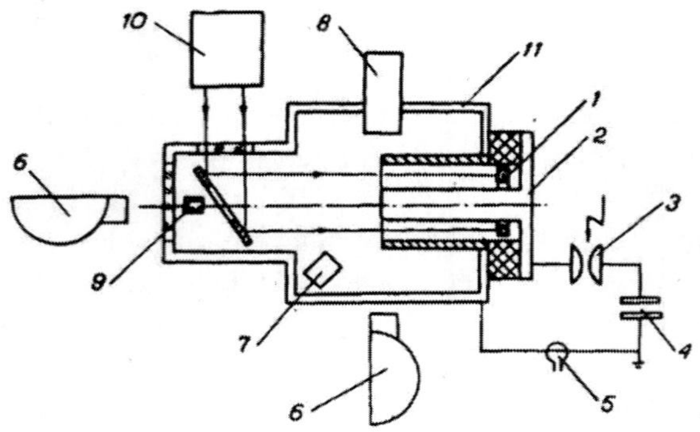

Figure 3.17. Scheme of the PF setup and location of the diagnostic equipment: (*1*) target, (*2*) injector, (*3*) discharge gap, (*4*) capacitor bank, (*5*) Rogowski coil, (*6*) SFR-2M photographic recorder, (*7*) pinhole camera, (*8*) thermoluminescent sensor, (*9*) Faraday cup, (*10*) CO_2 laser, (*11*) vacuum chamber.

Changes in the characteristics of the laser radiation and target make it possible to change the parameters of the plasma formed. The synchronization system, after a certain time delay, starts controlled discharge gap (*3*), short-circuiting the plates of the capacitor bank (*4*) to the PF electrodes of Meiser type. The current flows along the plasmoid and the thus produced magnetic field accelerates the plasmoid, usually called the plasma sheath, which, after reaching the end of the inner electrode, is compressed by the magnetic field, and from which a plasmoid is formed from a dense hot plasma – 'plasma focus.'

A peculiarity of the laser focus (LF) is the possibility of varying the parameters of the current sheath and its significant acceleration in the process of movement and collapse, as well as the increase in the efficiency of waist compression, in the absence of a gas medium.

Figure 3.18. LF development taken by the high-speed camera (frame exposure, 500ns; voltage, 29kV; pressure, 10Pa; target, polyethylene).

The research conducted allowed one to carry out operation modes, leading to the formation of pinch structures. Figure 3.18 shows a typical high-speed camera photograph.

Measurements performed by using X-pinhole cameras, led to the observation of X-ray sources with the characteristic region of 100μm in size and energy of 1keV. Using thermoluminescent sensors the energy spectrum of the radiation was recovered, which in the range 1–20keV was close to the spectrum of the plasma bremsstrahlung. Probe measurements showed that the laser plasma focus is the source of ions with very high energies.

Let us consider now the motion of the plasma sheath under simplified LF conditions when tracking is absent.

We obtain in the first nonzero approximation:

$$a = R - V_0 t + \frac{A}{12R} t^4 \text{ and if } V_0 = 0$$

$$a = R\left(1 - \frac{t^4}{12R^2 c^2 I^{-2} M_0 \tau^4} = R\left(1 - \frac{t^4}{12 t_1^4}\right)\right) \tag{3.3}$$

Thus, the law of the waist compression is characterized by the fourth degree of time and not the second, as in the case of the classical *Z*-pinch. In order to optimize the rate of energy input to the discharge, we ill change the parameters of the electrical circuit. The equation of the circuit has the form:

$$V_0 \sim \frac{1}{c}\int_0^t I(t)dt = (L_0 + L_1)I + L_1 I \tag{3.4}$$

where L_0 is the external inductance, and L_1 the discharge inductance.

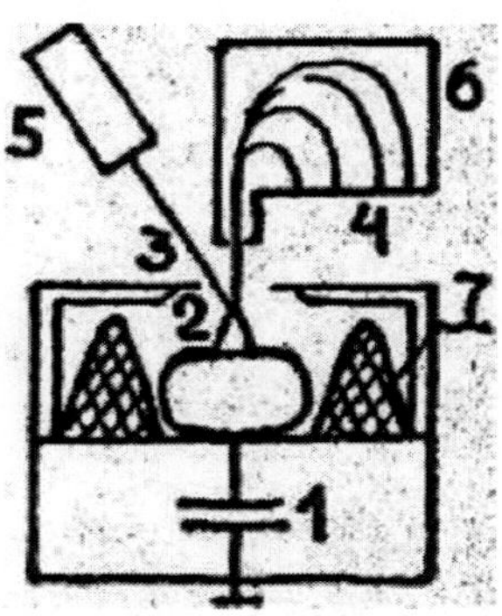

Figure 3.19. Experimental setup for studying the ‘fast’ LF: (*1*) capacitor, (*2*) and (*3*) electrodes, (*4*) film, (*5*) laser, (*6*) analyzer, (*7*) insulator.

For effective discharge development temporal coincidence of the current maximum flowing through the discharge is important (for the time of maximum compression). The problem is to match the discharge time to the dynamics of motion of the current shell. Since the initial velocity of the shell is large enough $V_0=10^6$–10^7cm/s, and the initial conductivity is high ($Q=T_e^{1.5}$), then the use of the usual parameters (C=1–10μF, L=0.1–1μG) is ineffective: the shell accelerates and collapses long before the current reaches a maximum, and so it is necessary to reduce the discharge time $T_{dis}=L_0^{0.5}/V_0$ at the same energy, by reducing the inductance and increasing the voltage across the discharge gap. This can also lead to an increase in the rate of the current rise:

$$d\,I(0)/dt = Vo/Lo \qquad (3.5)$$

The maximum achievable current is determined by:

$$max = (2Eo/L)^{0,5} \qquad (3.6)$$

On the other hand, the achievement of I_{max} requires the fulfillment of the ratio L_0=L_1. When L_0>L_1, too much power remains in the external circuit, and when L_0<L_1, too much energy is spent to accelerate the shell in the initial phase. Therefore, for the successful implementation of the LF idea, it is necessary to use sources with a fast energy input (such as Blumlein lines). This approach is implemented using a new original design [7] whose diagram is shown in Figure 3.19. It allows one to reduce to a minimum all parasitic inductances and ensures a minimum time of the energy input into the discharge, which is equal to (2–5)×10^{-8} s. With a small energy input E_0=0.1 J, the current up to 5 kA is provided. The results of experiments with this system are shown in Figure 3.20.

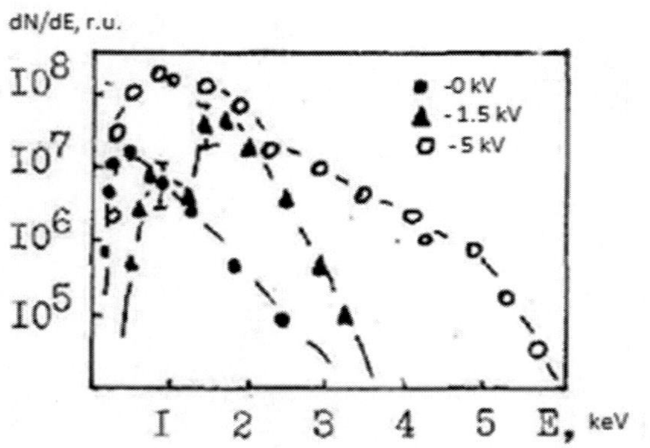

Figure 3.20. Energy distributions of aluminum as a function of the voltage at the capacitor V_0.

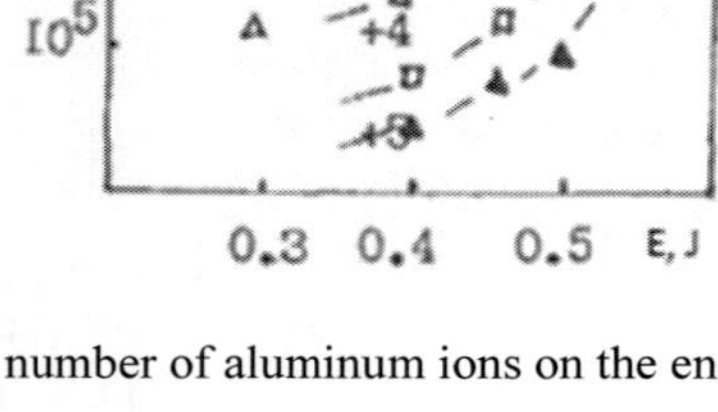

Figure 3.21. Dependence of the number of aluminum ions on the energy inputted into the LF.

Changes in the energy distribution of multiply charged ions with increasing voltage across the discharge gap from U_0=0 and U=5keV (aluminum anode, discharge gap d=5mm, C=0.01μF, angle 0° with respect to the device axis) testify about the increasing compression efficiency of the laser plasma by the magnetic field of the increasing discharge current. Figure 3.21 shows that this is accompanied by a sharp (almost 2 orders of magnitude) increase in the number of ions of the corresponding charge and in Z to 6 with increasing initial electric field up to E=1kV/cm.

Thus, the results obtained with the use of LF heating (LF-pinching) offer promising new opportunities not only for sources of multiply charged ions, but also for the study of processes in such plasma.

Apart from highly charged positive ions, LF expansion is accompanied by the formation of negative ions. The results of the mass spectrometry analysis of negative ions (anode material is graphite) are presented in Figure 3.22. The energy distribution of the negative carbon ions in the laser plasma (without the effect of the electric field) has a maximum at 15eV ion energy (2×10^9W/cm^2) and energy spread of 0.5keV. With increasing electric potential on the electrode system to V=5.6kV the maximum of the energy distribution shifts out in the energy range 50–100eV when the distribution amplitude grows by about 2 orders of magnitude. The number of negative carbon ions increased by 10 times with increasing energy input in the discharge up to 0.3J (Figure 3.23). Increasing the yield of negative ions can be attributed to the fact that there was an increase in the number of positively charged ions (Figure 3.21), and, therefore, an increase in the number of recombination neutral atoms. This recombination leads to an increase in negative ions.

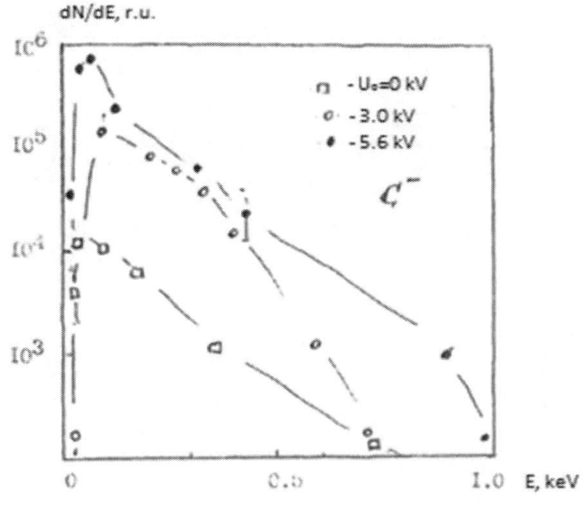

Figure 3.22. Dependence of the electrode potential on the energy distribution of the negative carbon ions.

The use of this type of LF heating [7] can greatly increase the possibility of laser-plasma medium as a source of positive and negative ions.

Further experiments on the use of the PF to heat the plasma were carried out on a modernized setup, which makes use of an injector made of cylindrical electrodes with diameters of 15 and 50mm, length of 150mm (the electrodes have inserts made of copper, graphite, polyethylene, acrylic glass, porcelain and other materials). As a radiation generator use is made of a pulsed CO_2 laser with a double transverse discharge. When employing an unstable resonator, an annular laser beam was directed to the target through the optical elements such as axicon, or through lenses located along a circle. The number of ions formed in the injector varied from 10^{14} to 10^{17}. In the experiments use was also made of additional diagnostic facilities (Figure 3.17): Rogowski loop, high-speed photo camera SFR-214, X-ray pinhole cameras, thermoluminescent detectors, grid probes and time-of-flight mass spectrometer. 10^9

The current oscillograms with the Rogowski loop made it possible to calculate a period of 5μs and an amplitude of up to 100kA for the discharge current injector. For targets of light elements (e.g., polyethylene) with a symmetric output of plasma to the end of the electrode system, on the anode we observed a spot of diameter 5 mm and length 20mm. The lifetime was equal to 3–4μs (Figure 3.18).

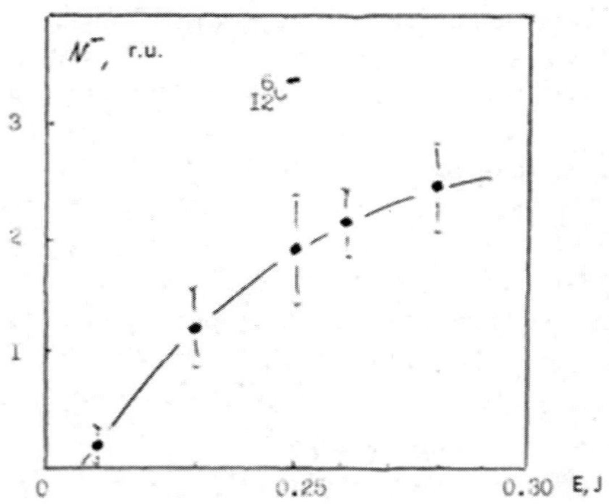

Figure 3.23. Dependence of the number of negative carbon ions on the additional inputted energy in the case of LF pinching.

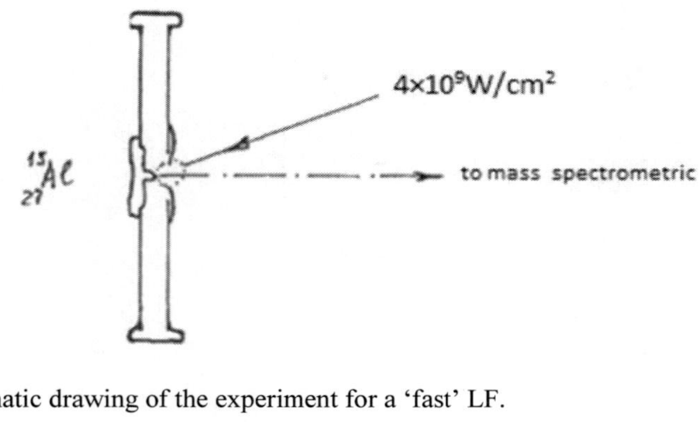

Figure 3.24. Schematic drawing of the experiment for a 'fast' LF.

The analysis of the results obtained and the qualitative peculiarities of the laser plasma pinching indicate the need to take the advantage of the 'fast' (10^{-7}s) energy input into the discharge (Figure 3.24). The dynamics of the formation of the charge and energy distribution of the output ion beam at a higher energy input and the electric field up to 10^3V/cm is determined by an increase in the compression efficiency of the laser plasma by the magnetic field of the increasing discharge current. The specific shape of the spectra of multicharged ions, to a certain extent, correlates with the spectra obtained in experiments with laser plasma at high laser intensities, which opens up new possibilities in the study of fundamental processes in laser plasma itself.

In the experiments on the LF formation the electron fluxes were investigated at the setup consisting of the electrode system; the discharge chamber; and the IK-25-12 storage capacitor, which is vacuum-packed to the discharge chamber to minimize the circuit inductance (Figure 3.24). This made it possible to carry out the process of pinching in the first quarter of the discharge period

Figure 3.25 shows the energy spectra obtained for inserts made of molybdenum and lead; these spectra make it possible to trace the effect of the changing polarity on the energy characteristics of electrons.

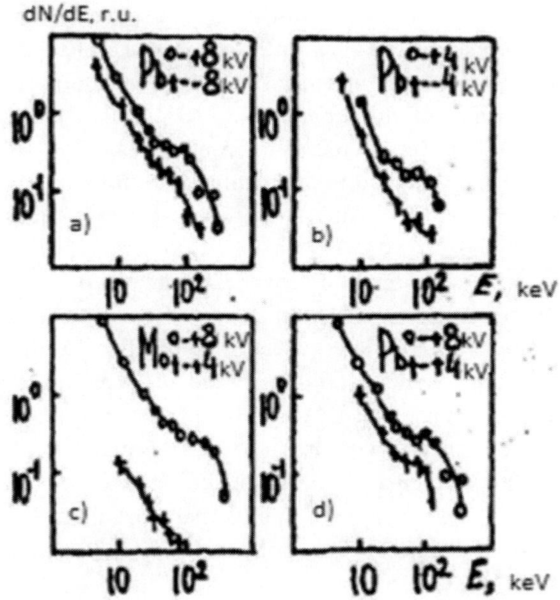

Figure 3.25. Energy spectra of electrons.

The traditional choice of polarity, when the central electrode is the anode, is characterized by an increase in the total number of high-energy electrons and the appearance of some features in the spectra at electron energies of 100keV. The presence of this feature is apparently due to the onset of an intense electron beam during micro-pinching of the discharge plasma. Therefore, we can assume that at this polarity compression is more effective. Reduction in the total number of registered electrons in the case of reverse polarity is explained by the spread of the electron beam during its expansion in front of the entrance slit of the analyzer. Also of interest is the dependence of the electron flux on the storage capacitor voltage (Figs. 3.25c, d).

The discharge current varies from 50 to 100kA with changing the voltage; in this case, the compression is readily achieved, as evidenced by the change in the shape of the energy distribution of electrons. One can see that no compression is observed for refractory molybdenum at an initiating laser energy of 0.1J and a voltage of 4kV due to the lack of seed laser plasma.

When the voltage is further increased up to 8kV, secondary processes after the breakdown of the discharge gap provide enough carriers for the regime, while the low-melting lead this regime already takes place at a voltage of 4kV, which corresponds to the discharge current of 50kA.

The results of the study of the X-ray polarization [8]. point not to the axial but the transverse (with respect to the discharge axis) motion of fast electrons in the micropinch. This should be expected because the Larmor radius of electrons with energies of 10–100keV is much less than the micropinch radius observed in the experiments [6]. Consequently, the electrons can move mainly along helical trajectories along magnetic field lines, moving in the longitudinal direction due to the drift in an inhomogeneous field. It is easily seen that the electrons outside the pinch will drift in the direction which is opposite to that of the current. Within the pinch the centrifugal drift is also directed against the current, and the diamagnetic drift – along the current. In our case, at a current of 10^5A and a microprinch radius of 10^{-3}cm, the drift velocity of electrons with energies of 10–100keV at the plasma boundary is of the order of 10^8cm/s, i.e., the time of micropinch 'acceleration' in the plasma will be10^{-10} s, which is obviously less than the lifetime of the micropinch itself [9]. Only a small fraction of the relativistic velocity electrons trapped in the axial region, for example, as a result of a drift in crossed electric E and magnetic B fields can rapidly, i.e., a rate determined by their kinetic energy, leave the discharge plasma [10].

Magnetized electrons in the pinch discharge plasma can be accelerated under the influence of some other mechanisms. In compression of the pinch plasma the collisionless electrons retain their adiabatic invariant and torque. Reducing the pinch radius and a corresponding increase in the magnetic field will lead to an increase in the transverse and longitudinal energy of the electrons [10].

For example, during the second compression of the micropinch [11] when its radius varies from 10^{-2} to 10^{-4}cm the energy of the electrons can increase by 10^2–10^4 times (compression time of 10^{-11}s) and is much smaller than the characteristic time of electron collisions. This mechanism of electron acceleration can explain the anticipatory appearance of radiation with energies of 100keV, compared with the radiation of multiply charged ions of the highest multiplicities appearing in the micropinch plasma at the stage of maximum compression [12].

The above analysis of a fairly wide range of experimental data allows us to conclude that the development of acceleration processes in the plasma of a multicharge ion source apparently plays an essential role in the phenomenon of emission of multiply charged ions, thereby affecting the plasma dynamics in the pinching process. In addition it should be noted that the fast electrons, leaving the region of the micropinch formation interact with the relatively cold peripheral plasma, thereby increasing the total contribution to the ion emission.

3.3.4. Sources of Neutral Atoms for Collective Heavy-Ion Accelerators (CHIAs)

The 1980 – 1990s saw a significant development of one of the collective methods – ion acceleration with electron rings [13].

This collective heavy-ion accelerator (CHIA) has a number of advantages over conventional accelerators, primarily due to the use of the electric eigenfield of the electron ring with a high intensity for ion acceleration. This ensures a significant gain in the rate of energy accumulation of particles, and hence the length and cost of collective accelerators is much less than that of conventional ones.

Ions are formed and stored in electronic rings of neutral atoms by electron impact ionization. By storing the ions inside the ring for a long time, on can obtain the ion beams of high intensity and degree of ionization. In addition, a collective accelerator is universal with respect to the type of accelerated ions. Such an accelerator makes it possible to obtain the ion beams with the elements throughout the periodic table and the high intensity of heavy elements up to uranium.

3.3.4.1. Requirements to Pulsed Atom Sources

To accelerate ions in a CHIA requires certain parameters of electron-ion rings. The number of electrons in the ring should be 5×10^{13}; otherwise, the strength of the electric eigenfield of the electrons will be insufficient for acceleration. On the other hand, the adgesator (an adiabatic generator of atoms) of the CHIA allows forming a ring with the number of electrons $N_e=1.5\times10^{13}$ [14]. The ions from a selected element should be accumulated from a pulsed flux of atoms crossing the electronic ring at the compression stage. This type of loading allows one to adjust the number of ions trapped by the ring and the loading time (in other words, the ion charge). For a pulse loading of the ring by ions, the initial flux of neutral particles must meet the following requirements:

1. The flux of atoms should be formed geometrically so that the maximum number of particles would cross the volume of the electron ring.
2. The interaction time (duration) of the atom flux with the electron ring should be optimal, i.e., longer than the characteristic time of ionization of neutral atoms (8×10^{-6}s for lead). The lower boundary is limited, apparently, by the design capabilities. The most optimal time is no more than a few tens of microseconds. The flow rate of neutral particles is selected as follows.

If the characteristic time of ionization of a neutral atom is much less than the time of flight through the ring, the jet is almost completely ionized at distances smaller than the smallest size of the ring, and a part of the electrons in the ring is not involved in the ionization process. In this case, increasing the jet velocity, we can increase the number of ions in the ring by making fuller use of its ionizing capacity. When tr<<to, a change in jet velocity does not affect the number of atoms in the ring.

The actual value of the velocity of neutral atoms should be 10^4–10^5cm/s.

1. The average number density of neutral particles in the jet must be such that, under the existing values of the duration t_0 and the specific geometry of the jet, to provide the required number of Ni ions in the ring. The value N_a atoms of different elements injected into the ring depend on the parameters of the electron rings in the accelerator and should be in the range from 10^9–10^{11}cm^{-3}.
2. The relative instability of the basic parameters of the pulsed fluxes of the atoms must be better than 5%. It should be noted that when the CHIA is used as a heavy ion accelerator injector [15] it is needed to have the relative stability no worse than 1%.

In addition to these requirements, the design of the sources must meet the following conditions:

1. atom fluxes generated should virtually overlap all the elements of the periodic table;
2. the source should be characterized by vacuum cleanness as it works under conditions of deep (10^{-7}–10^{-6}Pa) vacuum of the adgesator chamber;
3. the source should operate in a magnetic field (up to 2 T) and not to distort the generated magnetic field;
4. switching frequency of the source must be in the range 20–50Hz;
5. the source should have the life time of $>10^6$ operations;
6. the laser atomic source should be universal;

In the range of radiation flux densities 10^7–5×10^9W/cm^2, for a wide range of target materials the number of neutral atoms is greater than 95% of the total number of particles.

For lead, at a distance of 10cm from the plasma formation and a flux density of 4×10^8W/cm^2, the concentration of atoms is equal to 4×10^{11}cm^{-3}. The information obtained in our experiments on the atomic velocity distribution makes it possible to determine the flux duration at any distance from the target. Since the duration of the clusters of neutral atoms as a function of the flux density of the laser radiation and the target material at a distance of 10cm is 10–50μs, and the energy of the directional movement lies in the range from 1 to 1000eV, the velocities of the atoms are in the range 10^5–10^6cm/s.

3.3.4.2. Source Design and the Loading Scheme of Atoms

The characteristics of neutrals of plasmoids studied in detail made it possible to create the working versions of a laser source of atoms and the schemes of injection of such fluxes in the electron ring [16-20].

A laser source of atoms consists of a laser, a focusing lens and a target, on the surface of which laser radiation is focused. The target was made of a substance whose atoms it was necessary to obtain. The focusing lens and the target were placed in a vacuum chamber of the adgesator, and the laser was outside the chamber (radiation is coupled into the chamber through a window). The simplicity of the source design allowed one to position the target in the proximity to the region occupied by the electron ring, and thus, to effectively use the fluxes of atoms from the target.

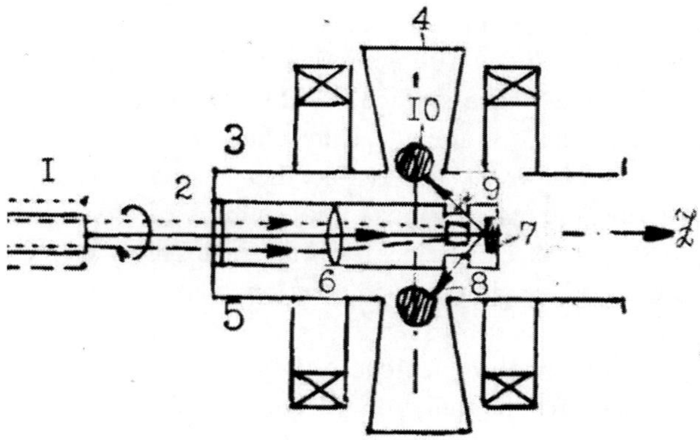

Figure 3.26. Scheme for loading a compressed electron ring (central version of a source of atoms).

One of the ways to load the ring (central source version) is shown in Figure 3.26. The laser (*1*), whose radiation (*2*) passes through the window (*3*), enters the adgesator chamber (*4*) and tube (*5*) (supporting structure) and is focused by the lens (*6*) on the target surface (*7*). The flux of atoms (*8*) through the holes in the collimating tube (*9*) propagates into the compressed electronic ring (*10*). In this arrangement, the flux of atoms should be directed at an angle of 45° to the Z axis. The azimuthal flux uniformity requirements and the possibility to inject atoms into the electronic ring at different moments of its compression (at different radii from 3.5 to 12cm) are achieved by delaying the flux of the atoms relative to the electron beam. The distance from the target to the compressed electron ring of radius 4cm is no more than 6cm, which allows one to load fairly dense fluxes stream of atoms.

The size of the partitions between the three collimating holes of the cylindrical surface of the tube (of diameter 20mm and wall thickness 0.5mm) is small (about 2mm), which provides good azimuthal uniformity of the ring loading. The angular distribution of the ions has the form of 'lobes' with the axis which coincides with the normal to the target, the higher charge ions having a narrower 'lobe'. When the power density is 10^7–10^9W/cm^2, one can observe in the plasmoid the ions with a charge of no more than +1, occupying an opening angle of no more than 45° from the normal. Thus, the tube of the source mostly screens the region of the compressed ring from laser plasma ions. The longitudinal magnetic field of the adgesator increases such a screening due to the 'narrowing' of the angular distribution of the ions.

Figure 3.27. Appearance of the scanner of the central version of the CHIA laser source.

The laser, as the outer part of the source, can be quickly replaced in the case malfunctioning (its life span included 106 operations). The element generally specifying the source parameter is the sample whose surface is subjected to erosion in the interaction with the laser light. The target is placed in the adgesator chamber; quick change of the sample for the central source version is almost impossible, because of the depressurization of the chamber. In such conditions, the maintenance of the stability of the atom flux parameters, i.e., adequate resourcing of the target can be achieved by changing the interaction point. In the laser source of atoms (central version) the interaction point changes by scanning the laser beam over the surface of a stationary target. Laser radiation is scanned by a special device providing a plane-parallel rotational movement of the laser around the *Z* axis, which is shown in the photo in Figure 3.27. Thus, the radiation falls onto the focusing lens at the points on the circle. The target lifetime increases with the size of the interaction spot. Changing the position of the interaction point on the sample within a few mm does not lead to a significant change of the atom flux parameters in the region of the compressed electron ring, because the angle between the normal and the axis, which is directed from the interaction point to the ring, as well as the distance between the interaction point of the ring change only slightly.

The scanning device is made on the basis of the optical table with the main and the alignment lasers, the system of plane-parallel motion of lasers driven by two RD-09 electric motors, providing the speed of 2.5rev/min. The rate of rotation and deviation of the radiation axis from the *Z* axis of the chamber can be controlled remotely.

To study the accumulation processes in the electron rings, it is needed to provide a pulsed flux of atoms into a ring of large radius (30 to 35cm). To implement such an operation mode, the second (peripheral) version of the laser source was designed. In this case the source is located in the lower vertical sleeve of the adgesator chamber. Radiation (*2*) from the laser (*1*) (Figure 3.28) is incident onto the total internal reflection prism (*3*), vertically passes through the cell window (*4*) of the adgesator chamber (*5*), and falls onto the spherical mirror (*6*), which focuses the radiation on the surface of the target (*7*). The atom flux (*8*) through the collimating aperture (*9*) enters the electronic ring (*10*). In this source the stability of the atom flux parameters and the necessary lifetime of the target are provided by changing the interaction point due to the sample rotation and the tilt of the spherical mirror. The target rotation is performed by a magnetic drive: AC magnetic field of the DSDR 1-2 motor penetrates through a thin-walled tube, which houses a rotor mounted on the axis of the target rotation.

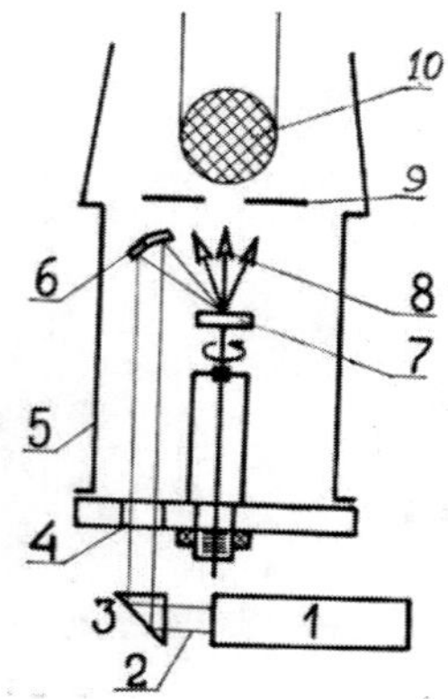

Figure 3.28. Arrangement of the peripheral source.

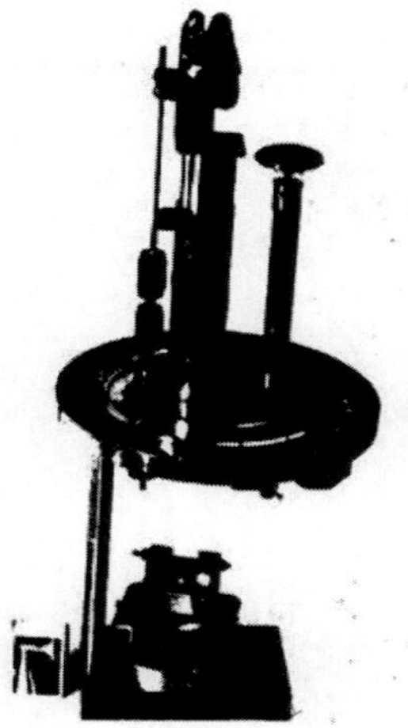

Figure 3.29. Appearance of the peripheral version of the CHIA laser source.

Figure 3.29 shows the appearance of the peripheral version of the laser source. A distinctive feature of this version is the fact that laser radiation falls onto the target at an angle of 40° to the normal. The magnetic field of the adgesator for fluxes of atoms is in this case transverse and should not significantly affect the characteristics of the flux of neutral particles. The magnetic field of the adgesator in the vacuum chamber of the bench was provided by a magnetic coil with the parameters: the average diameter, 6 cm; and the number of coils, 4×4=16. The current pulse in the form of a sine half-wave with duration of about 200μs and amplitude up to 10kA passed through the coil. The induction in the center of the coil in the air reached 3 T. Syncronization system provided the interaction of laser radiation with the target at the time corresponding to the maximum value of the field. The geometry of the target position with respect to the coil should lead to a change in the magnetic field; therefore, the target was mounted at a distance of 5 mm from the end of the coil. The real value of the magnetic field was at least 0.3T, which is sufficient to explain the effect of the magnetic field on the flux of the atoms. It was found that (i) the longitudinal magnetic field does not reduce the yield of neutral atoms from a laser plasma, and (ii) some increase in the number of atoms coming from the target along the normal is, apparently, due to the enhancement of the recombination processes.

In the laser source of atoms (in the adgasator chamber), when the flux of the particles comes from the target at an angle of 45° to its normal, there is a decrease in the number of ions.

Figure 3.30 shows the dependences of the average concentrations and partial pressure in the flux of the lead atoms loaded into the ring with a radius of 4cm on the radiation flux density (the measurement error is 20%). Changes in the concentration and flux density of lead atoms for different radii of the ring are shown in Figure 3.31. Figure 3.32 presents the dependences of the atom flux duration on the ring radii at different flux densities. It should be noted that the atomic flux parameters satisfy the requirements for loading rings with ions, obtained in the calculations and listed at the beginning of this section.

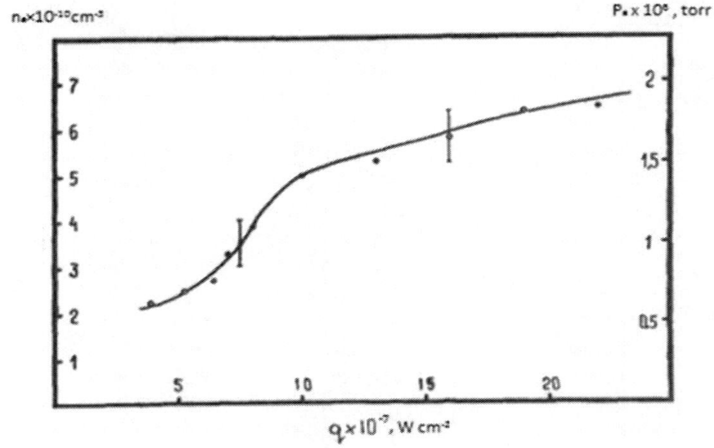

Figure 3.30. Dependence of the concentration and the partial pressure of lead atoms (R_{ring}=4cm) on the value of q.

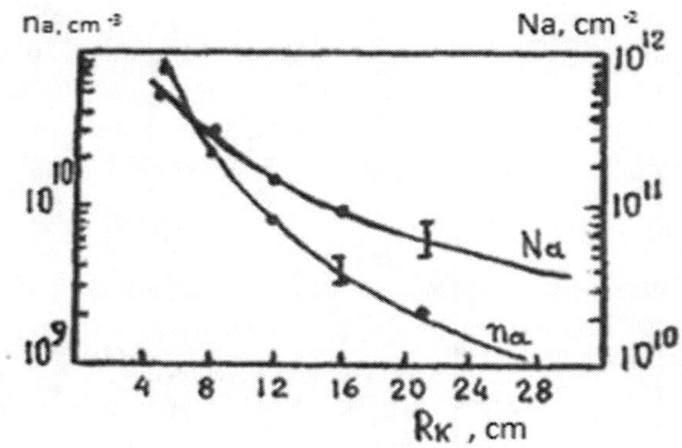

Figure 3.31. Dependences of the concentration and the number of atoms in the flux, N, on the ring radius at 2.5×10^8W/cm^2.

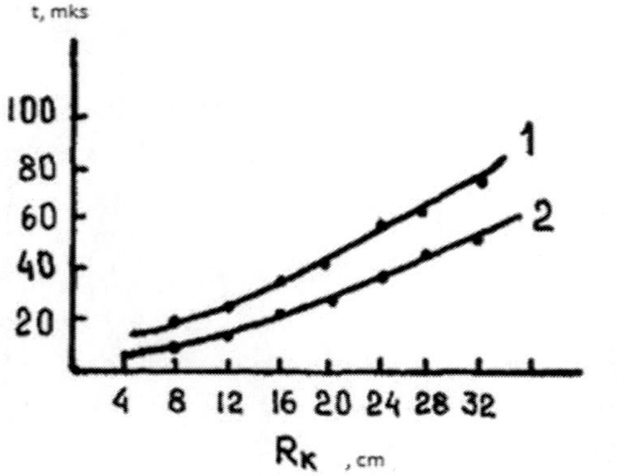

Figure 3.32. Dependences of the atom flux durations on the ring radius R_{ring} at (*1*) 2.5×10^8 and (*2*) 4.3×10^8W/cm^2.

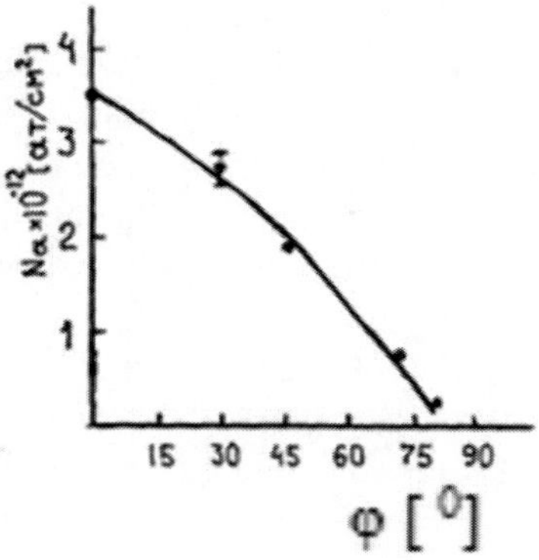

Figure 3.33. Angular dependence of the lead atom concentration.

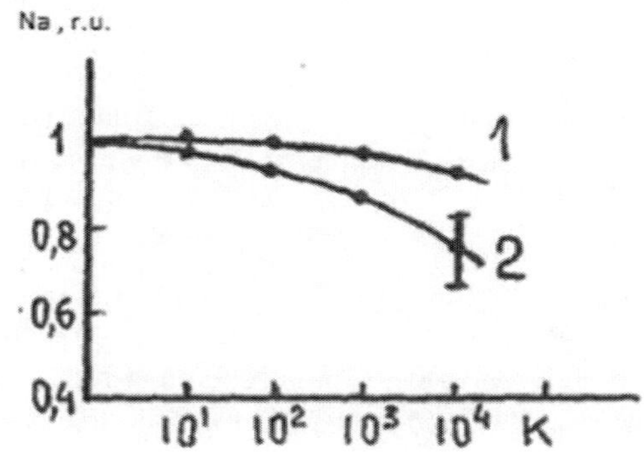

Figure 3.34. Dependences of the concentration of (*1*) iron and (*2*) lead atoms on the number of laser shots.

The experimental results showed that the tungsten samples have a very long life span, whereas the shortest life span (among the listed materials) is observed in indium targets. For other materials, we found a strong dependence of the target life span the on the size of the focal spot, which can be explained by a change in the number of atoms and retention of the atom flux duration in general.

Figure 3.34 on the dependences of the concentration of iron and lead atoms (in relative units) on the number of laser shots. A similar dependence was obtained for lead and carbon atoms (10^8W/cm^2). Under these conditions, the concentration of lead atoms will decrease by 10% when the same point of the target is exposed to no less than 10^4 laser pulses. Changing the interaction point in the working versions of laser sources can provide a life span of at least 10^7 laser shots. The atoms are injected into the ring near (a few centimeters) the stainless steel walls of the adgesator accelerator; therefore, a part of the flux of atoms can be reflected from the wall and fall into the ring. In addition, a local region with a high concentration of the working medium atoms can be formed, especially at high repetition rates of the source (50Hz). A relatively small reflection coefficient, a sharp drop in concentration with distance, and the accounting for the angular dependence of the concentration of the reflected atoms lead to the conclusion about their limited impact on the accumulation of ions in the electron ring. Indeed, the concentration of lead atoms in the flux injected into the electron ring of radius 6 cm (the number of electrons is 10^{13}) is equal to (5–10)×10^9cm^{-3}.

Given the above-mentioned factors, the concentration of the atoms reflected back into the ring does not exceed 5×10^8cm^{-3}, i.e., less than 5% of the concentration of atoms in the main flux. High velocities of the reflected atoms, 7×10^5cm/s, cause a sufficiently small (of the order of hundreds of microseconds) time during which the atoms non-ionized in the ring condense on the walls of the chamber.

This time is much less than that (20μs) between the individual shots of the accelerator at a frequency of 50Hz. Thus, the laser source of atoms does not produce a noticeable background of secondary particles for subsequent operation of the collective accelerator, and the influence of the atoms reflected from the chamber walls is negligible in the current process of accumulation of ions. To inject atoms in the electron rings use can be made of modified loading schemes. One of these schemes is shown in Figure 3.35. The laser radiation (*1*) propagating parallel to the *Z* axis passes through the optical system (*6*), which consists of an axial cone (axicon) and a lens, and then is focused in the form of a ring on the surface of a cone-shaped target (*7*) with an opening angle up to 90°. The region of the plasma formation and the nozzle holes in the fore-chamber lie in the median plane of the adgesator chamber (*4*); therefore, the azimuthal uniform flux of atoms (*8*) from the laser plasma propagates along the compression axis of the electron ring (*10*). This version of the central source allows a ring of different radii to be loaded.

The second version of the central location of the laser source is shown in Figure 3.36, which also has new design elements that allow for the injection of atoms at any radii of the adgesator into compressed electron rings. In this scheme, a reflector (*4*), made in the form of a cone with an opening angle of 90° and an axial through-hole, is placed between the lens and the target.

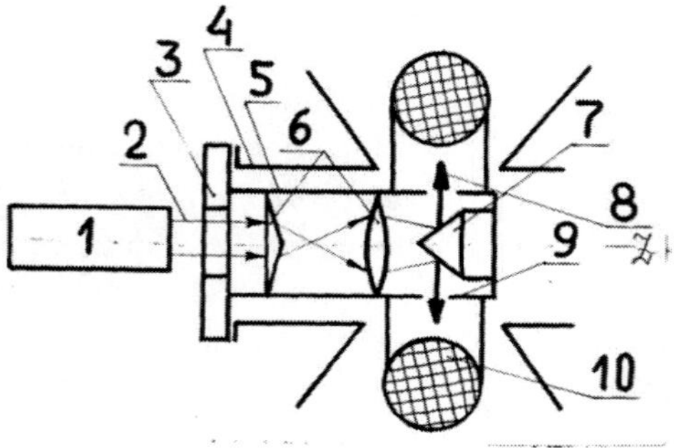

Figure 3.35. Scheme of the modified version of the central atom source.

Nozzle holes are also located in the medial plane of the adgesator chamber between the reflector and the target. The cone can be made of glass or ceramic. It is intended primarily for the effective reflection of the flux of atoms and formation of directed beams. In addition, charged particle fluxes of laser plasma can be neutralized on the cone surface. Studies have shown that the reflection coefficient of the deposited layer of the material in question (the reflective surface is in the form of a plate made of stainless steel) amounts to 30%. The particles are scattered in this case almost isotropically. Selecting the appropriate shape of the surface and having properly arranged nozzle holes, one can load the electron rings at selected radii of the adgesator.

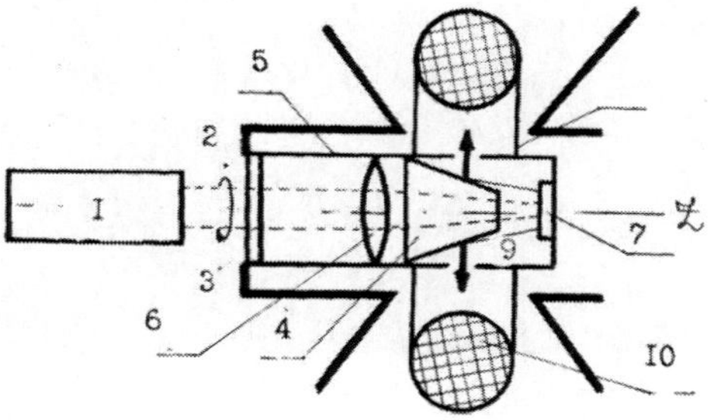

Figure 3.36. Scheme of the central version of the laser source CHIA in the adgasator chamber.

3.3.4.3. Use of Laser Sources of Neutral Particles in a Collective Accelerator at the Joint Institute for Nuclear Research

Experiments on loading an electron ring with ions by using a laser source were mainly carried out in a vacuum at 10^{-6}–10^{-5}Pa. It was found that the laser source operation in the adgesator chamber did not change significantly the vacuum in it. The first one – two hundred shots from the laser source led to a decrease in vacuum down to 2×10^{-5}. This process was associated, apparently, with the removal (during the first shots) of oxide films of the target material and gases adsorbed on the surface of the target. Thus, after the source installation and the depressurization of the chamber it is needed to periodically switch on the device in order to 'clean' the target surface. The source placed in the adgesator chamber had no effect on the pulsed magnetic field of the adgesator, i.e., the dynamics of the compression of the electron rings. This confirms the correctness of the choice of the source materials and design, sufficient accuracy in the nozzle tube adjustment. The relative stability of the radiation power (as well as its value over time) in real operating conditions in the accelerator could be efficiently monitored by the photodiode, the spread in these values not exceeding 5%.

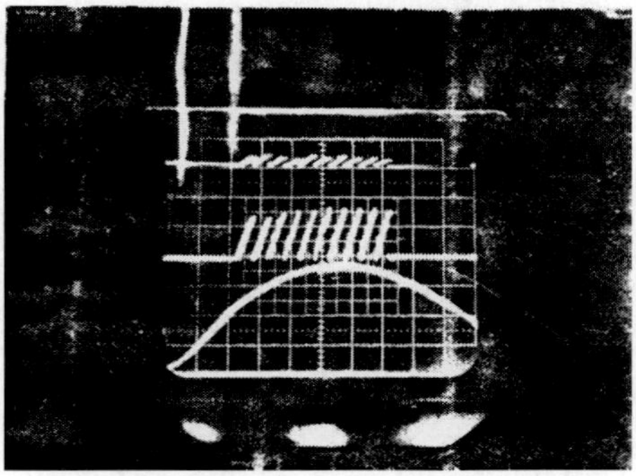

Figure 3.37. Oscillograms (top to bottom): bremsstrahlung from the ring with radius R_{ring}=3cm without triggering the laser source; in the case of injection of lead atoms into the ring with R_{ring}=35cm; the shape of the current at the last stage of compression (200μs/div).

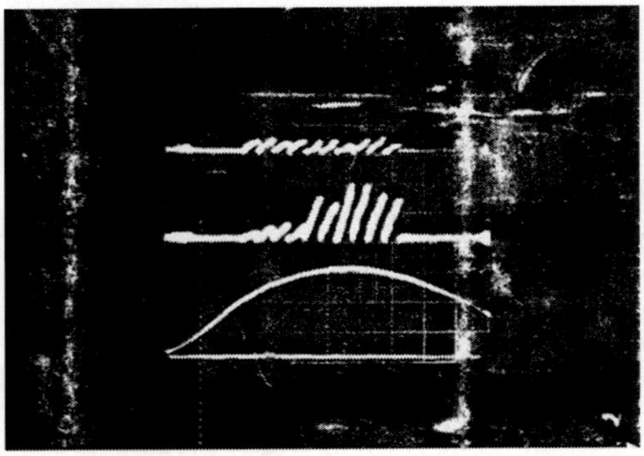

Figure 3.38. Oscillograms (top to bottom): bremsstrahlung from the ring with radius R_{ring}=3cm without triggering the laser source; in the case of injection of lead atoms into the ring with R_{ring}=3cm; the shape of the current at the last stage of compression (200μs/div).

A system of laser radiation alignment and a beam scanning mechanism used for scanning a target showed good results (after 3×10^4 laser shots); the area of erosion represented a ring with the size, corresponding to the calculated one. After the first series of experiments on the Kuti-20 accelerator, the parameters of the working version of the source were tested on the bench. Measurements showed that the atom flux parameters did not change, with an accuracy up to 10%.

The experiments conducted allowed us to estimate the duration of the atom flux and the rate of its propagation by the bremsstrahlung of the electrons in the ring. Figure 3.39 shows the oscillogram of the bremsstrahlung during the injection of the atoms into the ring with R_{ring}=35cm (in a constant magnetic field, without the ring compression). In this case, the depth of the potential well in the ring is not sufficient for trapping the ions, i.e., the duration of the electron bremsstrahlung corresponds to the time of their interaction with the injected atoms. It can be seen that the duration of the flux of atoms (80μs) is consistent with the data of Figure 3.32. At a distance of 35 cm a package of atoms 'spreads' due to the initial spread in their velocity. Figure 3.40 presents the oscillogram of the bremsstrahlung from the ring with a radius of about 3.5cm, which is injected into a flux of atoms. After an atom flux of duration 10 -15 μs passes through the ring, which is in agreement with Figure 3.32, one can observe the signals of the electron bremsstrahlung on the accumulated ions. Using the delay (in Figs. 3.39 and 3.40) between the laser photodiode signal and the bremsstrahlung signal, and taking into account the known target – ring distance, the velocity of lead atoms has been experimentally determined. The obtained values of the velocity well agree with those obtained by the methods described in previous chapters. Experiments on the loading of the rings with lead ions were held at an estimated number of electrons in the ring $(5–7)\times10^{12}$. The cross section of the atom flux by the median plane is a few centimeters in size (due to the relatively large size of the collimating holes in the tube), but the atom flux duration (a few tens of microseconds) is much less than the time of the ring compression (about 2.5μs). Therefore, taking into account the time of the flux propagation, it was possible to inject the atoms into the electron rings of a certain radius. Figure 3.41 shows the current oscillogram in the coil of the third compression stage and the oscillogram of the bremsstrahlung from the ring after the laser source triggering. The source was triggered 600μs before the current reached the maximum. The photomultiplier operated in the gamma photon counting mode in the gate of pulses. It can be seen that after the atom flux passes through the ring, the intensity of gamma radiation remains the same. Thus, lead ions were accumulated in the electron ring. The ratio of gamma photons in the case of a laser source to gamma photons in the case of another source of ions indicates the magnitude of the accumulated ions.

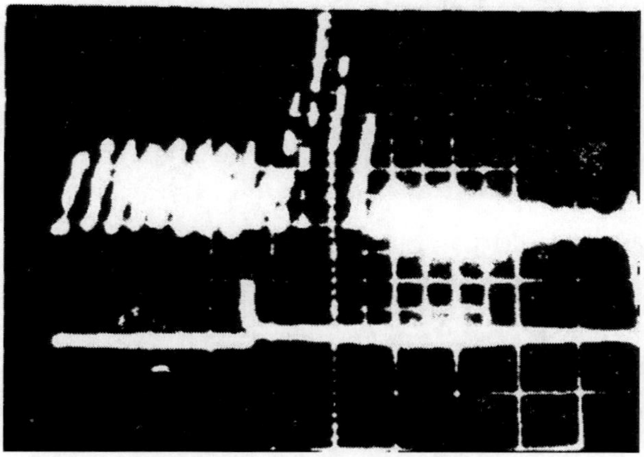

Figure 3.39. Oscillograms of the signals from the laser photodiode (bottom) and the detector (top) during the injection of the atoms in the ring with R_{ring}=35cm (50μs/div), T=30μs, 850W.

As noted above, by changing the operating mode of the source one can adjust the concentration of the atoms in the flux passing through the electron ring. The range of changes was defined by preset adjustment of the laser source. The laser radiation flux density on the target could be varied by changing the voltage on the capacitors of the discharge pump lamp and with the help of radiation attenuation filters. The entire range of working concentrations of atoms was almost overlapped. At a flux density of 2.5×10^8W/cm^2 ($N_a=5\times10^{10}$cm^{-3}) the ring totally or partially collapses (Figure 3.38). Figure 3.42a shows the oscillogram from the collector.

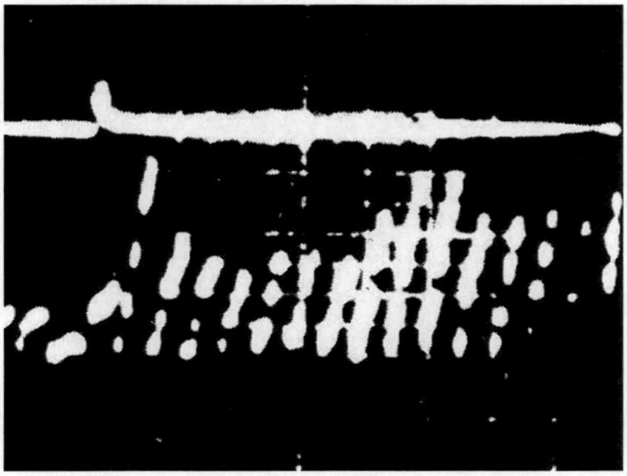

Figure 3.40. Oscillograms of the signals from the laser photodiode (top) and the detector (bottom) during the injection of the atoms in the ring with R_{ring}=4cm (20μs/div), T=10μs, 680W.

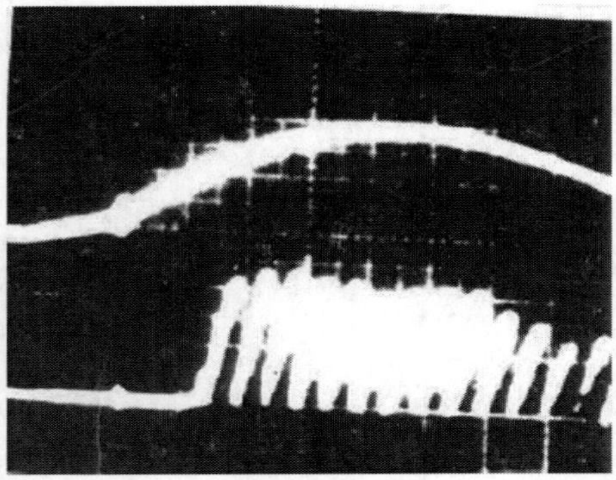

Figure 3.41. Oscillograms of the current signals in the third stage of compression (top) and the detector (bottom); 200μs/div.

It can be seen that the flux with a high concentration of atoms is injected into the ring, there appears a large bremsstrahlung signal, preceding the signal of 'reverse' injection). At lower densities of atoms the first signal is absent. The signal from the second detector operating in current mode is shown in the oscillogram of Figure 3.42b. The absence of the bremsstrahlung signal after the first pulse demonstrates the termination of ion accumulation in the ring. At 10^7W/cm^2 (N_a=5×10^{8}cm^{-3}) the bremsstrahlung from the ring with the detector sensitivity taken into account was not observed. Thus, it is possible to optimize the process of the ring loading with the ions when the laser source is properly adjusted. When the atoms are injected into the ring of a large radius (up to 35cm), which is then compressed, one can observe stable bremsstrahlung signals that existed until the destruction of the ring after compression. This experimentally recorded fact indicates the possible usage of the time of the loaded ring compression to increase the average charge of accumulated ions (Figure 3.37).

The experimental results show that the laser source of atoms is suitable for loading electron rings with ions in a collective accelerator.

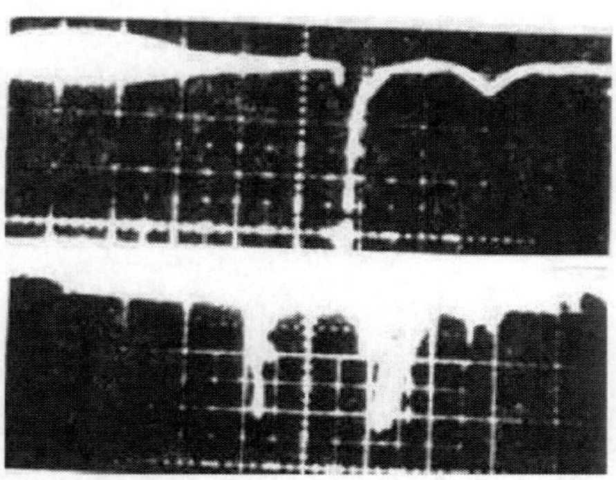

Figure 3.42. Oscillograms of the bremsstrahlung signals in the case of destruction of the ring (500μs/div and 850W) from (a) the surveillance detector and (b) the detector of the chamber.

The source operation showed the correctness of the design, the materials used as well as the high reliability. Control of the process of loading of the ring with the help of the laser source allows the accumulation of ions of different elements (up to uranium) to be optimized in the experiments.

3.3.5. Negative Ions for Acceleration Technology

3.3.5.1. Negative Ion Sources

Various fields of science and technology urgently require the creation of intense negative ion sources for use in high-energy accelerators in nuclear, atomic and applied research to study the fundamental processes of classical nuclear physics, as well as in programs on controlled thermonuclear fusion (CTF). The development of such sources has given an impetus to the design and construction of large tandem accelerator, which together with other facilities allow one to accelerate the heavy nuclei of some elements above the Coulomb barrier, thereby opening up new possibilities for the study of natural phenomena at the nuclear level.

In addition to this traditionally developed program, new applications of negative ion beams are currently experiencing rapid growth in chemical physics, geophysics, and surface physics. For example, the tandem-accelerator mass spectrometry is now being rapidly developed to study the precise isotopic composition of elements. Because of the relatively low energy of the electron affinity of atoms, the negative ions can easily donate electrons in collisions, thereby forming neutral beams. Accelerated negative ions are widely used to impregnate ions into semiconductor materials, in radiation damage, surface analysis based on Rutherford scattering, secondary ion mass spectrometry, X-ray fluorescence, depth profiling and many other areas.

Deeper understanding of the physical processes occurring with the direct participation of negative ions favored an increase in the amounts of information about their structure. This refers to the effects of radiation absorption in the stellar, including solar, atmosphere, to the electrical discharge and breakdown phenomena in electronegative gases, as well as to the processes in the Earth's atmosphere. Operation of most sensitive detectors of halogen compounds, such as many types of pesticides investigated in connection with the problem of environmental protection relies on the formation of negative ions. Note that the use of negative ions requires the creation of a universal source, capable of producing beams of negative ions that contain the maximum number of elements of the periodic table. The review of the currently available sources shows that the efficiency and versatility of such sources depends to a large extent on the mechanisms of ion generation and the values of the electron affinity of these elements.

3.3.5.2. Laser Plasma As a Source of Negative Ions

High efficiency of negative ions in different plasma and accelerator facilities has stimulated the development of intense sources. The currently available hydrogen ion currents are equal to 1 A in the pulse of the plasma sources. With increasing ion mass and decreasing energy of the electron affinity the intensity of negative ion beams decreases significantly. The obtained maximum intensities range from a few to hundreds of microamperes (Table 3.2.)

Table 3.2. Comparative characteristics of the sources of negative ions

Ion	Sample	$I^{max}{}_{\mu A}$	$I^{USSR}{}_{\mu A}$	$I^{review}{}_{\mu A}$	$I^{cal}{}_{\mu A}$ 100 Hz, μA
Li^-	LiF	3	2	3	10
C^-	C	40	2	40	200
O^-	SiO_2	100	50	100	100
F^-	LiF	100	100	100	500
Al^-	Al	3	3	2,5	5
Si^-	Si	20	1	20	100
Cl^-	NaCl	100	50	100	500
Cu^-	Cu	50	10	50	100
Ge^-	Ge	0,5	0,5	-	10
Br^-	Br^-	30	30	10	300
Y^-	NaY	50	50	-	400

A^- are the ions from a laser plasma (laser pulse repetition rate of 100 Hz).

Apart from the advantages of conventional sources, it is necessary to mention some drawbacks, namely:

1. Low emittance.
2. Use of secondary emission from the surface of a solid body; a number of serious shortcomings are usually noted.
3. Current density of negative ion emission is limited by the dielectric strength of the 'plasma – extracting electrode' gap and decreases with the size of the emission hole.
4. In the region of the emission hole it is necessary to provide the high gradient of the working material concentration for the efficient generation of negative ions and for reduction of their losses on the way to the emission hole.
5. Limitation of the beam intensity due to an increase in the permissible thermal loads on the anode in the emission hole.
6. Negative ion beam dispersion due to the action of the space charge.
7. Unavoidable presence of electronegative impurities in discharges.

When using a laser-produced plasma as a source of negative ions, these drawbacks are automatically removed or significantly restricted. When a laser-plasma spreads in a vacuum, there appear concentration gradients at the plasma boundary and the space charge of negative ions is compensated for up to large distances from the place of their formation, which contributes to simplification of their transportation to the destination point. Moreover, laser-plasma sources of negative ions have a number of practical advantages:

1. flexibility (the ability to obtain negative ions of any element of the periodic table);
2. sterility of selection of ions using a 'dry' vacuum (10^{-6} Torr);
3. wide variation in the location of the source with respect to external devices;
4. simple and compact design.

Optimization of the negative ion source operation requires information about the characteristics of ions, electrons and atoms, the concentration of particles and their local distribution in the region of generation, which is a very tedious task.

The validity of these estimates is confirmed by the existence of an optimum for formation of negative ions for 10.6μm radiation ($2–5\times10^9$W/cm^2) and by the corresponding dynamics in q for 1.06μ. The integral emission characteristics of negative ions as a function of q are in the range from 10^{-3}–10^{-4} for aluminum ions and 10^{-1} for carbon of the total yield of positively charged ions.

Table 3.2 presents comparative characteristics of the existing sources of negative ions of heavy elements with integral data output.

3.3.6. References

[1] Aglitskii E.V. et al. *Fiz. Plazmy,* 11 (3), 285 (1985).

[2] Bykovskii Yu.A. et al. The influence of the electrode field on the characteristics of the laser plasma. *Preprint MEPhI,* 060-90, 1990.

[3] Bykovskii Yu.A., Sil'nov S.M., Sheroziya G.A. 'The influence of a transverse magnetic field on the expansion of laser plasma,' *Fiz. Plazmy*, 12 (2), ??? (1986).

[4] Vatazhin A.B., Lyubimov G.A., Gegirer S.A. Magnitogidrodinamicheskie techeniya v kanalakh (Magnetic and Hydrodynamic Flows in Channels) (Moscow: *Nauka,* 1970).

[5] Kadomtsev B.B. Kollektivnye fenomeny v plazme (Collective Phenomena in Plasma) (Moscow: *Nauka,* 1988).

[6] Longmire C. Elementary Plasma Physics (New York: Interscience Publishers, 1963).

[7] Bykovskii Yu.A., Dolgov A.N., Lanchava B.M., Savelov A.S., *Sil'nov S.M. Preprint MEPhI*, 025-89, 1989.

[8] Bykovskii Yu.A., Romanyuk V.I., Silnov S.M. A Laser-plasma Ion Source, 1987, *Patent of USSR,* No. 1503595.

[9] Veretennikov V.A., Dolgov A.N., et al. *Preprint FIAN* No. 8, 1988.

[10] Aglitskii E.V. et al. *Fiz. Plazmy*, 11 (3), 285 (1985).

[11] Trubnikov B.A. *Fiz. Plazmy,* 12 (4), 468 (1986).

[12] Vikhrev V.V. et al. *Fiz. Plazmy*, 8 (6), 1211 (1982).

[13] Veretennikov V.A., Krokhin O.N. *Preprint FIAN* No. 59, 1983.

[14] Sarantsev V.P., Perel'shtein E.A. Kollektivnoe uskorenie ionov elektronnymi kol'tsami (Collective Ion Acceleration by Electron Rings) (Moscow: *Atomizdat*, 1979).

[15] Shirkov G.D. Accumulation of Ions in Electron Rings of Collective Accelerators. Thesis. Dubna, 1982.

[16] Aleksandrov V.S. et al. *Heavy Ion Accelerator Complex at JINR. – Preprint/JINR*, Dubna, 1983, p. 9-83-613.

[17] Mironov V.E., Sil'nov S.M., Sotnichenko E.A., Shestakov B.A. On the Interaction of the Neutral Component of the Laser Plasma with a Metal Target. – *Preprint/JINR*, Dubna, 1983, p. 9-83-536.

[18] Bykovskii Yu.A., Sil'nov S.M., et al. The Laser Source of Neutral Particles of a Collective Accelerator. - In. Proceedings of the Meeting on the Collective Acceleration Method: *Dubna,* 1982, D9-82-664, p.27-30.

[19] Bykovskii Yu.A., Sil'nov S.M., et al. About the Creation of a Laser Source of Neutral Atoms for a Collective Accelerator. – *Zh. Tekh. Fiz.*, 54-63, 527 (1984).
[20] Mironov V.E., Sil'nov S.M., et al. Operational testing of the laser source of atoms on KUTI-20 and PKUTI facilities. – *Preprint/JINR, Dubna*, 1983, p. 983-837.
[21] Bykovskii Yu.A., Sil'nov S.M., Sotnichenko E.A., Shestakov B.A. Zh. *Eksp. Teor. Fiz.*, 93 (8), 500 (1987).

Chapter 13

3.4. Simulation of Conducting Channels in the Atmosphere

3.4.1. Introduction

At present, many laboratories still continue the search for an efficient laser-based lightning protection system [1–3] and for producing a controlled high-conductivity channel in the atmosphere [4–8], capable of conducting a short-circuit current [9] in a natural or artificial electrical circuit. Zvorykin et al. [7] demonstrated a 1.5-fold increase in the length of the discharge gap broken under laser irradiation when switching extended (~1m) high-voltage (up to 390kV) electric discharges by 100-ns UV pulses of a KrF laser. Laser-based lightning protection systems, as is known, rely on the so-called long laser spark, which provides the conditions for connecting a thunderstorm cloud with a grounded metal rod, i.e., a classical lightning rod. Maximum lengths (~16m) of the laser-spark-controlled electric discharge at a voltage of 3000kV were obtained in Russia and Japan [3] using a 0.5-kJ pulsed CO_2 laser with spherical optics. Such control conductivity channels can be used in energy transmission, overvoltage protection systems, transport of charged particle beams, plasma antennas, etc. [8].

3.4.2. Lasers for Producing Sparks in the Atmosphere

It was shown in [10, 11] that a laser spark produced using conical optics demonstrates much better characteristics from the point of view of formation of conductivity channels in the atmosphere. Currently, among the huge variety of different lasers, only two types are being actively studied to be used in the formation of laser-spark-controlled high conductivity channels in the atmosphere: submicrosecond pulsed gas and chemical lasers (CO_2, DF) and femtosecond solid-state lasers [4, 6, 12].

The main advantage of femtosecond lasers is the ability of producing superlong ionized channels (so-called filaments) with a characteristic diameter of about 100μm in atmosphere along the laser beam propagation direction. With an estimated electron density of 10^{16}cm^{-3} in these filaments and the laser wavelength in the range of 0.5 – 1.0μm, the plasma hardly absorbs the laser light. In this case, the length of the track consisting of many filaments is determined by the laser intensity and can reach many miles at a femtosecond pulse energy of ~100mJ. However, these lasers could not be used to form long high-conductivity channels in atmosphere. The ohmic resistance of the thus formed conducting channels turned out to be very high, and the gas in the channels could not be strongly heated (the femtosecond laser energy is less than ≤1J). In particular, an electric breakdown (EB) controlled by radiation of a femtosecond solid-state laser was implemented in [5, 6] only at a length of 3 m (with a voltage of 2MV across the discharge gap).

As was shown in [13, 14], to form a high-conductivity channel in the atmosphere, which could commute an artificial or natural discharge (lightning), the gas must be heated to a temperature from 6000 to 8000°K, when thermal ionization begins. This condition is crucial, because the high initial electron density in the atmosphere is a necessary but not sufficient condition for realizing high conductivity in a long atmospheric channel. As was mentioned in [13], it is important not only to produce but also to maintain the electron density in an atmospheric channel that is sufficient for the breakdown. During gas heating thermal ionization becomes the main mechanism of electron 'production' in the channel. Specifically this is the main reason why the streamer mechanism cannot lead directly to an electric breakdown in the presence of high electron density (filaments) in a cold channel [13].

In the case of submicrosecond CO_2 or DF lasers with a high energy per pulse (several hundreds of joules), long conducting laser-plasma channels can be formed with conical optics. In contrast to spherical optics, one can use methods of lens dynamic correction to increase aberrations and elongate the focal segment by passing to a conical phase front [11]. For example, an axicon transforms the phase front of a plane wave so that it becomes conical rather than spherical and converges to the z axis at an angle γ. For small angles ($\gamma \ll 1°$), it is simply expressed in terms of the angle γ at the axicon base and the refractive index N of the axicon material : $\gamma=(N-1)\alpha$. Each ring element of a conical wave front with a radius R and width δR is focused into an axial element of length $\delta z=\delta R/\gamma$ and the entire focal segment has the length $L=R/tg\gamma=R/\gamma$ [11]. This length can be made significant. For example, at R = 10cm and γ =0.5°, it amounts to L~11m. The cylindrical symmetry of focusing suggests that the type of transverse intensity distribution is independent of the coordinate z, a situation corresponding to a 'diffractionless' wave beam. In reality, such beams also undergo diffraction; however, the diffraction energy loss from the central part of the beam is compensated for by the distributed lateral supply of radiation. Note that the theoretical length of the focal segment formed by the axicon is independent of the laser beam energy and power and is determined by only the beam diameter. At R = 100 cm and γ = 0.5°, it amounts to L~110 m. According to our estimates, when using CO_2 lasers with an energy of ~5 kJ per pulse, the length of these channels for wide laser beams and conical optics can be fairly large (much more than 100 m).

When laser radiation is focused by an axicon in a real experiment, the longitudinal intensity distribution $I(z)$ along the focal length L depends on z, since the transverse intensity distribution $I(R)$ of the input beam is transformed by the axicon into the longitudinal distribution $I(z)$ of the Bessel beam. When the radiation intensity in the beam reaches the threshold, there occurs a breakdown in the medium and an extended plasma channel is formed along the focal length of the axicon [11]. Initially, the channel has a diameter of 20–100μm and a length that is on order of the focal length L of the axicon. If the breakdown is stable, the plasma channel, in contrast to the case of spherical optics, becomes continuous for few nanoseconds. The density of the plasma, depending on the parameters of the medium, the wavelength λ, and other experimental conditions, varies from 10^{17} to 3см$^{-3}$, the temperature reaches 30–40eV, and specific conductivity is 10 (Ohm m)$^{-1}$. A few microseconds later the channel expands to 10 mm. Accordingly, the temperature drops to 1eV and retains at this level for about 100 μs, followed by a relaxation of the plasma [11].

Figure 4.1a shows a photograph of a plasma channel in scattered heating radiation with a wavelength λ =1.06μm at the breakdown onset ($\tau i \approx 50$ns, exposure time 5ns, delay time t_d=0 ns) [11]. The channel structure is affected by the combination of such parameters as the pulse width and the angle γ. For example, for a long pulse, laser-supported detonation may arise at inhomogeneities of the channel structure [15], as a result of which the structure becomes herringbone [11]. The photograph of a continuous laserspark (CLS) in Figure 4.1b was obtained by focusing CO_2-laser radiation in the atmosphere by a conical mirror with an opening angle γ = 177.8° [8]. One can see that a laser spark looks like a continuous cord, composed of bright tapered elements. There are no discontinuities that are characteristic of sparks produced by focusing radiation by spherical optics. A similar pattern was observed at laser energies below 300 J [8]. In the photograph of a CLS fragment in Figure 4.1c the longitudinal structure of the spark looks like a set of bright tapered elements, which are in contact and equally spaced. Figure 4.1d shows a photograph of the electric breakdown in the atmosphere for a channel formed by a CLS. When studying the spark between two rods immersed in its plasma and spaced by d=80cm [8], an electric breakdown occurred at a laser energy of 140J and minimum average electric field strength $E_{min} \approx 77$ V/cm.

The up-to-date experiments aimed at implementing laser-spark-guided electric discharges shows another fundamental difference in the case of femtosecond solid-state and long-wavelength lasers. In the presence of a laser-plasma channel composed of filaments, the breakdown voltage decreased by 30% [5]. The use of CO_2 and DF lasers made it possible to reduce it by a factor of more than 10.

As was pointed out in [16], the guiding properties of CLS (with a high voltage delayed by 10-20 μs) are similar to the action of a chain of metal balls. After an optimal time of ~10-20μs, the expansion of individual breakdown 'zones' leads to the formation of a continuous quasi-cylindrical conducting channel with a low gas density (~0.1 of the normal density), high equilibrium temperature (2700–2900°K), and electron density in the range from 10^7 to 10^8cm^{-3}.

3.4.3. USE OF A PULSE-PERIODIC LASER

Another approach to the formation of a conducting channel relies on the application of a pulse-periodic laser. It was shown in [17] that electric-discharge and gas-dynamic lasers with a high peak power can operate in a high-frequency pulse-periodic regime. This lasing regime allows one to produce a long conducting channel in the atmosphere, which exists for a long time. To this end, it is necessary to align a laser beam with an inverted Cassegrain telescope system, which consists of two spherical mirrors. The first (main) mirror of large diameter is immobile, while the second mirror can move according to a certain law. When moving the second mirror, the common real focus of the system changes its position in time from infinity to some minimum value, which is determined by the optical scheme of the telescope. Calculations show that, at a certain repetition frequency of laser pulses and a certain motion law for the second mirror, continuous extended plasma channels can be formed. However, practical implementation of this scheme with a 'running' focus meets a number of fundamental difficulties. The light distribution in the focal plane, yielded by a well correlated lens, is in essence due to the Fraunhofer diffraction. The specific features of the out-of-focus 3D monochromatic images of a point source, obtained with a round hole, were considered for the first time by Lommel. Along the optical z axis, the length of the focal segment that determines the longitudinal size of the energy-accumulation region depends quadratically on the focal length of the system. The transverse size of this region is proportional to the focal length and the angular divergence of laser radiation; therefore, the volume within which energy is concentrated is proportional to the focal length in the fourth power. For this reason, to form a homogeneous plasma channel several kilometres long, one must use a pulse-periodic laser with a shaped pulse energy and the second telescope mirror moving according to a complex program. A practical implementation of this method with a pulse-periodic CO_2 laser with an average power of 1 kW yielded a continuous plasma channel ~1m long [18].

3.4.4. Formation of a Current-Conducting Channel According to the Impulsar Program

A completely different approach to the formation of a current-conducting channel can be realized using laser firing according to the Impulsar program [19]. Motion of an Impulsar laser engine under irradiation by a high-frequency pulse-periodic laser with a pulse energy sufficient for producing a breakdown in the focus of the optical system of a laser reactive engine (LRE) gives rise to a continuous conducting channel in air as a result of the formation of a mixture of aerosol [20] with atmospheric air and its subsequent ionization [19]. In this case, according to the calculations, the aforementioned range of laser pulse repetition rates provides continuity of the conducting channel at optimal velocities of the focusing system. Indeed, each laser pulse focused by the optical system forms some extended plasma region with a high conductivity, which occupies a relatively small segment of the focusing system trajectory. If the laser pulses have a low repetition rate [21], these regions will form somewhat like a dashed line at some velocities of the focusing system. If the pulse repetition rate exceeds 10 kHz and the motion velocities are optimal for solving the problem stated, these high-conductivity regions will have no discontinuities, and a continuous conducting channel will be formed. However, when the moving optical system enters rarefied atmospheric layers, a problem of medium deficit arises. To solve it, the displaced optical system must have an efficient source of an appropriate medium (a material containing microscopic metal particles, which can easily be sublimated under laser irradiation) in the vicinity of the focus. In particular, one can use copper oxide (CuO) nanopowder [22], which provides synthesis of conducting aerosol copper particles upon laser heating. The presence of aerosol particles of heavy metals in the sublimated material makes it possible to increase the specific impulse of the LRE thrust, simultaneously with increasing the channel conductivity.

To implement a long conducting channel, one must solve several problems:

1. development of a high-frequency pulse-periodic CO_2 laser with an average power P~100kW and a peak power sufficient for initiating a breakdown in the focus of the LRE optical system;

2. development of a system of dynamic correction for the wave front of a wide-aperture laser beam with a diameter *D*#10 cm in the presence of small-scale turbulence in atmosphere;
3. consideration of the laser radiation absorption and scattering by aerosol particles in the LRE exhaust;
4. alignment of the starting position of LRE with the laser and the pulsed high-voltage (up to several MV) source;
5. obtainment of maximally possible specific impulse of the LRE thrust to reduce the time of aerosol channel formation;
6. choice of the material of aerosol heavy-metal particles and their dispersion composition that could provide high channel conductivity upon heating to the sublimation temperature in the LRE;
7. determination of the lower limit for the size of dispersion aerosol particles to ignite electric breakdown of a channel capable of transmitting a current, corresponding to a short-circuit current in natural or artificial electric circuits at a minimally possible average strength of electric field.

To address the latter two problems, we conducted a series of experiments. The results obtained are reported below. We consider urgent the problems of determining the dynamics of change in the channel conductivity at long delay times and its maintenance at the optimal level by introducing additional high-frequency generators into a chain; they will be analyzed in the next stage of the Impulsar program.

3.4.5. Formation of an Electrical Breakdown in the Channel Formed by an Exploding Thin Wire

The formation of a conducting channel based on the Impulsar technology was simulated using the results of the laboratory experiments on a controlled electric breakdown in an atmospheric channel containing aerosols [22, 23] of hot copper and copper oxide (CuO, Cu_2O) particles, formed as a result of wire explosion. Note that the proposed method for generating particles of variable size with a necessary concentration appears to be fairly efficient in terms of the price – quality criterion when one must carry out many experiments aimed at modeling the optimal conditions for high conductivity. There are a number of factors facilitating the electric breakdown (including the streamer stage) of long gaps in an atmospheric channel formed as a result of wire explosion at average external electric fields.

As was noted in [14], the main problem in the analysis of the formation of a spark channel in a discharge gap in atmosphere using a field of arbitrary configuration is to determine the mechanism of gas heating, which provides thermal ionization of the gas in the spark channel. The heating in the discharge channel was generally believed to be caused by the increase in the concentration of the charged particles supplied by shock gas ionization. However, the concentration of charged particles in the range of $10^{11} - 10^{13}\text{cm}^{-3}$ is limiting for shock ionization. The development rate of thermal processes in a gas is limited by the velocity of the electrons that transform the electric field energy to the thermal energy of gas molecules. The thermal inertia of the gas excludes the possibility of step changes in its temperature and therefore, its conductivity, both during heating and cooling the channel.

It was noted in [13] that ionization in a highly ionized plasma differs significantly from the similar process in a weakly ionized plasma, where molecules are ionized by the electrons gaining energy directly from the electric field. A field in a highly ionized plasma supplies energy to all gas electrons. The latter are thermalized as a result of collisions and obey Maxwell's distribution. The gas is ionized by the electrons that acquire sufficient energy during the energy exchange with other particles rather than directly from the field. It is much easier to maintain a highly ionized equilibrium plasma than a weakly ionized nonequilibrium plasma, because the former situation requires a much weaker field. For example [13], in an equilibrium plasma column of nitrogen arc at atmospheric pressure, which burns in a cooled tube of radius r=1.5cm at a current i = 10A, the field supporting the arc column has a strength E=10V cm^{-1}. Under these conditions, T = 8000 ^{0}K, the gas density N_a=10^{18}cm^{-3}, n_e=2×10^{15}cm^{-3}, the degree of ionization x_1=n_e/N = 2×10^{-3}, and E/N_a = 10^{-17}Vcm^2. In a nonequilibrium glow discharge column in nitrogen, the strength of the discharge-supporting field is larger by an order of magnitude: E=200Vcm^{-1}. Such a sharp difference is caused by the difference in the nature of ionization processes. In the case of weakly ionized plasma, the field must be sufficiently strong to accelerate electrons to an energy of 13–15eV (ionization threshold). At the same time, for highly ionized equilibrium plasma, the field can be much weaker, because in this case it is sufficient to accelerate electrons to only kT ~ 1eV. As was indicated in [13], electrons supply atoms with energy, and, during the energy exchange, all particles jointly concentrate energy in separate electrons, which serve as ionization sources.

Our experiments on the formation of a high-conductivity channel in atmosphere was performed with a PEV-2 copper wire in an enamel shell; the wire was 90 – 155cm long and had a diameter of 90mm. Figure 4.2a shows a schematic of the experimental setup; the arrangement of the sensors for measuring current by a Rogowski loop and voltage on the capacitor battery using a voltage divider; and the position of the end face of polished quartz fiber, which transfers radiation from the discharge channel to a high-speed pin-photodiode. The end face of the polished quartz fiber was located at a distance of 2–10cm from the wire axis near the discharge cathode or anode. The shape of the current pulse was measured using a differential screened Rogowski loop with a time resolution of ~30ns [24]. The shape of the voltage pulse across the discharge gap was determined with the aid of a screened voltage divider of mixed type (connected to the capacitor battery anode), which yields a measurement error of ~1 % at a width of the current-pulse leading edge >1μs [25]. Recording was performed using a Tektronix TDS 220 digital oscilloscope. The capacitor with a capacitance C=2μF was charged from a voltage source to U_0=7–10kV. When switched on by an external triggering pulse from a controlled discharger (IRT-3 ignitron), the capacitor was discharged through a copper wire 90μm in diameter (with allowance for the enamel shell, the wire diameter was ~110μm). As the experiments showed, the presence of an enamel shell facilitates the formation of a thin straight-line electric breakdown channel with a conducting region 4 – 8 mm in diameter in atmosphere and promotes a unique mechanism of aerosol formation in the form of balls 330–20μm in diameter with a black surface (oxidized during wire explosion). Large pieces of the enamel shell (up to 10cm long) remain integer, with burn-through areas on the surface. During the wire explosion and electric breakdown, the enamel shell stabilizes the spatial arrangement of the balls in the channel, which are centers of ionization plasma formations.

Figure 4.2b shows a photograph of electric breakdown in a channel formed during explosion of a horizontally oriented wire (*D*90 μm, length 105cm). This photograph was recorded through an FS-1 blue filter and a neutral NS-9 filter by an OLYMPUS C-5050 ZOOM digital camera. At an exposure of 1.6 s the photographs have an integral character; i.e., the exposure during photographing the discharge was determined by the discharge duration. The digital camera matrix operated with a maximum resolution. When taking pictures of the entire discharge channel, the distance *l* from the wire centre to the digital camera objective was ~2m. To reveal the electric breakdown details: sizes of individual hot particles formed during the wire explosion and the diameter of the channel and the inhomogeneities in the structure of channel luminescence, we took pictures in the macroscopic regime, in which the digital camera objective was located rather close (at 25–50cm) to the wire. When taking pictures of the wire explosion products, *l* was chosen to be 3 cm (supermacroscopic regime). The digital camera was used in the ZOOM regime. The spherical shape of the wire explosion products (*D*330–20μm) is primarily explained by the effect of surface tension forces on the molten copper drops during their formation. The large spread of particle diameters is explained by the fragmentation and aggregation of metal drops during wire explosion [26]. The black color of the particles is due to the formation of copper oxide on their surface as a result of the interaction of copper drops with air oxygen. The oxidation is reversible; i.e., upon heating to ~1200 ^{0}K, copper oxide is reduced to copper [22]. During the wire explosion, the oxygen concentration in the channel formed should sharply decrease as a result of the oxidation. It was shown in this experiment that, by the instant of electric breakdown, the voltage across the discharge gap U_a =7kV provides the electric field strength in the channel $E_c=U_a/d\approx 67$ Vcm^{-1}.

At the storage capacitance *C*=2μF, depending on the charging voltage and wire length, wire explosion may occur without electric breakdown. Figure 4.2c shows a photograph of a channel (*D*90 μm, length 115 cm) formed as a result of explosion in the absence of electric breakdown. It can be seen that the channel structure is intermittent: it consists of individual isolated plasma zones of superheated drops of wire material [27 – 29]. At a charging voltage U_0 = 10kV and storage capacitance *C* = 2 μF, the voltage across the capacitor battery decreases to 7.6 kV by the end of the wire explosion. At a constant charging voltage U_0, the maximum pulse current during wire explosion is mainly determined by the initial wire resistance R_0 and amounts to 1.25 kA. When changing the length of 90-μm copper wire within 90 – 155 cm, the measured resistance ranged from 2.7 to 4.2 W. At U_0 = 10 kV and *C* = 2 – 4 μF, the maximum amplitude of the pulse current was in the range of 1 – 1.4 kA. After the wire explosion, a current $i\approx$10 A flows for some time through the hot channel and then decays to zero. As a result, the residual voltage across the capacitor battery decreases to 7 kV. Despite the presence of residual current after the wire explosion, which heats additionally the channel, the electric breakdown does not occur at these parameters of the discharge circuit. The amount of energy *W* spent on wire explosion, calculated from the current and voltage characteristics, was found to be ≈35 J.

In the series of experiments performed on a horizontally oriented wire, the capacitance of the capacitor battery was taken to be constant: *C* = 2 μF; the only variable parameters were the wire length (90 – 155 cm) and charging voltage (7 – 10 kV). It was found experimentally that, at a wire length of 100 cm, the minimum charging voltage at which there is no electric breakdown in the gap does not exceed 7 kV; a value corresponding to the minimum accumulated energy: $W_0 = 49$ J.

In the next series of experiments on the electric breakdown in a channel, we considered a vertically oriented wire. This version is preferred, because it takes into account the possible influence of the vertical component of electric field at the Earth's surface on the electric breakdown in the channel formed by wire explosion.

Figure 4.3a shows a photograph of electric breakdown in a channel formed by a vertically oriented wire (*D* 90 μm, length 143 cm) at the initial stored energy W_0 = 190 J, which corresponds to the charging voltage U_0 = 10 kV and the storage capacitance C = 3.8 μF. The electric breakdown can be seen in the oscillograms of discharge channel current, voltage, and luminescence, which were recorded by a high-speed pin-photodiode with a time resolution of ≈ 1 ns. The light arrived at the photosensitive area of pin photodiode through a silica fiber ≈2 m long, whose polished receiving end face was located near the discharge anode at a distance of 1.5 – 4 cm from the discharge channel. The radiation pulse width at half maximum was ≈30 μs and hardly differed from the corresponding current pulse width during electric breakdown (the light-pulse oscillogram in Figure 4.3e shows only the arc phase of the breakdown). The oscillogram of the discharge current (Figs. 4.3b, 4.3d) indicates that it is close to zero during the current pause [30, 31], whose duration is ΔT≈140 μs. The results of our studies showed that the current during the pause amounted to 10 – 50 A (depending on the discharge-circuit parameters).

There are many theories explaining this phenomenon; they will be considered elsewhere. As was shown in [31], the electric breakdown delay for a copper wire depends on the average electric field strength E_c in the channel at the instant when the arc phase of the breakdown begins. The voltage oscillogram demonstrates that the breakdown occurred at a voltage of ~8.4 kV. Using the current and voltage oscillograms, one can also determine the channel resistance during the electric breakdown at the instant corresponding to I_{max} = 360 A. In this case, R_{min} = 17 W. Here, in contrast to the data of [32, 33], the discharge-channel diameter is of little importance, because the discharge is slow, and the inductive component can be neglected when estimating the voltage drop across the discharge channel. In Figure 4.3b, which illustrates the case of the electric breakdown in a channel formed by wire explosion at C = 3.8 μF and inductance L= 2 μH, the wave resistance $\rho = (L/C)^{0,5} \approx$ 0.7ohm. Judging by the volt–ampere characteristics of the discharge gap, L≈143 cm in the case of the electric breakdown at Rp≈18 Ohm, Rp>2ρ (18>1.4 Ohm), which corresponds to the aperiodic shape of the discharge.

It was noted in [9] that the channel resistance is affected by the amount of energy released in the channel rather than the current through the latter. At a wire length of 143 – 152 cm and the maximum capacitance C_{max} = 4.5 μF, a breakdown in a channel formed by wire explosion occurred with a probability close to 100 %. Figure 4.3f shows a photograph of a controlled electric breakdown at C = 4.5 μF, U_0 = 10 kV, and an interelectrode distance of 155 cm. In this experiment, an electric breakdown was ignited at a voltage of ≈8.3 kV. The duration of the current pause was ΔT≈200 μs. At a wire length of 152 – 159 cm, an electric breakdown occurred with a probability of ~30 %. For wires longer than 159 cm the breakdown probability (at the same parameters of the scheme) was close to zero. Thus, the average minimum electric field strength in the channel formed by wire explosion, at which breakdown can occur, is Emin≈ 52 V/cm. The suggestion about the possibility of long-term support of current breakdown with an amount of accumulated energy increased by an order of magnitude and the same charging voltage was partially confirmed.

The channel for recording radiation in the visible wavelength range has a low sensitivity, because it detects only the radiation entering the angular aperture of the polished receiving end face of the quartz fiber. If the distance between the end face and the discharge channel is reduced to minimum (≈1.5 cm), one can detect the following phases in the oscillogram of a visible-light pulse. During the first 16 μs after the wire explosion, the visible light intensity is low (the wire explosion gives rise to radiation with wavelengths in the range of $\lambda \approx 2.5 - 0.5$ μm) [29]. Then the oscillogram of visible light pulse exhibits the radiation corresponding to the streamer (weak), leader (weak), and arc phases of the breakdown.

An analysis of many photographs indicates that the electric breakdown trajectory (broken line, circle, etc.), which is set by spanned copper wire, is highly controllable and contains no branchings. The experiments revealed that the vertical orientation of exploded wire is indeed preferred in the case of streamer breakdown of maximally long gaps ($L_{max} \approx 159$ cm) at $U_0 =$ 10 kV and $C_{max} = 4.5$ μF.

Let us analyze the requirements to an ionized formation that can initiate electric breakdown. The minimal conditions that are necessary for initiating a streamer were reported in [13]. We will consider a uniform conducting sphere with a radius R_s, placed in a uniform field of strength E_0. To form a self-propagating plasma channel, the external field must be displaced from the plasma formation and amplified beyond it. To displace the external field from the entire volume of the sphere, charges must be nonuniformly distributed over the surface, and there must be no less than $N=3\varepsilon 0\pi R^2 Eo/e$ electrons in the field. Taking into account that the radius of this formation should be no less than α^{-1} (α is the effective ionization coefficient), we can estimate the lower limit of the number of electrons at which a streamer may arise [13] as follows: $Ne\ min \approx 3\varepsilon 0\pi E0/e\alpha^2$ As an example, we will consider air at atmospheric pressure. At the breakdown field strength E_0 = 31.4 kV/cm and effective ionization coefficient α= 12.4 cm^{-1}, we have Ne min =2 10^9. The smaller the plasma formation radius, the more severe the limitation on the plasma density is. At the minimally allowable value $R \sim \alpha^{-1}$ the electron concentration in air at atmospheric pressure should be ne min ≈ 4 10^{11} cm^{-3}. The plasma formation size R also must be no smaller than the ionization length α^{-1}, in a field with a strength equal to the external field strength or even somewhat higher. In air at atmospheric pressure the ionization length is α^{-1}~0.8 mm. If the initial size of plasma formation in a homogeneous field satisfies the aforementioned requirements, this formation should generate ionization waves to both sides along the field, thus forming a plasma channel. The plasma channel development is accompanied by amplification of the field at both leader heads [13].

Our experiments showed that all above requirements to an ionized formation [13], which must be fulfilled to ignite an electric breakdown, are implemented in the channel of the discharge gap formed by wire explosion with the aforementioned parameters. The characteristics of separate plasma formations ($R_s \approx 0.5 - 4$ mm, average distances ~ 3 – 5 mm, and temperature $T \approx 3000 - 5000$ °C) with clusters of superheated drops and their aggregates at the centre, as well as low oxygen content in the wire-explosion channel, reduce significantly the average field strength that is necessary for electric breakdown. Note that the size of individual plasma formations exceeds that of superheated drops of wire material by one to two orders of magnitude. The average density of the wire-explosion products in the channel is $\rho \approx 0.1$ kg/m^3. The electric field in separate spherical plasma fractions can be amplified by a factor of no more than 3. Further studies should reveal new factors leading to electric breakdown in the channel produced by wire explosion at $E_c = 52$ V/cm. We should note the original work by Komel'kov [34], who analyzed the leader-discharge channel. We showed that the specific resistance per unit length in the channel of leader positive retarded discharge is 11.8 W/cm, while the longitudinal gradient along the channel in the end of the leader phase is 55.5 V/cm. The main parameters of the leader channel turned out to be the same as in the case of arc discharge. The result obtained in [34] was interpreted in [13] as not quite correct because of incorrect analysis of the discharge-gap voltage oscillograms. Our results disprove the conclusion of [13].

To check the influence of the size of individual plasma formations in the channel on the mechanism of electric breakdown, we performed experiments with explosion of a copper wire L~30 – 36 cm long and Ø 90 μm in diameter at $U_0 = 10$ kV and $C_{max} = 4.5$ μF, i.e., at a much higher specific energy contribution. The channel formed by wire explosion with these parameters contains a fairly large number of wire-material nanoparticles [22], whereas large particles (individual plasma formations) are completely absent. In this case, the second requirement to the development of streamer mechanism of breakdown in the channel is violated. As a consequence, a fivefold decrease in the gap length did not lead to breakdown at the corresponding parameters of the discharge circuit. It occurred only for wires not shorter than L≥60 cm, a minimum value at which the size of plasma formations in the wire explosion channel became sufficiently large to implement a streamer breakdown.

3.4.6. Formation of Electrical Discharges in a Plasma Channel Produced by a Solid-State Laser

We also investigated the regime of electric breakdown in a plasma channel containing evaporated aerosol particles of metals and their compounds, which is formed by focusing radiation of a solid-state laser, operating in the modified spike regime, onto targets made of different materials [35]. To form a plasma plume on a target, we used a phosphate glass laser with a wavelength $\lambda = 1.054$ μm, operating in a modified lasing regime with a peak power of 100 – 500 kW. Photographs of lasers operating in this regime are shown in Figs. 4.4a and 4.4b. Solid-state lasers of this type are unique. Actually, they are laser hyperboloids (concerning the way of extracting radiation from a totally reflecting flat cavity through a round hole of large diameter, which can be seen in the upper photograph). Lasers of this type can produce large plasma plumes in both pulsed and pulse-periodic regimes. A laser pulse with a width up to 700 μs consisted of many (approximately 150) high-contrast pulses 150 – 300 ns wide with an average repetition rate of 200 kHz (Figure 4.4c). The radiation divergence at the laser output $\alpha \sim 4\ 10^{-4}$ rad, was close to the diffraction one. The total energy per laser burst reached ~30 J (see the laser in Figure 4.4a). Graphite, vinyl plastic, textolite, ferrite, permalloy, and carbonyl core were used as targets. Their exposure to focused laser radiation led to the formation of a cloud of rapidly expanding plasma, which contained aerosol particles of evaporated target material ~ 50 – 1000 nm in size. No large particles were found in the plasma cloud. Note that a copper target cannot be used to form an electric breakdown in this plume, because evaporated large copper drops have velocities of several hundred meters per second and cannot form a quasi-stationary channel for electric breakdown. The conductivity of this plasma has been studied little, while the dynamics of changes in its properties is of great interest to search for the ways of forming (according to the Impulsar program) superlong high-conductivity channels in order to transfer energy at large distances. The experiments were performed with a system of electrodes, between which an electric discharge was ignited (Figure 4.5). An anode in the form of a copper cylinder was connected to the high-voltage output of the capacitor battery with a capacitance C = 2 μF, which was charged to a voltage of U0=10 kV. A duraluminum cathode(K) had a hole Ø 6 mm in diameter, through which laser radiation was guided to the target using a lens with a focal length of F= 300 mm. The distance between the electrodes was L~~80 mm. The photo of the plasma plume on the target made of graphite is shown in panel (b). The results of image processing on the PC with the use of the HSB color model are presented in panel (c).

At a laser pulse energy below 30 J the gap was switched using the plasma formed on the target. Typical photographs of the electric breakdown in atmosphere on a graphite target (at a laser burst energy of 12 J) for a gap ~70 mm wide, obtained using an FS-1 filter, a neutral NS-9 filter, and an OLYMPUS C-5050 ZOOM digital camera, are shown in Figure 4.6. In highly inhomogeneous fields, streamers are formed primarily in the regions of maximum field strength near the electrodes and, depending on the voltage polarity, are directed either to the anode or to the cathode [13]

In the photographs in Figure 4.6 the streamer directed to the anode develops more rapidly than that directed to the cathode. Streamers overlap at a distance of ~25 mm from the anode, and an electric breakdown of the gap occurs. An analysis of this process shows that the breakdown has a random trajectory, as well as a natural lightning. In addition, the results of experiments with different targets show that electric breakdown occurs at a minimum average electric field strength of ~300 V/cm. The experiments revealed also that the electric breakdown delay (which is shown in the oscillograms of discharge-circuit current and luminescence) for the graphite target, could exceed 300 μs after the laser irradiation.

In Figure 4.6c, the capacitance C = 2 μF. The current oscillogram indicates that the damping oscillation period T = 8 μs; therefore, the discharge inductance is 0.8 μH. The contour wave resistance ρ≈ 0.6 ohm. According to the current – voltage characteristics of the L≈70 mm discharge gap for the graphite target, the discharge gap resistance is Rp≈0.5 <2ρ ohm, a value corresponding to an oscillatory discharge with a high damping decrement. The long breakdown delay with respect to the laser irradiation onset is apparently caused by the closing of the discharge gap by the expanding vapor of the target material and discharge-gap preionization by UV radiation of the plasma plume.

The targets made of other materials were characterized by a shorter possible delay. Changing the delay for the voltage pulse applied to the capacitor, one can measure the plasma characteristics at different instants and thus estimate the conductivity of the channel at different stages of its cooling and expansion. In our opinion, the strong electric field in these experiments is due to the absence of large particles in the erosion-plasma cloud. To carry out successfully the Impulsar program, we believe it necessary to study the dynamics of plasma conductivity and the residual aerosol products after cooling the plasma.

3.4.7. Experimental Results

Below we report the estimated values of the channel conductivity, which were obtained both in our previous experiments on the breakdown of atmospheric air using long- and short-wavelength radiation and the experiments with aerosol plasma (this study).

1. For the plasma channels obtained on graphite, ferrite, Permalloy, and carbonyl core targets using a phosphate glass laser, the gap specific resistance was 10 ohm/m and the specific breakdown voltage was ~300 V/cm.
2. For the plasma containing a set of dispersion and condensation aerosols, which was obtained by wire explosion, the gap specific resistance was~5 ohm/m and the specific breakdown voltage was ~50 V/cm.
3. For the plasma channel formed by a pulsed CO_2 laser with a conical focusing optics, the gap specific resistance was 100 – 400 ohm/cm and the specific breakdown voltage was ~80 V/cm.
4. For the plasma channel obtained with the aid of a femtosecond pulsed laser, the gap specific resistance was ~0.1 MW/m and the specific breakdown voltage was ~7.5 kV/cm [4].

These estimates, based on the results of the electric-breakdown experiments, show that a voltage of 7 MB, at a total introduced energy of~300 kJ, is sufficient to form a plasma channel ~1 km long and 8 mm in diameter.

Note that the minimum specific resistance of the channel depends strongly on the specific introduced energy; the transverse size of plasma formations; the concentration and size of aerosol particles; the repetition rate of laser pulses and their energy; and the humidity, temperature, and many other parameters of the medium where a conducting channel is formed.

The results of the studies show that the discharge plasma has a high conductivity and can efficiently be used to form a conducting channel in order to transfer energy over long distances.

3.4.8. Conclusion

We determined the range of sizes of copper and copper oxide particles (from 50 nm to 300 μm) and their minimum volume density (~ 0.1 kg/m^3) that are necessary to provide an electric breakdown in order to form a plasma channel according to the "Impulsar" program. The results of the model experiments showed that nanoparticles in the plasma channel can merge into larger aggregates during the LRE operation.

The experiments on electric breakdown showed also that a high conductivity of a plasma channel with a diameter of ~8 mm and a minimum specific resistance of ~5 ohm/m is implemented at the maximum introduced specific energy: ~200 J/m. At a larger channel diameter, the specific energy is proportional to the squared ratio of diameters.

The minimum average electric field strength (~52 V/cm) in the channel, at which a streamer-leader electric breakdown of the discharge gap begins, is due to the following factors:

1. the presence of dispersion and condensation aerosols of hot particles of copper oxide (CuO, Cu_2O) and copper with oxidized surface (smaller than 300 μm in diameter) in the channel for a long time (1 – 10 μs) due to their low mobility;
2. the presence of superheated metal drops at the centre of plasma formations, which are in the metastable (unstable) state at temperatures $T \sim 3000 - 5000$ °C and play a role of hot dots [31] during thermionic emission;
3. formation of large linear aggregates of aerosol particles [20], which are formed during wire explosion;
4. low oxygen content in the channel as a result of fast oxidation of aerosol particles formed during wire explosion.

3.4.9. FIGURES AND PHOTOS

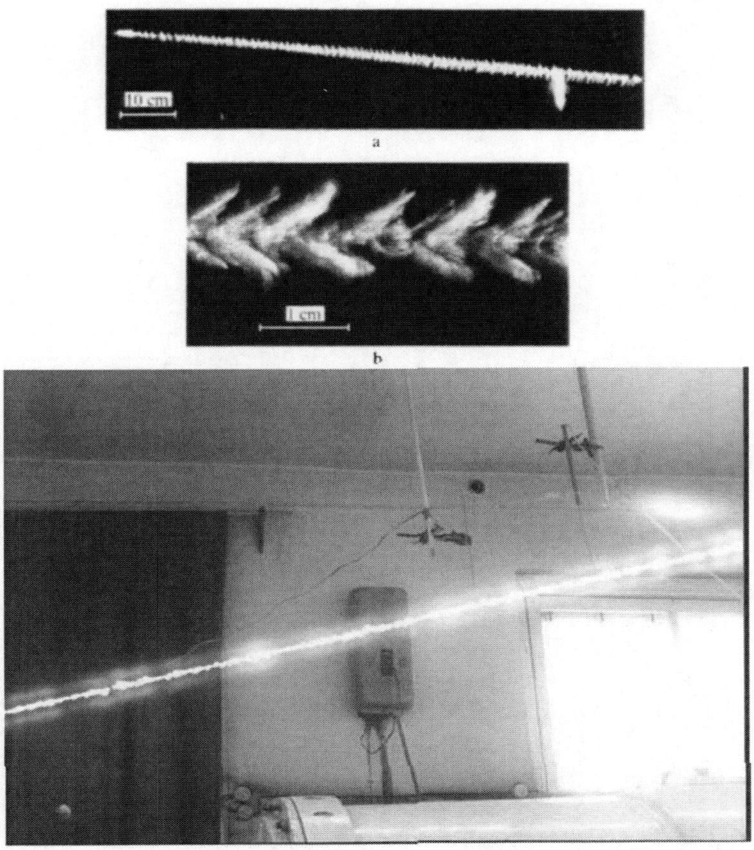

c

Figure 4.1. (Continued).

d)

Figure 4.1. (a) Schematics of focusing laser radiation by a conical lens (CL) and a photograph of the channel in scattered heating laser radiation (LR) with a wavelength 1.06 μm at the breakdown onset (50 ns, exposure time 5 ns, delay time = 0) [11]; (b, c) photographs of a continuous laser spark for CO_2-laser radiation focused by a conical mirror with an opening angle = 177.8°; and (d) a photograph of electric breakdown in atmosphere in a CLS-induced channel.

a

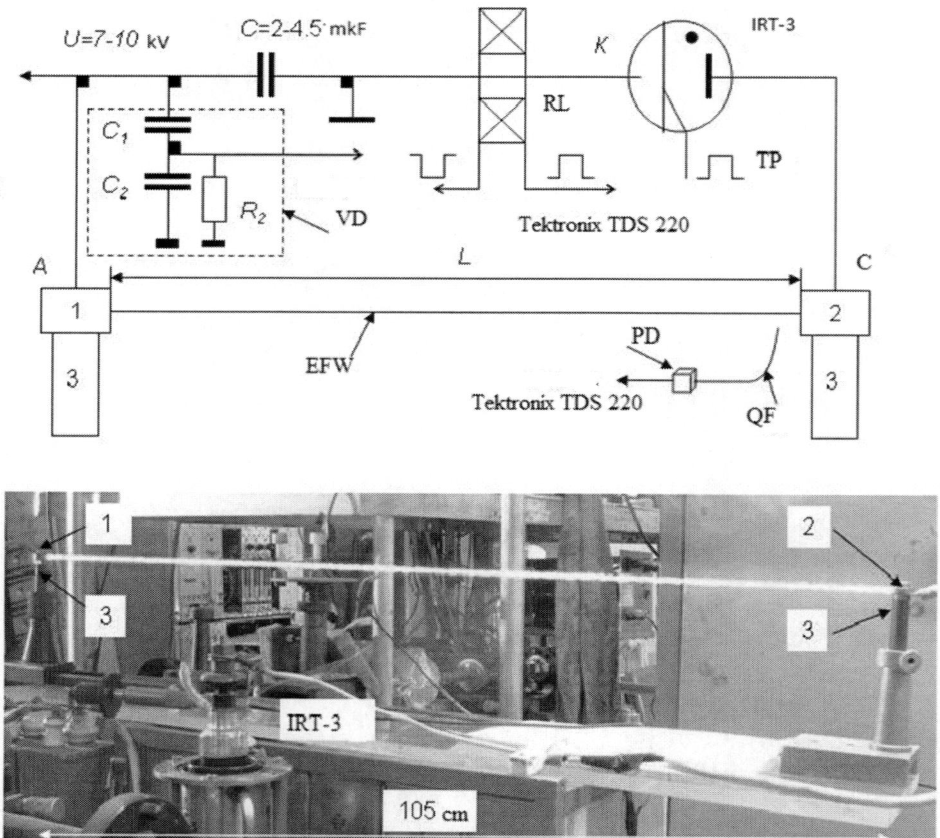

b)

Figure 4.2. (Continued).

Figure 4.2. (a) Schematics of the experiment on igniting an electric breakdown in a wire-explosion channel for a horizontally oriented wire: (*1*, *2*) brass electrodes, (*3*) ebonite insulators, (IRT-3) ignitron switcher, (RL) Rogowski loop, (VD) voltage divider, (TP) triggering pulse, (EFW) exploding fine wire, (PD) photodiode, (QF) silica fiber, (A) anode, and (C) cathode; (b) a photograph of controlled electric breakdown; and (c) a photograph of wire explosion in the absence of breakdown).

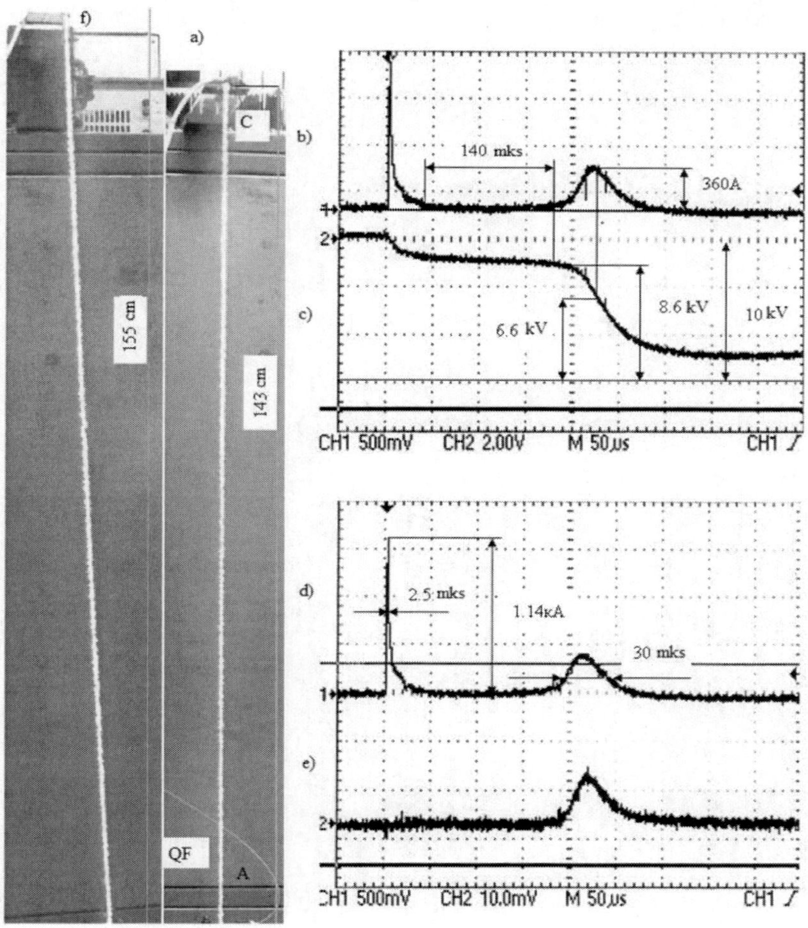

Figure 4.3. (a) Photograph of controlled electric breakdown for explosion of a vertically oriented wire (the interelectrode length is 143 cm); (b – e) oscillograms of the discharge channel (b, d) current, (c) voltage, and (e) luminescence at a capacitance $C = 3.8$ μF and charging voltage $U_0 = 10$ kV; and (f) a photograph of controlled electric breakdown at $C = 4.5$ μF, charging voltage $U_0 = 10$ kV, and an interelectrode distance of 155 cm.

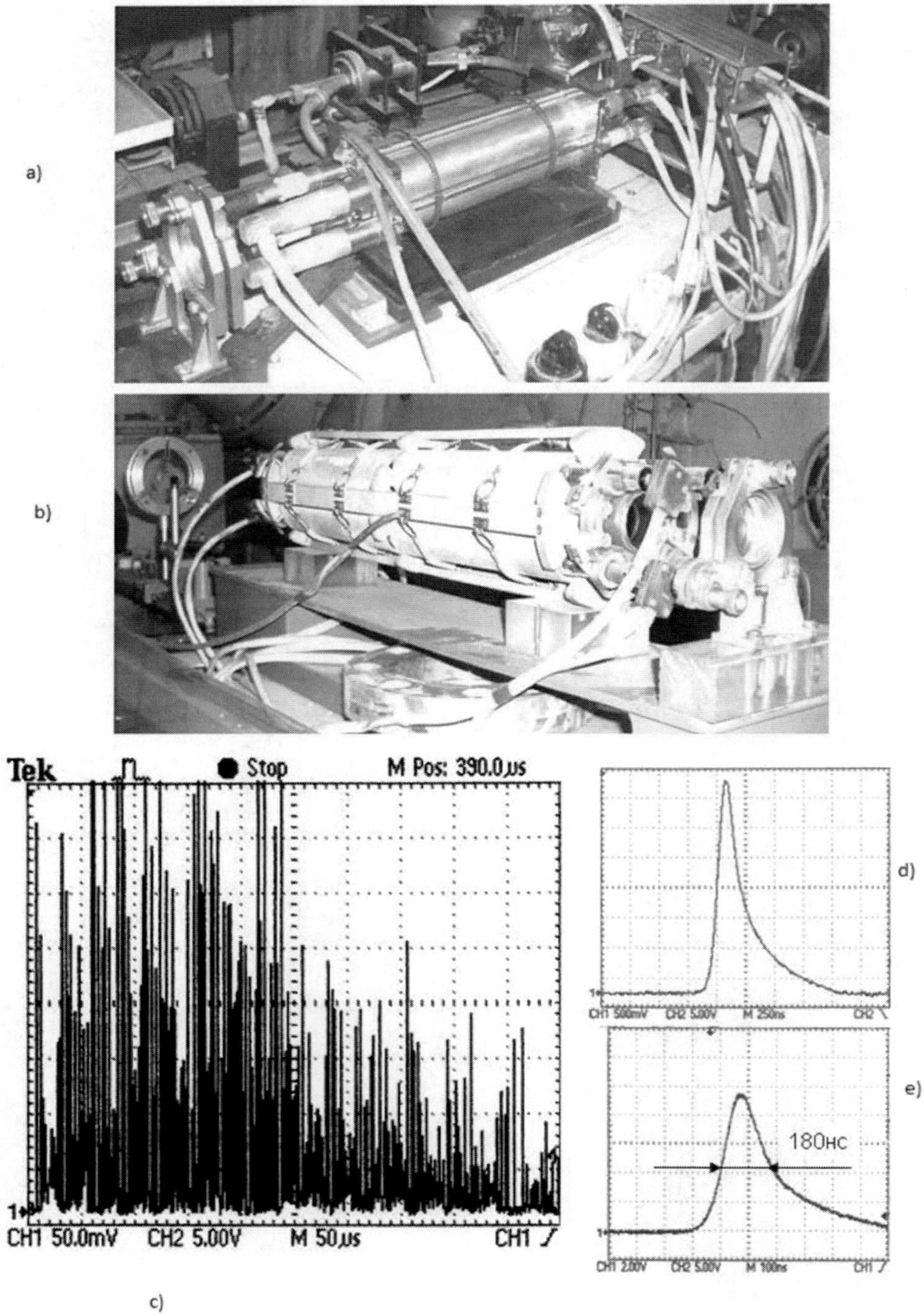

Figure 4.4. (a, b) General view of solid-state lasers and (c – e) oscillograms of (c) the laser radiation and (d, e) separate laser pulses.

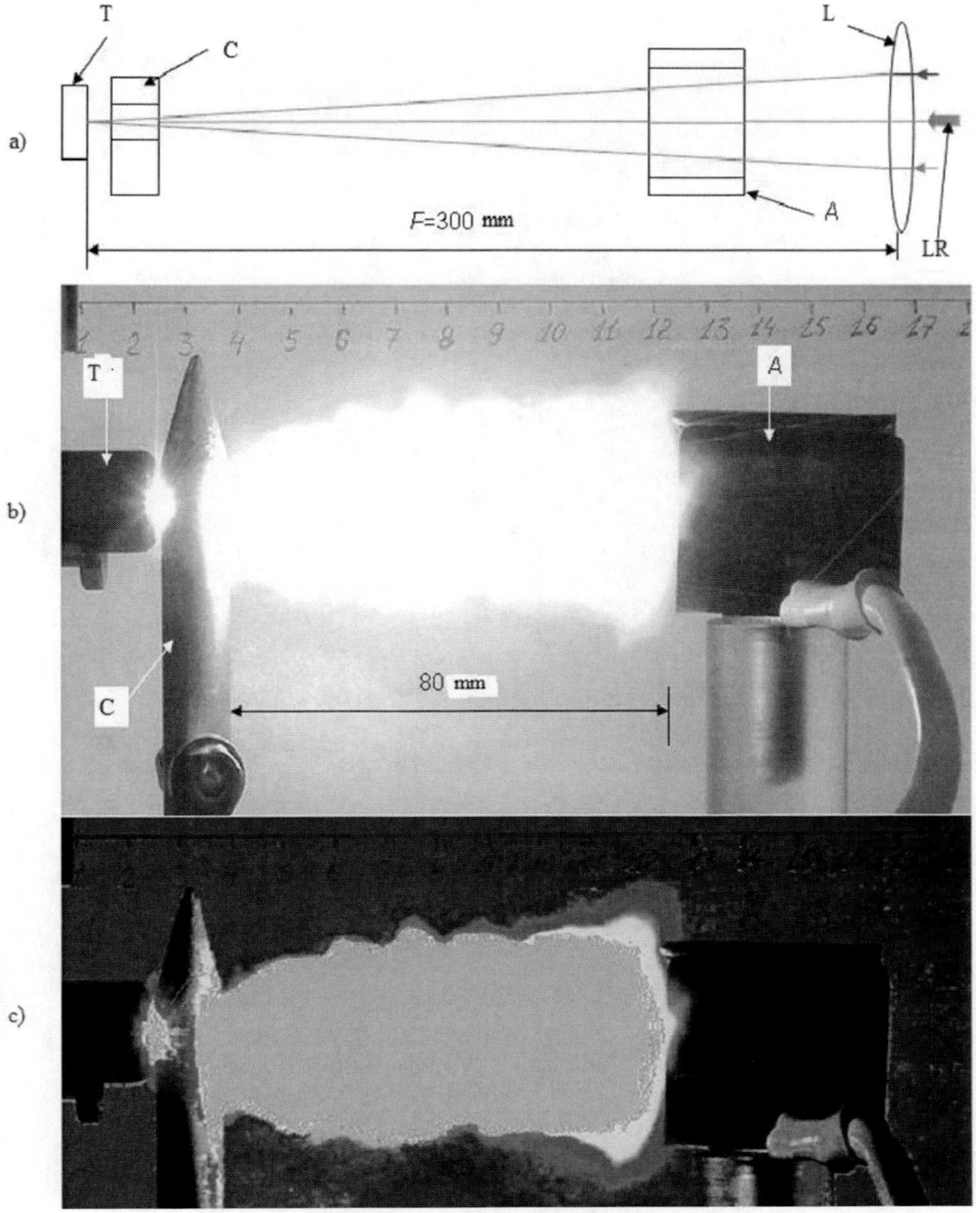

Figure 4.5. (a) Schematics of the system for studying the conductivity of plasma plume products (T indicates a target), (b) a photograph of a plasma plume on a graphite target, and (c) the results of image processing in the system of HSB color coordinates.

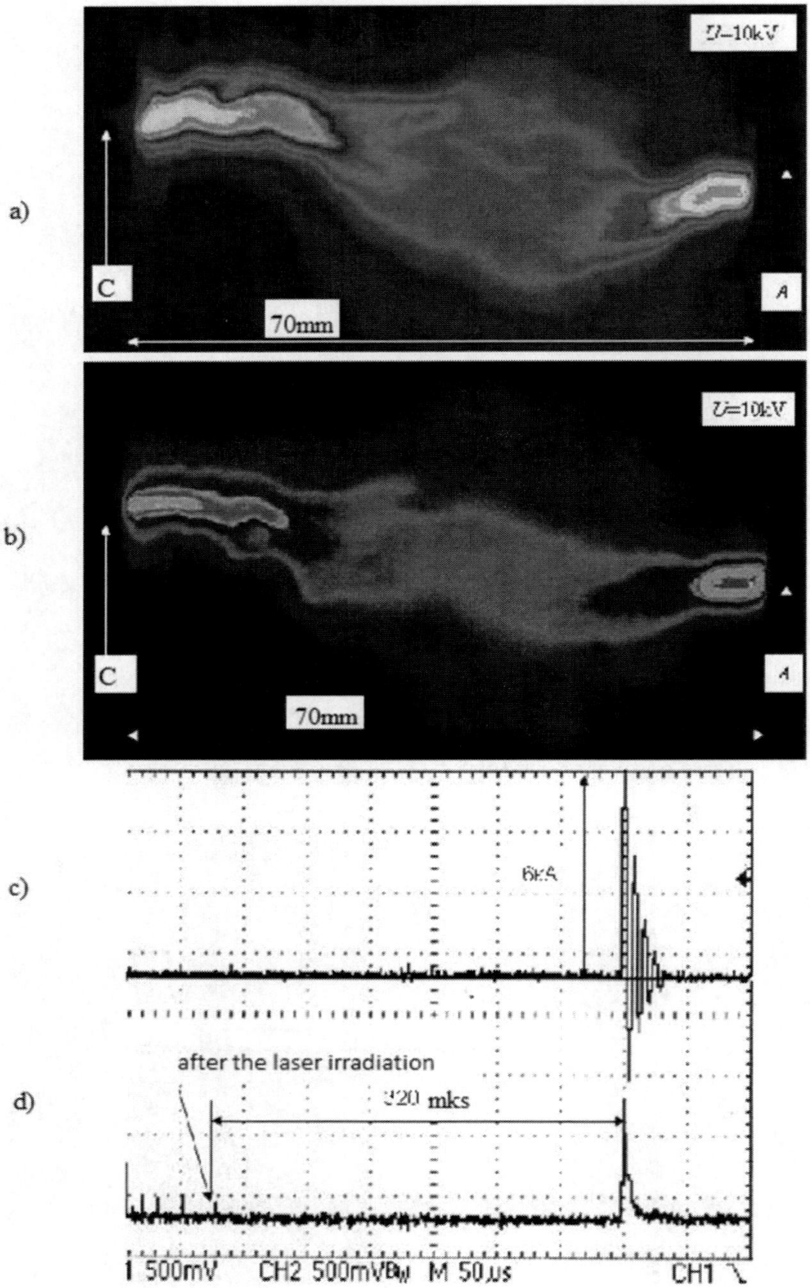

Figure 4.6. (a, b) Photographs of electric breakdown of a 70-mm gap in atmosphere in plasma-plume products (photographing through an FS-1 filter and a neutral NS-9 light filter) and the results of image processing in the system of HSB color coordinates, (c) a current oscillogram at electric breakdown of the gap (voltage $U_0 = 10$ kV, storage capacitance $C = 2$ μF) in plasma plume, and (d) a luminescence oscillogram for the discharge channel in aerosol plasma.

3.4.10. References

[1] Apollonov V.V. et al. *J. Opt. Soc. Am. B*, 8, 220 (1991).

[2] Alexandrov G.N. et al. *Elektrichestvo*, 12, 47 (1980).

[3] Kinoshita F., Morooka Y., Uchiumi M., et al. *Proc. XII Intern. Conf. on Gas Discharge and their Applications* (Greifswald, Germany, 1997) p. 475.

[4] Diels J.-C., Rudolph W. *Ultrashort Laser Pulse Phenomena: Fundamentals, Techniques, and Applications on Femtosecond Time Scale* (Burlington: Acad. Press, 2006).

[5] Kasparian J., Rodriquez M., M'ejean G., et al. *Science*, 301, 61 (2003).

[6] Berge L., Skupin S., Nuter R., Wolf J.-P. *Rep. Prog. Phys.*, 70, 1633 (2007).

[7] Zvorykin V.D., Levchenko A.O., Ustinovskii N.N. *Kvantovaya Elektron.*, 41, 227 (2011) [*Quantum Electron.*, 41, 227 (2011)].

[8] Apollonov V.V., Vasilyak L.M., Kazantsev S. Yu., et al. *Kvantovaya Elektron.*, 32, 115 (2002) [*Quantum Electron.*, 32, 115 (2002)].

[9] Bazelyan E.M., Raizer Yu.P. *Fizika molnii i molniezashchita* (Physics of Lightning and Lightning Protection) (Moscow: Fizmatlit, 2001).

[10] Apollonov V.V. *Opt. Eng.*, 44, 1 (2005).

[11] Pyatnitskii L.N., Korobkin V.V. *Trudy IOFAN*, 57, 59 (2000).

[12] http://www.teramobile.org/publis.html.

[13] Bazelyan E.M., Raizer Yu.P. *Iskrovyi razryad* (Spark Discharge) (Moscow: Izd-vo μFTI, 1997).

[14] Aleksandrov G.N. *Zh. Tekh. Fiz.*, 37, 288 (1967).

[15] Raizer Yu.P. *Lazernaya iskra i rasprostranenie razryadov* (Laser Spark and Propagation of Discharges) (Moscow: Nauka, 1974).

[16] Asinovskii E.I., Vasilyak L.M. *Entsiklopediya nizkotemperaturnoi plazmy. Vvodnyi tom. Kniga II* (Encyclopedia of Low-Temperature Plasma: Introduction. Book II) (Moscow: Nauka, 2000) p. 234.

[17] Tret'yakov P.K., Grachev G.N., Ivanchenko A.I., Krainev V.L., Ponomarenko A.G., Tishchenko V.N. *Dokl. Akad. Nauk.*, 336, 466 (1994); Apollonov V.V., Kiiko V.V., Kislov V.I., Suzdal'tsev A.G., Egorov A.B. *Kvantovaya Elektron.*, 33, 753 (2003) [*Quantum Electron.*, 33, 753 (2003)].

[18] Grachev G.N., Ponomarenko A.G., Smirnov A.L., Statsenko P.A., Tishchenko V. N., et al. *Kvantovaya Elektron.*, 35, 973 (2005)

[19] [*Quantum Electron.*, 35, 973 (2005)]; Apollonov V.V. *Proc. X Intern. Conf. on Photonics and Optoelectronics* (Wuhan, China, 2009) p. 13.

[20] Apollonov V.V., et al. *Sposob sozdaniya tokoprovodyashchikh kanalov v neprovodyashchei srede* (Method for the Formation of Conducting Channels in a Nonconducting Medium), RF Patent No. 2400005 (May 20, 2009).

[21] Fuks N.A. *Mekhanika aerozolei* (Mechanics of Aerosols) (Moscow: Izd-vo AN SSSR, 1955).

[22] Ageev V.P., Barchukov A.I., Bunkin F.V., et al. *Kvantovaya Elektron.*, 4, 2501 (1977) [*Sov. J. Quantum Electron*, 7, 1430 (1977)].

[23] http://www.nanosized-powders.com.

[24] Rukhadze A.A., Shpigel' N.S. *Elektricheskii vzryv provodnikov* (Electric Explosion of Conductors) (Moscow: Mir, 1965) p. 341.

[25] Gerasimenko N.I., Grashina N.A., Medvedkov A.G., Meshcheryakov A.B., Pletnev N.V. *Prib. Tekh. Eksp.*, No. 1, 212 (1988).

[26] Goncharenko G.M. *Trudy MEI. Vyp. 45. Elektroenergetika* (Proceedings of MEI. Issue. 45: Electric Power Engineering) (Moscow: Izd-vo MEI, 1963) p. 146.

[27] Gavrilov V.N., Litvinov E.A. *Zh. Prikl. Mekh. Teor. Fiz.*, 34, 28 (1993).

[28] Kvartskhava I.F., Plyutto A.A., Chernov A.A., Bondarenko V.V. *Zh. Eksp. Teor. Fiz.*, 30, 42 (1956).

[29] Martynyuk M.M. *Zh. Tekh. Fiz.*, 44, 1262 (1974).

[30] Peregud B.P., Abramova K.B. *Dokl. Akad. Nauk SSSR*, 157, 837 (1964); Abramova K.B., Valitskii V.P., Vandakurov Yu.V., Zlatin N.A., Peregud B.P. *Dokl. Akad. Nauk SSSR*, 167, 778 (1966).

[31] Protopopov N.A., Kul'gavchuk V.M. *Zh. Tekh. Fiz.*, 31, 557 (1961).

[32] Vlasto's A.E. *J. Appl. Phys.*, 38, 4993 (1967); Vlasto's A.E. *J. Appl. Phys.*, 39, 3081 (1968).

[33] Aleksandrov A.F., Zosimov V.V., Kurdyumov S.P., Popov Yu.P., Rukhadze A.A., Timofeev I.B. *Zh. Eksp. Teor. Fiz.*, 61, 1841 (1971).

[34] Aleksandrov A.F., Rukhadze A.A. *Fizika sil'notochnykh elektrorazryadnykh istochnikov sveta* (Physics of High-Current Electric-Discharge Light Sources) (Moscow: Atomizdat, 1976).

[35] Komel'kov V.S. *Dokl. Akad. Nauk SSSR*, 58, 57 (1947).

[36] Apollonov V.V., Pletnev N.V. *Sposob neodnorodnogo vyvoda energii svobodnoi generatsii vysshikh poperechnykh tipov kolebanii iz lazera i lazer* (Method for the Inhomogeneous Extraction of the Energy of Free Higher Transverse Lasing Modes from a Laser and a Laser), RF Patent No. 2239921 (November 10, 2004).

[37] Sinton R., van Herel R., Enright W., Bodger P. *J. Appl. Phys.*, 110, 093303 (2011).

Chapter 14

3.5. LASER PLASMA AS A SOURCE OF IONS FOR MASS SPECTROMETRY

INTRODUCTION

Laser mass spectrometry is widely used to determine the trace elements and study the distribution of impurities in inorganic and organic materials.

The advantages of mass spectrometric methods of elemental analysis, as is known, include: lack of restrictions on the physical form of the sample; ease of sample preparation; needelessness for prior information about the composition of the sample; determination of up to 60 elements in pure substances and up to 40 in the complex natural objects (in the case of simultaneous detection of all isotopes); wide dynamic range; low detection limit, which allows expending extremely small amounts of a sample, i.e., to make use of non-destructive analysis; bulk, local and stratified analysis with high surface ($<0.1\mu m$) and depth (3nm) resolution. These advantages favor the widespread use of the method for the analysis of natural objects and technical materials. Besides, more stringent requirements are now imposed on such methods of analysis in order to reduce the detection limits, increase the reproducibility and accuracy of analysis. Urgent is also the development of standardless quantitative analysis. One of the problems of mass spectrometry is to identify the factors that govern the relationship between ion compositions in mass spectra and elemental composition of the sample. In this connection, much has been made to understand the processes occurring during the formation of ions from the particles a solid body. On this basis, work is underway to elaborate standardless analysis techniques. Successful implementation of the methods should be based on theoretical and experimental studies of the mechanisms of ion formation, the relationship of the intensity of the ion currents with the concentrations of elements determined, their parameters, the composition and the basic properties of the matrix, and the conditions of the impact on the sample in the ion source. One of the main goals of the theory as applied to the elemental mass spectrometric analysis is to provide a method for calculating the relative sensitivity coefficients.

As seen in previous chapters, theoretical and experimental results have already been obtained in the field of laser mass spectrometry. However, the relative yield of ions of different elements is studied mainly for the case of simple substances (three component materials at best). The theoretical research in this area is extremely difficult due to non-equilibrium processes, spatial inhomogeneity of the plasma source, changes in its parameters during the target irradiation and plasma expansion. One can offer an approximate description of the processes leading to the formation of the mass spectrum. It is based on the representation of the process in two stages, each of which is assumed independent and successive:

1. material atomization, ionization of atoms, and expansion of plasma;
2. formation of an ion beam, mass analysis, and detection of ions.

The main idea here is that at each stage the behavior of atoms of different elements is determined by one of the main parameters characterizing these elements. This greatly simplifies the description of all the main processes. In this approach, the relative sensitivity coefficient can be represented as the product of the relative yield of ions and the instrumental factor. The relative yield of ions characterizes the discrimination of atoms of different elements in the first three stages of the formation of a mass spectrum, which are associated with the plasma processes. The instrumental factor takes into account the discrimination of ion composition, which is related to the final stages of the formation of the mass spectrum [21].

In laser mass spectrometry, in considering the first stage most applicable is a quasiequilibrium model [22]. Let us clarify the physical meaning of model temperatures: They characterize the degree of nonequilibrium processes under consideration. The higher the atomization temperature, the faster and more unbalanced the evaporation; the higher the ionization temperature, the stronger the effect of the non-equilibrium recombination process in the expansion of the plasma [23–24]. From the analytical viewpoint, the higher the model temperatures, the smaller the difference between the final ionization degree of atoms of different elements, despite the difference in their physical properties. In the limit, when the values of both temperatures tend to infinity, the coefficients of the relative yield of ions are close to unity. This means that discrimination in the stages of atomization, ionization and recombination are completely excluded, i.e., the ion composition of the plasma after its expansion is equivalent to the atomic composition of the sample. Thus, the previously proposed models, according to which the coefficients of the relative yield of ions under certain conditions is equal to unity in the laser mass spectrometry [25], are special cases of the quasi-equilibrium model (provided that the instrumental factor is the same for different elements).

The instrumental factor is found using the calibration with respect to pure elements or by statistically processing the results of a large number of analyzes of standard samples.

Statistical processing of the results of the experiments and the available set of literature data has shown that the quasi-equilibrium model allows one to interpret the relative sensitivity coefficients more accurately than any empirical formulas [26]. The main advantage of the proposed model is that it allows one to judge the nature of the processes occurring during the formation of ions, and the main reasons for the deviation of the ion current composition on the elemental composition of the sample. Naturally, the quasi-equilibrium approach is necessary only to simplify the calculations of the relative sensitivity coefficients and in any case does not replace the physical processes that occur during the formation of a plasma torch.

Research in this area offers ways to stabilize the conditions of formation of ions by controlling the irradiation parameters of a sample in the source. This improves the reproducibility and accuracy of analysis results using both standard and standardless analysis of the sample compositions. Control of the mass spectrum composition (compared to singly charged, multiply charged and complex ions) should lead to a significant reduction of influence of overlapping and thus to a decrease in the detection limits in concentrations of many elements. A further decrease in the detection limits in mass can be achieved by increasing the residual ionization degree of atoms and by decreasing the spread of ions in energy.

3.5.1. Laser Ion Source for Mass Spectrometers

Laser ion sources for mass spectrometric analysis of solids can be divided into two main types according to the purpose for which the laser is used: either as an evaporator for sampling, or for producing a dense high-temperature laser plasma, which is an emitter of ions. The first type of the laser source and analytical capabilities of combining a laser (free-running regime) and a mass spectrometer were studied by many authors. Although the characteristic dimensions of the structural elements of laser sources in these cases vary somewhat, the principle of their action is the same, and the main parameters are close. For the second type of ion sources a dense high-temperature plasma, which is an emitter of ions, is produced during laser irradiation.

One of the simplest designs of a laser ion source of second type–a time-of-flight mass spectrometer–was proposed and developed by a group of Yu.A. Bykovskogo [25]. The source had a target irradiated with a laser, a uniform electric field produced by the flat grid and a target, where the generated ion beam is accelerated.

Such a source design with a uniform electric field to accelerate ions in a time-of-flight mass spectrometer with an electrostatic analyzer was investigated in detail by these authors.

A laser ion source for time-of-flight mass spectrometry with an electrostatic analyzer using a single-potential lens was considered in [27]. If the laser beam is incident on the sample surface (thin film) from the side which is opposite to the analyzer, the range of the samples to be analyzed is very limited. Nevertheless, this source design is used in a LAMMA-500 laser mass spectrometer (Liebold-Heraues, Germany). Given the advantages and disadvantages of the previous sources, a design was proposed, which makes use of a set of apertures to reduce the region of the laser-plasma plume from which the ions enter the accelerating gap of a pre-injectors for subsequent transport to the electrostatic mass analyzer. To eliminate the effect of the backward expansion and possible breakdowns, and thus to change the accelerating voltage, a closed cycle expander was used to extract ions in an electrostatic ion mass analyzer. Consequently, the laser plasma ions arrive into the accelerating gap of the pre-injector only from the region bounded by the system aperture. In the ion beam extraction system use was made of a scheme fixing the boundary with the help of a flat fine-grained grid.

3.5.1.1. Temporal and Mass Resolution of the Devices with an Electrostatic Analyzer

The operation principle of mass spectrometric systems is based on the separation of ions with different ratio of their mass to charge (m/z). Regardless of all the principle of separation of ions, mass spectrometric systems consist of three main components: an ion analyzer, which is used to separate ions according to their ratio m/z; a detector, which permits receiving extracted ions; and a recording device, which retrieves information. The main parameters of a mass spectrometer are the resolution and light intensity. The resolution of a mass spectrometer is determined by its ability to separate ions with different masses. This value is usually expressed in relative units Rm=M/▲M, where ▲M=m1-m2 is minimum difference of ion masses, detected separately. The light intensity of a mass spectrometer in general is the ratio of the number of ions, which have reached a detecting device, to the total number of ions produced in the ion source. Thus, the light intensity of a mass spectrometer is characterized by the transmission coefficient of the ion beam through the analytical part of the mass spectrometer.

In order to improve the main characteristics of mass spectrometric devices, a special field is created in the analyzer, which is focused with respect to energy and direction. The sensitivity of a mass spectrometric device relies on a combination of characteristics of the ion source, analyzer, and detection system. First, the sensitivity of the analysis depends on the amount of a substance (in grams) required to detect a certain element of interest in this substance. Second, the sensitivity can correspond to the minimum relative concentration of the detected impurity element. This characteristic is called the relative detection limit. Both these characteristics are determined by a set of parameters of the ion source, analyzer, and detecting device as well as by the object to be analyzed, background noise in the device, etc.

To obtain a sufficiently high resolution, a time-of-flight analyzer (EMAL-1) utilizes an axially symmetric field whose ion-optical characteristics are calculated using the theory developed in [28].

When the object surface is irradiated by a *Q*-switched laser, the time of emission of laser plasma ions is determined by collisional processes that take place after the end of the laser action. The mean free path of ions depends on their energy and plasma density. During expansion the density decreases and there comes a time when the mean free path of the ion at a given energy exceeds the size of the plasmoid. The emission time depends on the ion energy and decreases faster than $1/E^{0.5}$. For example, for tungsten ions with an energy of 100eV, the emission time is equal to or less than 250ns. The emission region from which ions are detected, increases with the detection threshold. The resolution should drop down a level from which the pulse width is measured.

In the energy range of 50–500eV, the number of ions is approximately the same; in addition, in the analytical part of the mass spectrometer, all the ions have approximately the same energy E=Eo+U, where U is accelerating voltage of a pre-injector, because U»Eo, and as shown in paper [29] the maximum resolution that can be obtained in EMAL-1 devices is 400–500. It should be noted that when a pre-injector with an accelerating voltage Uy=1.5kV is used, the energy range of the ions detected during one laser shot is very small. Thus, the desire to increase the resolution of the device leads to an increased uncertainty in the determination of the concentration of elements. This is due to the fact that the energy transmission of the mass spectrometer is less than the energy spectrum of ions of different elements. Figure 5.1 shows the typical oscillograms of the signals measured by EMAL-1 for lead.

3.5.1.2. Limiting Concentration Sensitivity

The time-of-flight and electrostatic analyzer can be viewed as a system, which forms the ion beam. The flux density in the beam must not exceed a certain critical density of the ion current [29] for the heaviest masses. In the case of a laser ion source there are a number of additional functions performed by the analyzer. First, use is made of a pre-injector which separates the electrons and ions of the plasma and converts the pulse plasma flux into the ion current pulses. Second, the analyzer selects only a certain interval from the full energy spectrum of the plasma determined by time, and hence by the mass resolution of the device.

The emission region decreases with increasing laser plasma ion energy. Thus, by decreasing object slit (by stopping or reducing the slits S1 and S2), we artificially introduce discrimination for low-energy ions, and therefore the energy spectra of the laser torch, obtained by means of such devices differ from the energy distribution during the free expansion of the plasmoid.

The absolute sensitivity of the device is obviously related to the detector sensitivity and the ion loss factor in the formation of the ion beam. As a detector EMAL-1 utilizes secondary electron multipliers SEM-1 and SEM-2, which allow detecting individual ions and have a time resolution better than At>20ns. However, this sensitivity is not statistically justified, since in this case the error amounts to 100%. Thus, we can assume at least 10–100 ions detected by the SEM to be the limiting value of the absolute sensitivity; in this case, the higher the accuracy of the concentration determination, the lower the sensitivity or the higher the statistics. If Emax-Emin=1000eV, the absolute sensitivity is 10^{-12}g. This indicates that for EMAL-1, when the sampling of the analyte is realized by single laser shots, it is possible to achieve high absolute sensitivity of the analysis, and the concentration sensitivity is determined by the resolution of EMAL-1. Figure 5.2 shows the dependence on the concentration sensitivity on the mass resolution for ion energies E=400eV. Figure 5.3 presents the dependence of the limiting resolution on the energy of singly charged ions. Figures 5.4 and 5.5 show typical oscillograms of singly charged ions for different alloys. The main features of such devices, directly related to the concentration sensitivity and mass resolution, are: high mass resolution is limited primarily by two parameters–the size of the ion emission region (or the time of emission) and temporal resolution of a particle detector. The concentration sensitivity is unambiguously related with a dynamic operation range of a SEM, broadband amplifier, and oscilloscope. Thus, a correct quantitative analysis can be performed at the level of no more than 0.01%. For semi-quantitative analysis the concentration sensitivity range can be increased. When detecting the ions in individual acts of laser interaction with the analyte, the limiting concentration sensitivity is 10^{-4}%. The concentration sensitivity and mass resolution increase with the use of higher energy ions.

Thus, to achieve a high sensitivity concentration and high mass resolution requires an elemental analysis at a radiation flux density of 2 10^9W/cm^2 in the range of ion energies of 300–400eV.

3.5.1.3. Express Method for Quantitative Analysis of Alloys

In time-of-flight systems, as a rule, power transmission of the device is substantially less than the width of the energy spectrum of the ions of the laser plasma torch. Since the energy spectra of the ions of different elements in the region of high energies are, in general, different, the detection of the ions in a narrow energy range leads to errors during the elemental analysis of materials.

Figures 5.1–5.6 show the amplitudes dependences of the signals and the relative yield of niobium-based alloys. For 'pure' niobium the maximum energy distribution of singly charged ions corresponds to an energy of 100eV, whereas for tungsten alloys the maximum energy distribution of tungsten shifts to higher energies and is equal to 150eV. In this case, the power transmission of the mass spectrometer was about 2eV.

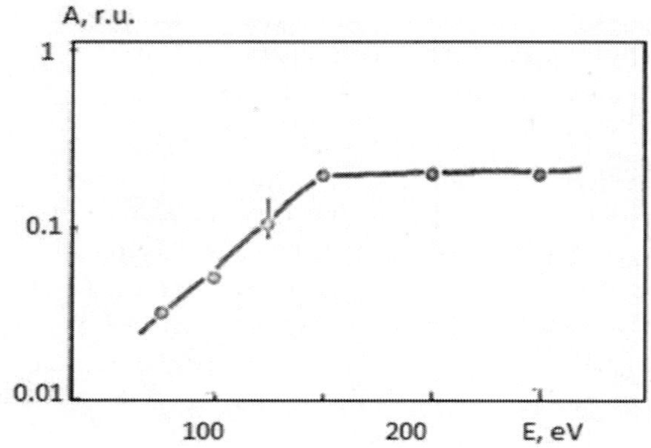

Figure 5.1. Dependence of the signal amplitude of tungsten on the energy of singly charged ions for the $Nb^{+}1.1\%W$ alloy.

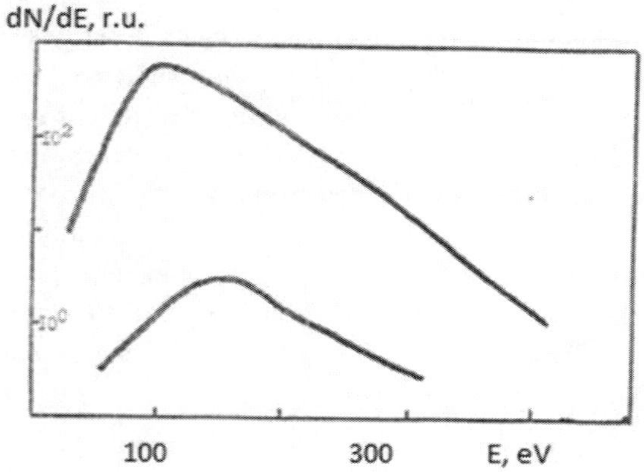

Figure 5.2. Dependence of the relative yield of singly charged ions on the detection energy for the $Nb^{+}1.1\%W$ alloy.

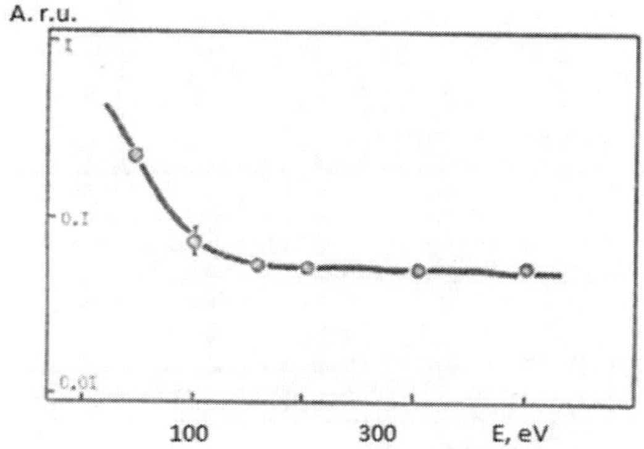

Figure 5.3. Dependence of the ratio of the signal amplitude of carbon to the signal amplitude of the niobium Nb^{+}1.1% C alloy on the energy of the singly charged ions.

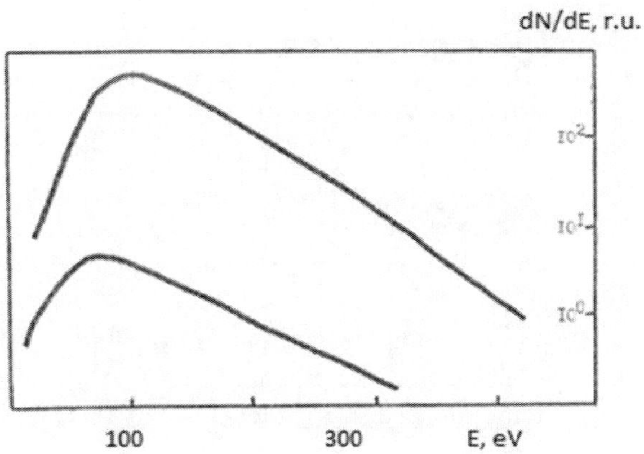

Figure 5.4. Dependence of the relative yield of singly charged ions on the detection energy for the Nb^{+}1.1% C alloy.

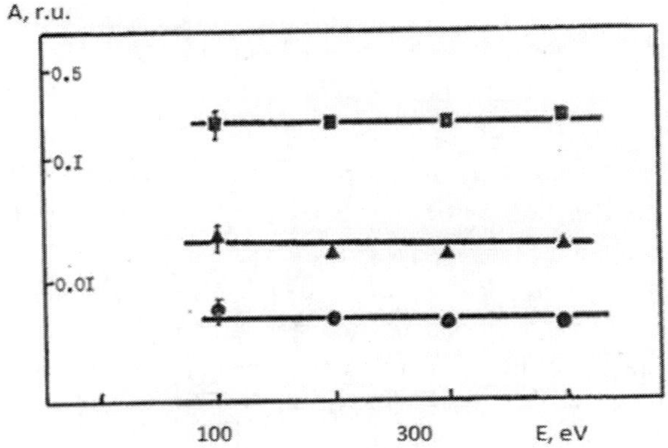

Figure 5.5. Dependences of the ratio of the signal amplitude of molybdenum to the signal amplitude of niobium for the allows: (■) Nb^{+}20% Mo, (▲) Nb^{+}2% Mo, (•) Nb^{+}0.5% Mo.

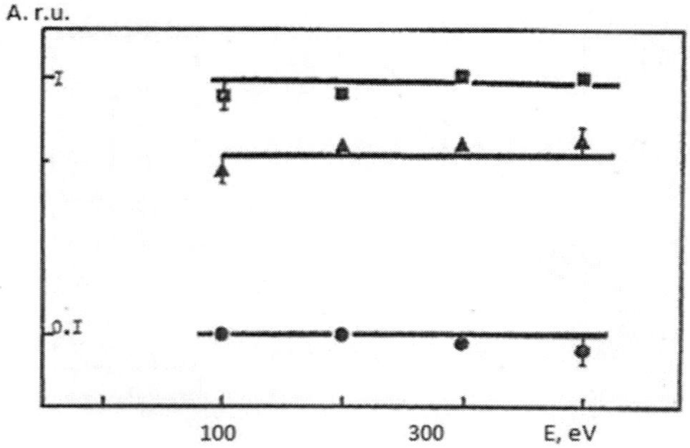

Figure 5.6. Dependences of the ratio of the signal amplitude of zirconium to the signal amplitude of niobium for the allows: (■)Nb^{+}1%Zr, (▲)Nb^{+}0.5%Zr, (•)Nb^{+}0.1%Zr.

Figures 5.5 and 5.6 show the relative yield of singly charged ions as a function of the detection energy for the two alloys. These graphs clearly explain the errors arising in determining the concentration of elements, if the power transmission of the mass spectrometer is much less than the energy spectrum of the ions. For alloys we failed to observe the differences in the energy distributions of singly charged ions.

Figures 5.5 and 5.6 show the dependences of the relative yield of singly charged ions on the detection energy for three concentrations of molybdenum (20%, 2% and 0.5%) and zirconium (1%, 0.5% and 0.1%).

Thus, for express analysis of complex niobium alloys with a concentration of carbon (0.1%–0.5%), molybdenum (2%–5%), zirconium (0.5%–1.5%), tungsten (5%–15%), such devices should mass resolution of at least 50 on a level of 0.0005 (Figs. 5.4, 5.5), the accuracy of determination of the concentration being at least 10%.

In these alloys, carbon and tungsten were detected by the amplitudes of the ion signals of recombination energy distributions. For quantitative analysis of the samples' composition, this method requires a high stability for the following parameters:

- flux density of the laser radiation, i.e., the laser energy and the size of the focal spot;
- accelerating voltage of the pre-injector and the bias voltage on the plates of a spherical capacitor;
- deviation of the center of the focal spot of laser radiation from the axis of the mass spectrometer path should not be significant (10% of the size of the focal spot).

Calibration curves (Figures 5.7–5.8) at 6×10^{9}W/cm^{2} and the diameter of the focal spot of~100μm for different concentrations of tungsten and carbon are presented for the ion energy of 400eV. The concentration of carbon and tungsten is determined with an accuracy of better than 10%. Table 3.1 shows the results of analysis of industrial alloys using EMAL-1 and compares them with the results of chemical analysis.

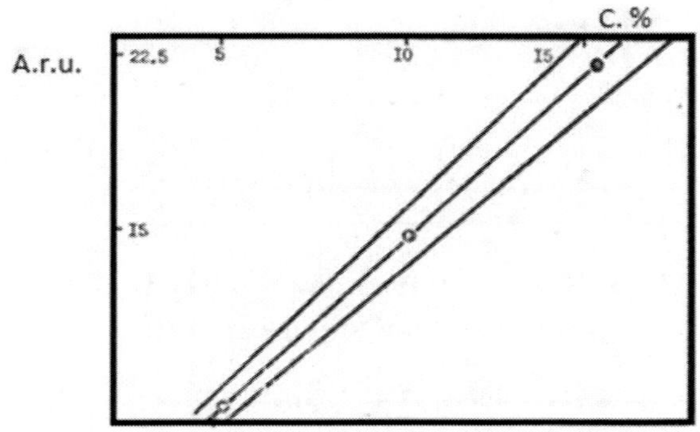

Figure 5.7. Dependence of the amplitude of the tungsten ion signal with *E*=400eV on the concentration in a binary alloy with niobium.

EMAL-1 was used for isotopic analysis of Zr, Mo, W, Pb in the alloys with niobium and pure materials. As seen from Tables 3.2–3.5 relative error does not exceed 10%, except zirconium (20%), when the isotope concentration differ significantly.

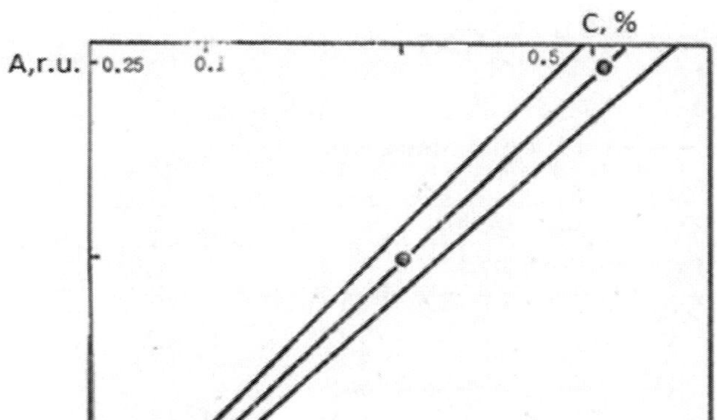

Figure 5.8. Dependence of the amplitude of the carbon ion signal with *E*=400eV on the concentration in a binary alloy with niobium.

Table 3.1. The results of analysis of industrial alloys using EMAL-1

	Alloy	210	213	231
W	Chemical analysis	7.6	10.5	4.7
	EMAL-1	7.3 ± 0.7	5.4 ± 0.9	5.1 ± 0.5
Zr	Chemical analysis	1.43	1.3	1.2
	EMAL-1	1.4 ± 0.1	1.3 ± 0.1	1.1 ± 0.1

Table 3.2. EMAL-1 was used for isotopic analysis of W in the alloys with niobium and pure materials

	W_{182}	W_{183}	W_{184}	W_{186}
EMAL-1	27.4 ± 2.1	15.9 ± 1.5	28.9 ± 2.0	27.8 ± 1.7
Isotope concentration, (%)	26.41	14.40	30.64	28.41

Table 3.3. EMAL-1 was used for isotopic analysis of Zr in the alloys with niobium and pure materials

	Zr_{90}	Zr_{91}	Zr_{92}	Zr_{94}	Zr_{96}
EMAL-1	50.8±3.1	11.9±0.7	17.8±1.5	17. ±1.5	3.5±0.7
Isotope concentration (%)	51.46	11.23	17. 11	17.40	2.8

Table 3.4. EMAL-1 was used for isotopic analysis of Pb in the alloys with niobium and pure materials

	Pb_{206}	Pb_{207}	Pb_{208}
EMAL-1	27.5±2.5	23.5±2.2	49.0±4.0
Isotope concentration (%)	25.1	21.2	52.3

Table 3.5. EMAL-1 was used for isotopic analysis of Mo in the alloys with niobium and pure materials

	Mo_{92}	Mo_{94}	Mo_{95}	Mo_{96}	Mo_{97}	Mo_{98}	Mo_{100}
EMAL-1	15.5±1.0	8.6±1.01	15.8±1.2	16.9±1.5	10.2±1.1	21.8±2.0	9.9±1.0
Isotope concentration (%)	15.84	9.04	15.72	16.53	9.46	23.78	9.63

These studies made it possible to improve the ion source for mass spectrometry by taking into account the features of ion emission from a laser-plasma torch; therefore, the following characteristics were obtained:

- mass resolution at half-height of ion peaks was at least 400;
- mass resolution at a level of 0.0005 of ion peaks was at least 50;
- stability of the arrival time of the ion in the SEM was at least their half-widths;
- relative concentration sensitivity at a mass resolution of 50 was better than 0.01%.

3.5.2. Problems of Quantitative Analysis of Laser Mass Spectra

Quantitative analysis using laser mass spectrometry is an urgent task. Improving reproducibility of the results is achieved by stabilizing the conditions of ion formation in the source and by automating the processing of mass spectra. However, only random errors are reduced in this case.

The problem of reducing systematic errors or their correct identification is still relevant. Usually this problem is solved by introducing the relative sensitivity coefficients of the elements. The relative sensitivity coefficients of different elements change over a wide range (up to the order of magnitude) in different experiments [22,26]. The reason probably lies in the fact that these coefficients often include discriminations of ions of different elements both in the formation of a plasmoid, and during transportation of the ion beam to a mass spectrometer. Purely physical factors influencing the formation of relative sensitivity coefficients for different elements are considered in this paper. Systematic errors arising due to instrumental factors are related to the influence of the scattered magnetic field and the space charge of the beam. Despite the fact that the reasons of the discrimination in the ion mass and charge of the mass analyzer, assembled under the Mattauch–Herzog scheme, are known, the ways of proper account of these factors are not considered in the literature.

The authors of paper [27] consider the influence of the refractive index of the ion beam at the boundary of the electric field on the width and shape of the analytical lines. This effect can be taken into account, for example, by integrating the line during photometry [13]. In paper [30] the correction for the scattered magnetic field is not accurate enough because of the uncertain degree of interaction of the beams of different masses and structural imperfections in the assembly of the analytical device. The data of paper [31] make it impossible to determine the systematic error in the absence of standard compositions, which are similar in composition to the sample. This is probably due to the fact that these papers focus on the physical side of the problem, with the analytical part being poorly developed.

The problem of achieving the best accuracy and maximum sensitivity of the analysis should be solved by careful consideration of all systematic errors, with optimization of parameters of incident radiation and ion-optical system. In addition, investigation of the processes of production of new elements by the obtained mass spectra is possible only in case of a complete separation of the physical and instrumental factors in determining the relative sensitivity coefficients of the elements. This is the reason to consider instrumental errors as components of errors arising during calibration.

3.5.2.1. Estimate of the Quality of Analytical Lines

A necessary condition for quantitative analysis on the devices with double-focusing (EMAL-2) is the absence of defects in the analytical lines of the mass spectrum, which arise due to ion-beam optical aberrations during its transport. In general, this is achieved by using the methods of adjusting the analytical system of the instrument [4]: the mass spectrometer was calibrated for a set of pure elements C, Si, Fe, Ge, Sn, Yb, and Pb by measuring the dependence of the photoemulsion sensitivity on the ion mass. The emulsion contrast was determined by measuring the isotopes of tin.

The quality of the analytical lines was assessed visually by the mass spectrum of a standard sample (Br669). Analytical lines used for photometry should have a strict rectangular shape, uniform intensity in height and approximately Gaussian distribution in width. After the analytical lines that satisfy the above conditions are obtained, it is necessary to optimize the parameters of the ion-optical system. The relative sensitivity coefficient is an integral parameter that takes into account the differences in the relative yield of ions of elements from laser plasma and the discrimination of ions during their transport to the mass analyzer. Using the conventional method of calibration, i.e., the dependence of the photoemulsion sensitivity on the ion mass, the obtained values of the relative sensitivity coefficients implicitly contain coefficients K. Since K values are determined by the masses of elements, their contribution to the relative sensitivity coefficients may be the cause of the general dependence of these coefficients on the mass. Thus, the use of the refined calibration characteristics allows one to consider the obtained values of the relative sensitivity coefficient as the values characterizing the relative contribution of ions of different elements of a plasmoid. To increase the light intensity of the mass spectrometer, it is needed to optimally combine the conditions of laser irradiation of the sample, the parameters of the ion beam formation system and the parameters of the mass analyzer.

Expansion of ion beams due to the own charge and disordered spread in transverse thermal velocities during their transport between the object and the aperture slits separated by a distance equal to L, on the one hand, and the finite size of the slits, on the other hand, determine the existence of possible discriminations in the ion masses at the edges.

The absence of significant discrimination in the mass analysis because of the action of the space charge of the ion beam during its formation and transport (for the given parameters of the ion-optical system: $0.05\text{mm}<d<0.6\text{mm}$ and $2\ 10^9\text{W/cm}^2$) is considered optimal for the most efficient ionization of atoms of trace elements in order to achieve the best accuracy and maximum sensitivity. Discriminations related to the scattered magnetic field and space charge are independent; therefore, to achieve maximal sensitivity, while maintaining the best accuracy of the results, it is needed in a reasonable manner to minimize the combined influence of these effects. To this end, the size of the object slit S is optimized. On the one hand, the increase in S to a value greater than the optimal value increases the number of ions necessary for the emergence of the analytical line and its broadening. On the other hand, reducing S below the optimal value significantly increases the exposure time needed for the emergence of the required analytical line. Accounting for the instrumental factor in the construction of the calibration curve allow one to consider the obtained values of the relative sensitivity coefficient as parameters characterizing the relative yield of ions of different elements from a plasmoid. To ensure the maximum sensitivity in the analysis of samples of different composition use, advantage is taken of the position of the object slit corresponding to the maximum ion current of the irradiated sample. In order to achieve the lowest detection limits of trace element contents, the size of the object slit is optimized.

3.5.3. RSC Dependence from the Analysis Starting Conditions

Recent years have been marked by a substantial reduction in the number of publications on sample composition analysis using laser mass-spectrometry. This is due to the fact that available mass-spectrometers with laser ionization do not fit some of the requirements of practical analysis [1,2]. The problem of validity of the results is of value not only in the case of a laser excitation source. This problem is connected with determination of relative sensitivity coefficients (RSC) for various impurities. It was shown in [3,4] that parameters of exciting radiation such as fluence q and spot diameter d influence the value of RSC of impurities in a copper matrix. The authors of [3,4] proposed an *a priori* approach to calculate RSC. However, variation of the matrix mass when analyzing other samples is expected to result in variation of main parameters of the plasma plume (for example, lifetime τ_0 [5]) and relative yield of ions of different impurities from laser-produced plasma. Such a "matrix effect" was reported by several authors [6,7]. The influence of the matrix mass on the energy spectrum and relative yield of ions of main components of the plasma plume was studied in detail in Refs. [5,8,9]. However, the behavior of the impurity component was not studied in the above papers.

We report here on experimental studies of the influence of the matrix component of a plasma plume on ionization efficiency of impurity atoms, processes of acceleration and recombination of impurity ions in laser-produced plasma. This work is a continuation of a series of experiments on studies of recombination processes in multicomponent laser plasma started in [3].

Experiments were made with a mass-spectrometer EMAL-2 [10] using a pulsed Nd: YAG laser with a repetition rate of 50Hz operated in the regime of minimal longtime instability of the output power [11]. The analytical part of the mass-spectrometer was calibrated according to recommendations of [12] to ensure proper accounting for the influence of the instrumental factor on the mass discrimination of ions along the flight path in the mass-spectrometer. The target material was a standard Al-Si-45 compound containing impurity elements (Si, Ti, Cr, Mn, Fe, Cu, Zn, As, Sn, Sb, Pb) at concentrations ranging from 1 to 10^{-4}at%. Experimental data were processed using the method described in [4,13]. Experimentally, we studied the yield of impurity ions per unit illuminated area N^{+z}/S as a function of laser spot diameter d ranging from 0.05mm to 1mm, at fixed values of $q=2\times10^8$ and 8×10^9Wcm^{-2}.

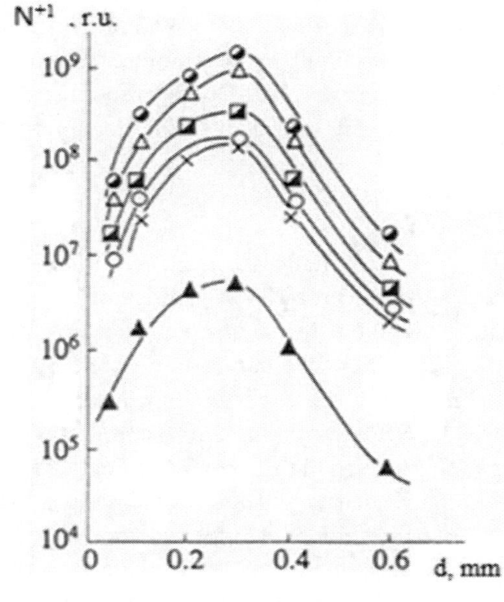

Figure 5.9. Yield of singly charged impurity ions per unit irradiated area of an aluminum matrix as a function of laser spot diameter at q=2x109Wcm $^{-2}$.(o-Cu, ^-Fe, п-Ti, o- Zn, x-Cr, A-As.

3.5.4. Influence of Plasma Torch Initial Size and Matrix Contents on the Impurities Analysis Result

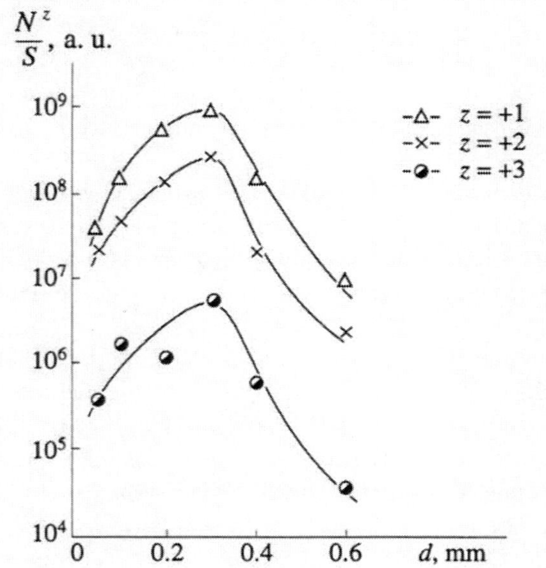

Figure 5.10. Yield of singly and multiply charged ions per unit irradiated area of an aluminum matrix as a function of laser spot diameter at q=2x10^9 Wcm^{-2}.

The yield of singly charged impurity ions as a function of the laser spot diameter at q=2 10^{9}Wcm^{-2} is shown in Figure 5.9. The dependence N^{+z}/S is similar for all impurity ions: this function rapidly grows as d increases from 0.05 to 0.3mm; in doing so at d>0.2mm the function slope with respect to the abscissa becomes more gradual; at d=0.3mm the function attains its maximum and starts to decrease steadily with further increase in d.

The yield of iron ions with z ranging from 1 to 4 as a function of d at

q=2 10^{9}Wcm^{-2} is shown in Figure 5.10.

The function N^{+z}/S for z=+1 - +3 attains the maxi mum at d =0.3mm, and for z=+4 it has the maximum value at d=0.4mm. From these profiles one can conclude that the rate of increase of the number of ions with the increase of d becomes lower at larger z, while the rate of decrease (the slope of the right side of the curves in Figure 5.10) becomes higher at larger z. Such dynamics of the charge spectra with variation of d are typical of all impurity ions.

Let us compare the yield of impurity ions from samples with copper and aluminum matrices. The yield of singly charged iron ions per unit illuminated area as a function of laser spot diameter for A1 and Cu matrices at q=2 10^{9}Wcm^{-2} is shown in Figure 5.11. Comparing the dependencies NFe^+/S on d one concludes that in the N^{+z}/S case of A1 matrix, the maximum of the function lies at larger d (0.3mm) than in the case of the Cu matrix (d=0.2mm). The slope of the left side of the curveis N^{+z}/S larger for the A1 matrix. The rate of decrease of the function N^{+z}/S is also higher for iron ions in the aluminum matrix. Note also that these distinctions are typical of all impurity ions.

Thus, the function N^{+z}/S behaves similarly for all impurity ions both in copper [3] and aluminum matrices. In doing so in the case of impurities in the aluminum matrix the function N^{+z}/S attains its maximum at larger spot diameters, and the rate of increase of the number of ions with increase in d is larger. The question arises of why the maximum of the function N^{+z}/S shifts for the same ions? It was shown earlier that me maximum of this function is governed by the competing processes of ionization and recombination and can be determined from a simple relation $\tau_i(z, A) \sim \tau_0(z, A) \sim \tau_r(z, A)$. For the same impurity we have $\tau_{0Al}(z, A) \approx \tau_{0Cu}(z, A)$, where the lifetime τ_0 can be expressed as $\tau_0 \sim d/v$. If one assumes that v is the local sound velocity then $v \sim A^{-0.5}$, $\tau_0 \sim dA^{0.5}$.

Consequently, we have

$$d_{\max Cu}\sqrt{A_{Cu}} \approx d_{\max Al}\sqrt{A_{Al}}, \quad (1)$$

in remarkable accord with the experimentally observed shift of the maximum from 0.2mm to 0.3mm.

From this analysis one concludes that the relation $\tau_0 \sim dA^{0.5}$ can be used for *a priori* calculation of impurity yield.

3.5.5. Influence of Matrix on Ionization Efficiency of Impurities

It was shown in Refs. [5,8,14] that the left branch of the dependence $\tau_0 \sim dA^{0.5}$ is governed predominantly by the processes of ionization, while the right branch of the dependence is governed by the processes of recombination in a plume; in doing so the efficiency of both processes increases at larger d. The maximum of the dependence is determined to a certain degree by competition of the processes of ionization and recombination.

Table 3.6. Ionization rate constants of impurity ions (in arb. units) in copper and aluminum matrices at q=2 10^9Wcm^{-2}

matrix \ element	P	Cr	Fe	Zn	As	Sn	Sb	Pb
Cu	1.37	6.15	2.94	2.44	2.3	3.0	2.7	3.3
Al	2.11	18.3	14.8	3.1	3.9	5.6	5.8	7.2

Let us compare ionization of impurity atoms in copper and aluminum matrices. The ratio of the number of ions formed at the maximum of the function N^{+z}/S to that formed at d =0.05mm and q=2 10^9 Wcnr^2 in these two matrices is given in Table 3.6. It is seen that efficiency of ionization characterized by the growth rate of the function N^{+z}/S is higher for impurities in the aluminum matrix compared to that in the copper one. This fact is also clearly seen in Figure 5.11.

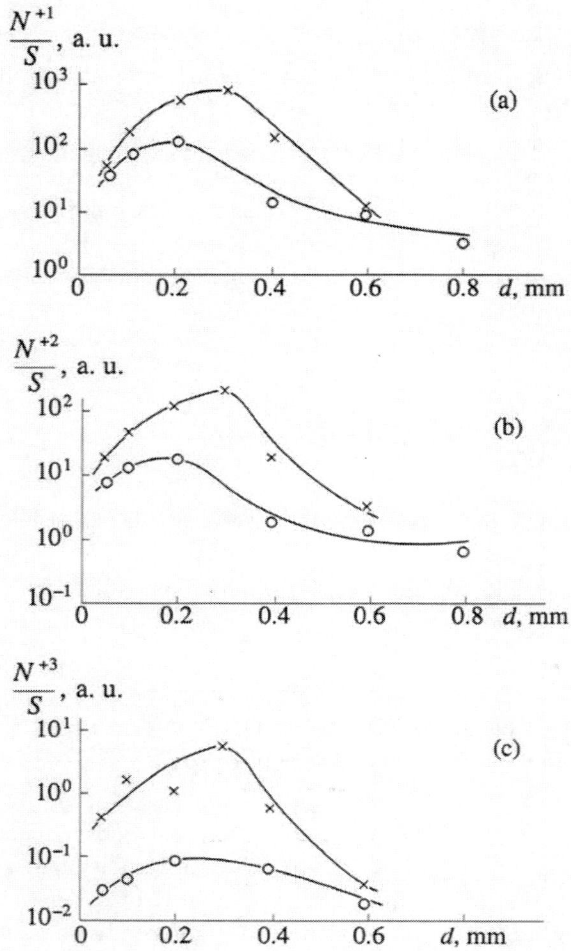

Figure 5.11. Yield of singly and multiply charged ions per unit irradiated surface area as a function of laser spot diameter for aluminum (x) and copper (o) matrices: +1(a), +2(b), +3(c).

The observed dependence can be explained as follows. Taking into account the fact that during plume expansion, ions of the matrix and impurity behave similarly, one can suppose that the difference in ionization efficiency is due to characteristics of the plasma plume determined by the matrix (the target excitation conditions are assumed to be the same). The lifetime τ_0 governed by the matrix component [5] has little influence on efficiency of ionization processes because we neglect the recombination process in the plume at $d<d_{opt}$. On the other hand, ionization of impurity elements is governed by Ref. [15].

$$\frac{dn_i}{dt} = n_e n_a \alpha_e, \tag{2}$$

where n_a, n_i, n_e are the densities of atoms, ions and electrons, respectively. Considering the fact that the quantity $\alpha_e n_e$ is the probability of two-particle collisions per unit time we see that increased ionization efficiency in the aluminum matrix compared to the copper one is either due to larger value of n_e, or due to higher T_e. In doing so the values n_e and T_e are coupled to each other, but increase of T_e results in increase of probability of electron collision with an atom, while increase of α_e results in increase of n_e for the corresponding impurity element because according to Eq. [16]

$$\alpha_e \sim \frac{\sqrt{T_e}}{\varphi^2} \exp\{-\varphi/T_e\}. \tag{3}$$

Coupling of n_e with T_e can be traced as follows. It was shown in [3] that for copper matrix

$$T_e \simeq 2\times 10^{-5} q^{2/3}\,(\text{eV}), \tag{4}$$

where the coefficient at q is a function of plasma density δ and mass of the matrix component M [17]. Assuming $\delta = n_e$ we have

$$T_e \simeq \{n_e/M\}^{2/9} q^{2/3}. \tag{5}$$

Consequently, the increase of the ionization rate of impurity atoms in the aluminum matrix compared to the copper one (provided q is constant) can be explained as follows. In the aluminum matrix the ionization potential of the matrix component φ_M is higher, resulting in increased ionization rate of the matrix atoms α_{eM}, in accordance with Eq. (3). The matrix atoms are ionized both by the laser field and by electron impact. They are the main source of electrons in plasma. Consequently, increase of the ionization rate of the matrix atoms results in increase of electronic density n_e and energy, in accordance with relation (5) and the relation

$$\frac{dn_e}{dt} = n_{aM} n_e \alpha_{eM}. \tag{6}$$

Here dn_i/dt was replaced by dn_e/dt, in compliance with assumptions made above. In turn, the ionization rate of atoms of a fixed impurity is determined by the increase of electronic energy according to relation (3), and the value of dn_i/dt is determined by n_e, in accordance with Eq. (2). It should be noted that the question as to which quantity, n_e or T_e, governs the increase of α_e, remains unanswered. Solution of this problem is rather difficult because the quantities n_e and T_e are closely coupled to each other, and Eq. (5) describes this coupling only approximately.

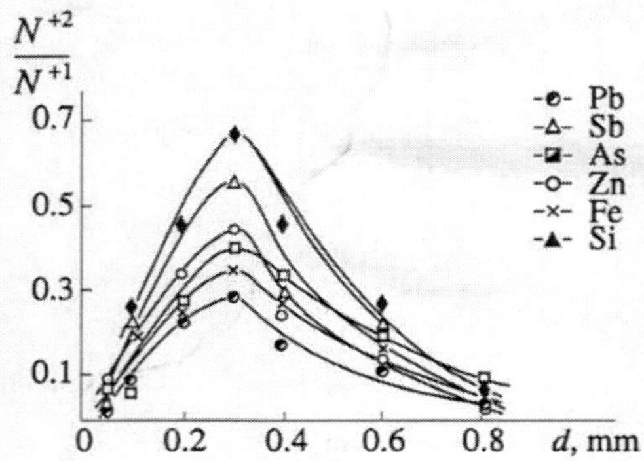

Figure 5.12. Ratio of the number of doubly charged ions to that of singly charged ones for impurities in an aluminum matrix, as a function of laser spot diameter at q=2 10^9Wcm^{-2}.

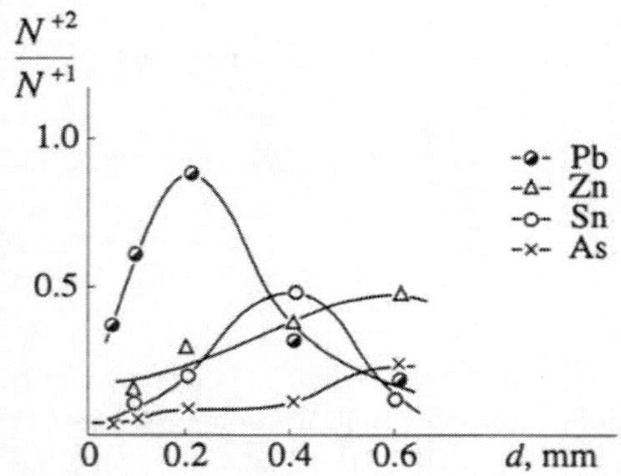

Figure 5.13. Ratio of the number of doubly charged ions to that of singly charged ones for impurities in a copper matrix, as a function of laser spot diameter at q=2 10^9Wcm^{-2}.

3.5.6. Acceleration and Recombination of Impurity Ions in Presence of Different Matrix Components of the Plasma Plume

Plasma plume expansion is accompanied by ion acceleration due to pressure gradient and presence of an electric double layer determined by areas with noncompensated charge [18]. The process of multicomponent plasma expansion is determined by the matrix component [8]. The acceleration efficiency of matrix and impurity ions is proportional to the ratio of the ion charge to its mass [19]. It was shown in [19] that recombination complicates consideration of ion acceleration, because accelerated ions gain additional velocity with respect to neutral particles of the plasma plume, thus increasing probability of recharging.

We will describe ion acceleration in a plasma plume in terms of charge ratios of ions during expansion, as was done above. The ratio of the number of doubly charged ions to the number of singly charged ones as a function of laser spot diameter d at q=2 10^9Wcm^{-2} is shown in Figs 5.12 and 5.13 for aluminum and copper matrices, respectively. A characteristic feature of the dependencies $\frac{N^{+2}}{S}(d)$ is the maximum at d=0.3mm, the height of the latter increasing with the increase of ion mass. Similar increase of the ratio $\frac{N^{+2}}{N^{+}}$ at the maximum with increase of M is also observed for ions in a copper matrix, though the maximum shifts to smaller values of the spot diameter d with increase of M. For example, for Zn and As the maximum value of the ratio $\frac{N^{+2}}{N^{+}}$ is achieved at d=0.6mm, for Sn - at d=0.4mm, and for Pb the maximum is achieved at d=0.2mm. Such difference of the plots can be attributed to the fact that the mass of aluminum is essentially smaller than that of all impurities. As a result, the maximum of the ratio $\frac{N^{+2}}{N^{+}}$ is achieved at the optimal value of the laser spot diameter from the point of view of the maximum yield of singly and doubly charged ions. Moreover, a larger value of n_e for aluminum plasma (as was pointed out above) results in more effective ion acceleration. In the case of the copper matrix, masses of copper and a majority of impurities are close, and the maximum of the ratio $\frac{N^{+2}}{N^{+}}$ only shifts towards d_{opt} (d=0.2 mm). This behavior is clearly seen for Sn and Pb ions having masses considerably different from that of copper. Obviously, with increase of n_e in a copper plasma the maxima of the ratio $\frac{N^{+2}}{N^{+}}$ for all impurities will tend to d_{opt}=0.2mm.

It should be noted that differences in dependencies for copper and aluminum matrices are not connected with the peculiarities of ionization processes.

This conclusion is illustrated by Figure 5.14 which shows the ratio of the number of doubly charged ions to that of singly charged ones as a function of inverse ratio of the corresponding ionization rate constants α_{el}/α_{e2} of impurity atoms in $\frac{N^{+2}}{N^{+}}(d)$ and aluminum at q=8 10^8Wcm^{-2}. The value $\frac{N^{+2}}{N^{+}}$ decreases with increase of the ratio α_{el}/α_{e2} with the same rate in copper and aluminum. Larger absolute values $\frac{N^{+2}}{N^{+}}$ of in aluminum plasma can be attributed to more effective processes of ionization, in particular due to shift of d_{mix} (z=1) to larger values *(d_{max}*=0.3mm).

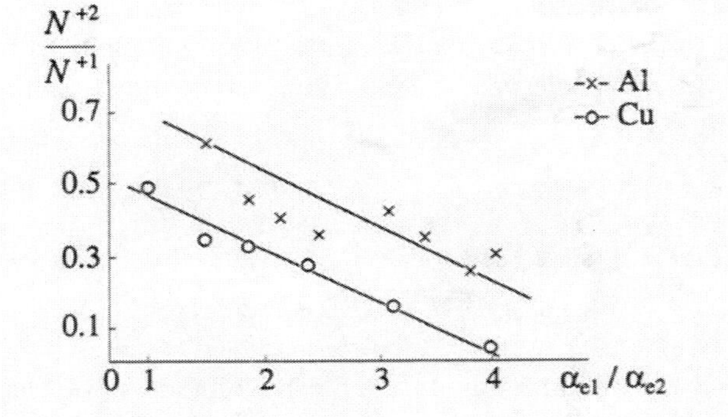

Figure 5.14. Ratio of the number of doubly charged ions to that of singly charged ones as a function of inverse ratio of the corresponding ionization rate constants of impurity atoms in aluminum and copper matrices.

Table 3.7. Comparison of the experimental and calculated values of RSC of impurity ions in an aluminum matrix at q=8 10^8Wcm^{-2}

element	Si	Ti	Cu	Cr	Zn	As	Sn
RSC_{exp}	1.10 ± 0.07	1.30 ± 0.16	1.05 ± 0.04	1.30 ± 0.12	0.71 ± 0.05	0.65 ± 0.07	1.30 ± 0.13
RSC_{calc}	0.90	1.48	1.05	1.51	0.60	0.52	1.21
Δ, %	18	14	0	16	15	20	7

It was pointed out above that the right branch of the function $\frac{N_+^{+z}(d)}{S}$ is governed by dominating processes of recombination. Let us determine the recombination rate constant α_p similar to α_e, namely from the rate of decrease of brightness as d increases. The experimental data on recombination rate constants of singly charged ions as functions of their mass at various values of q are shown in Figure 5.15. It is seen that α_p increases with increase of M from 1.8 for P to 2.3 for Pb at q=8 10^8Wcm^{-2}. In doing so the character of the dependence of α_p on M does not change and remains linear: α_p~M. As q increases from 2x10^8 to 2x10^9Wcm^{-2}, α_p increases for all impurities though this increase is not significant and amounts to less than 30%. This circumstance can be attributed to the fact that increase of q results in increase of the number of "cold" electrons and consequently in increased efficiency of recombination processes in the plasma plume. Additionally this simultaneously raises the yield of doubly charged ions. As these ions recombine, they contribute to the singly charged plasma component. Consequently, the variation of α_p with increase of q is not so considerable as one might expect from consideration of the influence of Ne only at the stage of ion recombination. It was shown in [3] that in the process of copper plasma expansion $\boldsymbol{\alpha_p}$ does not depend on $\boldsymbol{M}$ for fixed z. Strong correlation between the recombination rate constant and ion mass of the corresponding impurity in aluminum is due to variation of the plasma plume parameters. Increase of n_e results in more effective recombination as is seen in Figure 5.11 in the case of singly charged ions of iron in copper and aluminum. Variation of τ_0 in accordance with Eq. (1) results in increase of α_p for impurity ions with a mass larger than that of the matrix component. This is due to faster expansion of aluminum plasma compared to the copper plasma.

Thus efficiency of acceleration of impurity ions and their recombination increases in aluminum compared to copper due to larger values of n_e in aluminum plasma. In doing so, faster expansion of aluminum plasma results in increase of the recombination rate constant with increase of the impurity ion mass.

3.5.7. Influence of a Plasma Plume Lifetime on Relative Yield Of Impurity Ions

It was shown in [5] that the lifetime of the plasma plume can be changed by changing a laser spot diameter at a target or by changing the mass of the matrix component. The influence of d variation on the relative yield of singly charged impurity ions (or RSC) in a fixed (copper) matrix was discussed in detail in [3,4]. Let us briefly summarize the results of previous studies.

Three intervals of variation of d (or τ_0) were distinguished from the point of view of processes in the laser plasma which influence the relative yield of ions. At small values of *d* (≤0.2mm) or τ_0, ionization processes dominate. At *d* ≤0.2mm the degree of ionization of impurity atoms is determined by their ionization rate constants.

In the second interval of the laser spot diameter variation (≤0.2mm ≤ 0.6mm), ion acceleration plays an important role and leads to variation of a relative yield of ions because the efficiency of ion acceleration is proportional to the ratio Z/M. In this interval the relative yield of ions of different elements can be equal, however this equality takes place only for several chemical elements. The relative yield of ions of other elements is other than unity and is determined by efficiency of the processes of ionization, ion acceleration and recombination for ions with different values of M and φ. The character of this dependence is determined by experimental conditions. Because in this interval the relative ion yield is controlled by competing processes of acceleration and recombination, calculating this yield for different elements is highly complicated.

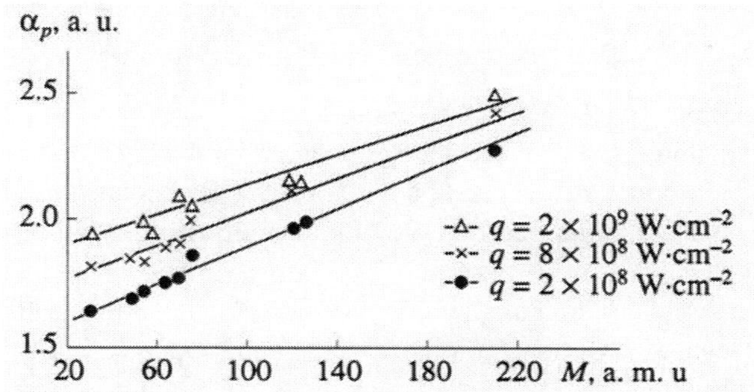

Figure 5.15. Recombination rate constant of impurity ions as a function of mass at various laser fluences for an aluminum matrix.

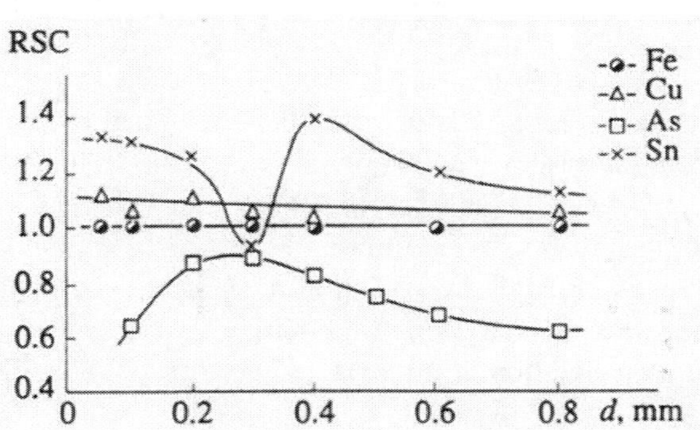

Figure 5.16. RSC of impurity ions in an aluminum matrix as a function of laser spot diameter at q=2 10^9Wcm^{-2}.

At *d*≥0.6mm the recombination plays the decisive role. In this range of *d* the absolute value of ionic yield is governed by the nature of the recombination process and by efficiency of ionization and ion acceleration processes (and competition of acceleration and ionization processes). Consider variation of the relative yield of ions at different masses of the matrix component of the plasma plume. Figure 5.16 shows RSC of impurities in a sample with aluminum matrix as a function of *d* at q=8 10^8Wcm^{-2}. Note that only singly charged ions were taken into account when calculating RSC because their yield amounted to more than 90% of the total number of ions. At d<0.3mm ionization processes dominate. As the laser spot diameter *d* increases from 0.05 to 0.3mm the value of RSC for Sn drops from 1.30 to 0.91, and the value of RSC for As grows from 0.65 to 0.88. Such behavior of RSC as a function of *d* for these elements is due to the fact that the first ionization potential of Sn is lower, and that of As is higher, than the first ionization potential of the internal reference element (iron, in the case under consideration). In addition, it is seen that the most striking decrease of RSC for Sn takes place at *d* ranging from 0.2 to 0.3mm. Variation of RSC with increase of *d* in this interval is governed by the fact that increase of the plume lifetime results in more effective interaction of electrons with evaporated atoms, which is described by the relation obtained in [3-5]

$$N_i \sim \frac{1}{\tau_i} \exp(\tau_0/\tau_i). \tag{7}$$

However, variation of *d* results in simultaneous variation of τ_0 and $\tau_{I.}$ Thus, at 0.05≤d ≤0.3mm the number of ions created is governed by the exponential in Eq. (7).

In the case of admixture of Cu, RSC does not vary as the spot diameter *d* varies from 0.05mm to 0.3mm. This is explained by equal ionization rate constants (or ionization times) for Cu and Fe atoms.

Let us compare the experimental results with *a priori* calculation of RSC using a formula obtained in [3,4]. It was shown that for small $d<d_{max}$ and moderate values of *q*, that is, in the case of the dominant yield of singly charged ions in the laser plume, we have:

$$RSC_{st}^{x} \approx \left(\frac{\varphi_{st}}{\varphi_x}\right)^2 \exp\left\{\frac{\varphi_{st} - \varphi_x}{T_e}\right\}. \tag{8}$$

It is seen from Table 3.7 *(d=0.05*mm, q=1 10^9Wcm^{-2}) that relation (8) well describes the relative yield of ions in the case of the aluminum. matrix, as well. The largest discrepancy ▲ obtained in the case, of As does not exceed 20%. In our calculations we took into account dependence of T_e on plasma density and mass of the matrix component, in accordance with Eq. (5). In doing so we assumed that variation of n_e as the copper matrix is exchanged for the aluminuft one is determined by variation of the ionization rate constant^ of the matrix atoms, so that as follows from the solution of Eq. (6).

$$\frac{n_{e\,Cu}}{n_{e\,Al}} \approx \exp\{(\alpha_{e\,Cu} - \alpha_{e\,Al})\, n_a^0 \tau_0\} \times \left\{\frac{1 + \exp(\alpha_{e\,Al} n_a^0 \tau_0)}{1 + \exp(\alpha_{e\,Cu} n_a^0 \tau_0)}\right\} \tag{9}$$

At d>0.3mm the process of ion acceleration is increased. It was shown earlier in [19] that efficiency of acceleration of ions is inversely proportional to their mass. In doing so, accelerated ions gain an excess velocity with respect to the matrix atoms, which can result in effective recharging [20]. Thus recharging of Fe^+ ions on A1 atoms is more effective than that of Sn^+ ions, due to larger relative velocity gained under acceleration in a double electrical layer. As a result, RSC of Sn increases as d is increased from 0.3mm to 0.4mm. For Cu, RSC does not vary because the difference between the masses of Cu and Fe is small. Thus, in this range of d the relative yield of impurity ions is governed by the competing processes of acceleration and ion recharging. Obviously, the efficiency of both processes depends on the matrix component of the plasma plume. It was shown above that the ionization rate constant of the matrix atoms governs N_e, which in turn influences the efficiency of ion acceleration. In doing so the more effective is the interaction of ions with electrons in the electric double layer, the larger is the relative velocity of ions with respect to that of matrix atoms, leading to increase of the recharging cross-section [20].

At d>0.4mm, ion recombination becomes the dominant process. Contrary to the results obtained for copper, the recombination rate constant is proportional to the mass of an ion (under expansion of the aluminum plasma). Consequently RSC for Sn and As decreases, because M_{Sn} and M_{As} are larger than The essential dependence of α_p on M is due to the fact that the matrix mass (Al) is considerably less than the atomic masses of all impurity elements, so that one can speak of partial separation of ions with respect to their masses as the plasma plume in a "nonfrozen" ionization state expands. Weak dependence of α_p on M can be explained by friction which results in matrix ions carrying away a fraction of the impurity ions. In doing so the character of the recombination process is governed by the plume parameters and its matrix component [5].

3.5.8. Conclusion

At fixed parameters of target irradiation, efficiency of main processes which occur under multicomponent plasma plume formation and expansion is governed by the matrix.

If the laser spot diameter is small, so that processes of ionization dominate in the plasma plume, the absolute yield of impurity ions is determined by the ionization rate constant of the matrix atoms (this constant governs the number of electrons in the plume). In this case the relative yield of impurity ions is determined by the conditions of excitation, and characteristics of the corresponding impurity atoms (their first ionization potentials) and atoms of the matrix (their mass and ionization rate constant), and can be calculated *a priori.*

The density of electrons n_e, which depends on a_{eM}, determines the efficiency of ion acceleration in an electric double layer, provided other conditions are the same. At $d>d_{max}$ impurity ions effectively recharge on matrix atoms in the plasma plume, the efficiency of recharging increasing for impurity ions with smaller mass.

At $d>d_{max}$ the recombination of created ions dominates in the plasma plume. The character of recombination depends on the kind of matrix. The recombination rate constant of impurity ions depends on their mass, provided the mass of the matrix ion is largely different from that of the impurity ion, resulting in partial separation of ions with respect to their mass. The efficiency of recombination is also governed by characteristics of the matrix component which determine the time of plasma expansion and the number of "cold" electrons in the plume.

Thus, one can control the absolute yield of impurity ions by changing the matrix component of the plume so that parameters necessary for multicomponent laser mass-spectrometry are achieved.

3.5.9. Hydrocarbon Background of Laser Sources

In solving various materials problems in microelectronics, optoelectronics, chemical industry, development of new technologies, etc., there inevitably arises the question of the mass-spectrometric control of the elemental composition of very pure substances, which allows simultaneous detection of all the elements of the periodic table with a detection limit of no worse than 10^{-6}–10^{-7} at. % and accuracy of 30% without the use of standards.

A very important problem is the detection of gas-forming impurities such as H, N, C, K, and O in pure substances. The presence of background lines of residual gases, such as hydrocarbons, in the laser mass spectra often leads to significant errors during the laser-mass spectroscopic (LMS) analysis in determining the gas-forming impurities and identification of analytical lines in multi-component samples. The LMS analysis of the hydrocarbon background was performed using a JEOL JMS-01-BM-2 double-focusing spark source mass spectrometer (Japan) [32]. The vacuum in the spark source of ions was 2-10^{-1} Pa, in an electrostatic analyzer — 2-10^{-5} Pa.

The most intense lines are formed by the ions CH_3^+, $C_2H_3^+$, $C_3H_3^+$, $C_3H_5^+$, and $C_3H_7^+$ containing an odd number of hydrogen atoms. The intensity of each of these lines in terms of the concentration may in some cases be as high as 0.8–2 10^{-3} at %. Figure 5.17 shows a fragment of the mass spectrum of an osmium sample with an exposure of the 3 nC. One can see lines of hydrocarbon ions from $C_2H_2^+$ to C_3H^+ with their content from 10^{-4} to 2 10^{-3} at %. At an exposure of 20 nC. The mass spectrum of a multicomponent biological object (Figure 5.18) exhibits lines of hydrocarbon ions from $C_4H_2^+$ to $C_5H_{12}^+$ with their content from 10^{-6} to 10^{-4} at %. In a series of hydrocarbon ions with the number of hydrogen atoms $n = 2$ to 12, one can observe more intense lines of ions with odd masses in comparison with even masses from 1.5 to 25 times. This phenomenon can be explained by the difference of the ionization potentials of the corresponding molecules and radicals: for odd masses they are lower than the even ones (Figure 5.19).

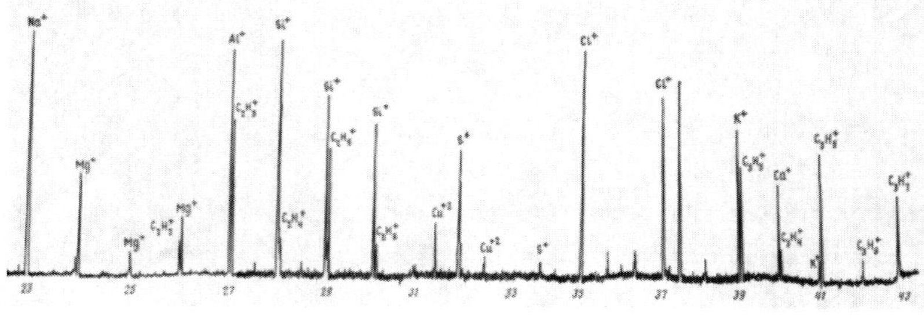

Figure 5.17. Fragment of a mass spectrum of osmium (exposure of 20 nC and double lines near the impurity elements belong hydrocarbon ions from $C_2H_2^+$ to $C_3H_{12}^+$).

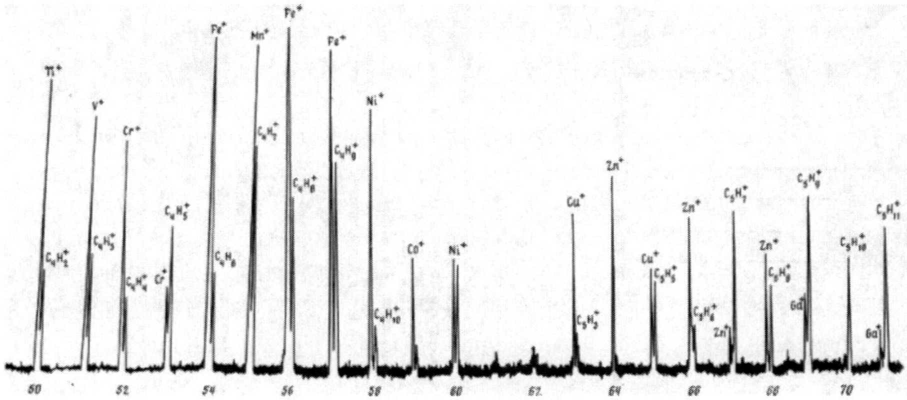

Figure 5.18. Fragment of a mass spectrum of osmium (exposure of 3 nC and double lines near the impurity elements belong hydrocarbon ions from $C_4H_2^+$ to $C_5H_{11}^+$).

With increasing exposure of 10 to 30 nC in the mass spectra exhibit heavier hydrocarbon ions with $n = 7$. However, their total number is one or two orders of magnitude lower than that of hydrocarbon ions with $n = 2$.

The total content of hydrocarbons in the mass spectrum amounts to 10^{-2} at %. To determine the content of oxygen ions of types CnHmOp$^+$ and to detect them among the hydrocarbon ions require an enhancement of the spectrometer resolution (on the left wing of the analytical lines of heavy hydrocarbons C_6H8^+–$C_7H_9^+$ (in the form of a small step) the corresponding oxygen ions can be present).

When the vacuum in the electrostatic analyzer increases to 5 10^{-6} Pa and in the ion source of the mass spectrometer to 2 10^{-5} Pa, the hydrocarbons in the mass spectra at these levels of sensitivity are not detected.

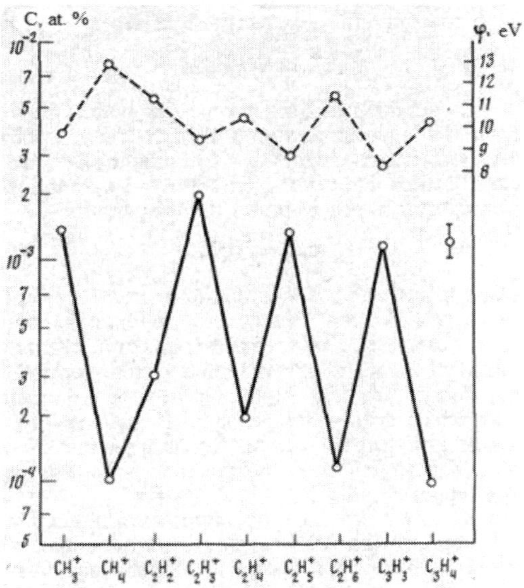

Figure 5.19. Relative content of hydrocarbon ions CH_3^+–$C_3H_4^+$ in laser plasma (the upper curve shows the distribution of the ionization potentials of hydrocarbon molecules for even and odd masses).

3.5.10. Low-Background Ion Source for a 'LIDIA' Ultrasensitive Mass Spectrometer

The search for small amounts of superheavy atoms in nature requires the use of a highly sensitive and rapid analysis. This can be done by the mass spectroscopy with direct detection of ions. To this end, an ultrasensitive mass spectrometer 'LIDIA' [33] was developed, the scheme of which is shown in Figure 5.20. This mass spectrometer includes a laser-plasma ion source, a vapor or laser target, an electrical system of selection of the ion component, and a multi-split system to reduce the background of rescattered ions. The relative sensitivity of the mass spectrometer is 10^{-7}–10^{-9} at %, the sample weight is 100 mg, and the analysis time is 2-3 hours.

The choice of a laser-plasma ion source is justified by a number of advantages: in the first place, because of the minimal yield of ionized molecular compounds that can result in background mass lines in the region of the atomic mass $A = 250$. The laser-plasma ion source is also preferable because of the low flow of matter, the possibility of local analysis, the lack of selectivity, and the absence of ionic components not related to the target material.

3.5.10.1. Laser-Plasma Source (LPSI) Operation Characteristics

The laser source was an LTI-PCh-8 solid-state electro-optical Q-switched quantum generator. The energy generated by the laser energy was 60 mJ and the pulse duration was 12 ns. The radiation was focused on the sample surface to a 300-μm-diameter spot. The plasma produced on the target surface expanded inside a 100-mm-long drift tube and reached the module of beam formation and acceleration. The ion beam with an energy of 40 keV was analyzed by a sector magnet in the focal plane of which a Faraday cup was placed. The current pulse signal was observed on an oscilloscope.

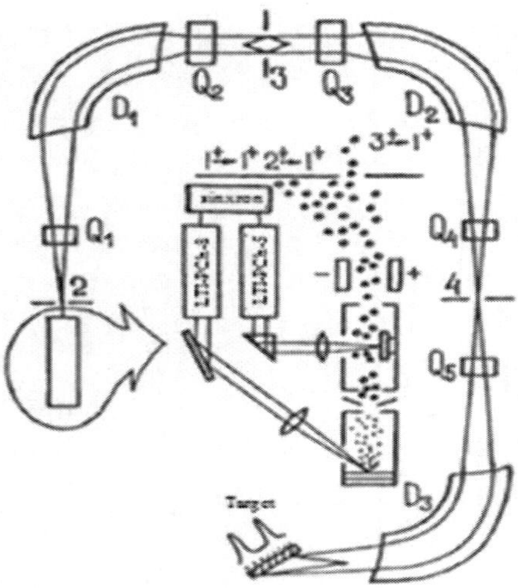

Figure 5.20. 'LIDIA' ultrasensitive mass spectrometer.

In designing the 'LIDIA', an additional charge exchange of the ion component is envisaged because the number of molecular ions with increasing ion charge decreases by several orders and the charge exchange process causes an additional decrease in the ionized centers in the flow. Simple evaluations in our experiments show that with the charge exchange $3^{+}\rightarrow 1^{+}$, the relative sensitivity of 10^{-9} at % is achieved during 3 hours of the LTIPCh-8 laser operation at a frequency of 50 Hz. In designing a source for the 'LIDIA', operation with singly charged ions is provided against the reduced background of molecular compounds. As was shown, there is a large difference in the energy spectra of atomic and molecular ions, which is effectively used in low-background laser-plasma sources. The simplest way is pre-selection of the ion energy by a decelerating electric field.

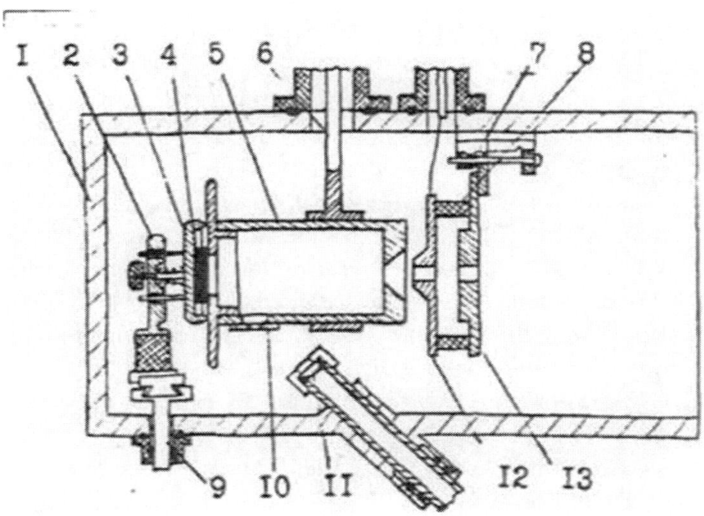

Figure 5.21. Scheme of a LPSI (Laser-plasma source ions)device: 1 – housing, 2 – pressure foot, 3 – cup, 4 – target, 5 – expander, 6 – high-voltage input, 7 – bridge moving electrodes, 8 – insulating glass, 9 – rod of the vertical scan, 10 – glass plate; 11 – focusing lens, 12 – second electrode, 13 – third electrode.

Consider the design of a LPSI (Figure 5.21). The expander is made in the form of two electrically isolated parts (6) and (9), connected by a fluoroplastic insulator. On the surface of the insulator there is a ridge that prevents the surface of the evaporated material sample from dusting. The rod-holder with an additional electric input connects the various parts of the expander to the low voltage power poles. The 'plus' is connected to the end of the expander (6), located near the extracting electrode (12). Inside the expander an electric field is produced and the ion energy can be extracted. The rectified voltage is adjusted through the primary circuit of the transformer. On both sides of the expander a high voltage of up to +40 kV is applied via the ceramic insulator. The sample is mounted on a movable cup (3), and the ions are extracted by the ion optics (12, 13).

After the first series of experiments we found some design defects of this source. In particular, the background of charged molecules is eliminated at a supply voltage of 200 V, which is probably due to the geometry of a locking field (e/d – 1.0) and to the phenomenon of the accelerating voltage drop in the expander volume. To address these disadvantages, we placed additionally in the expander a grid (1×1 mm, 90% transparency), which serves to produce a locking field with the configuration of a plane capacitor. With such a source we studied the yield of charged molecules for specially prepared oxide samples, at a confining potential in the range of 10 – 20 V. The ability to remotely change the confining potential without turning off the high accelerating voltage allows a direct comparison of the oxide yield with and without a confining potential.

At a level of 5 10^{-6} from the total ion current, we failed to observe charged molecules at 20 V, while at 0 their yield was (1 – 3)10^{-3} for different types of samples. Thus, a LPSI with the energy selection reduces the background level of the molecular ions by 500 times on average, and is one of the main parts of the 'LIDIA' facility.

3.5.11. References

[1] Liebich, F,, Ramendik, G.I., Blokin, A.G., et al., 1987, *Zh. Anal. Khim.,* 42, 1783.

[2] Ramendik, G.I., Kiyuchkova, O.I., and Kaviladze, M. Sh., 1983, *Zh. Anal. Khim.*, 38, 1749.

[3] Bykovskii, Yu.A., Oksenoid, K.G., Ramendik, G.I., et al., The Role of Impurity Atom Ionization, Ions Acceleration and Recombination in Laser Plasma Source Mass-Spectrometry, *Preprint of Moscow Engineering Physics Institute,* Moscow.

[4] Oksenoid, K.G., Ramendik, G.I., Sil'nov, S.M., et al., Zh. Anal. Khim., 45, *858.*

[5] Bykovskii, Yu.A. and Sil'nov, S.M., 1987, Ions Recombination in Laser Plasma, *Preprint of Moscow Engineering Physics Institute,* Moscow.

[6] Bykovskii, Yu. A., Zhuravlev, G.I., and Belousov, V.I., 1978, *Plasma Physics,* 323.

[7] Belousov V. I. Vysokochist, Veschestva. 1987, 213.

[8] Bykovskii, Yu.A. and Nevolin, V.N., 1985, *Laser Mass- Spectrometry* (Moscow: Energoatomizdat) (in Russian).

[9] Sil'nov, S. M., 1988, *The Last Stages of Laser Plasma Expansion* (Moscow: Moscow Engineering Physics Institute) (in Russian).

[10] Boriskin, A.I., Bryukhanov, A.S., Bykovskii, Yu.A., et al., 1981, *Nauchnye Pribory,* 28.

[11] Ramendik, G.I., Sotnichenko, E.A., Oksenoid, K.G., etal., 1989, *Zh. Anal. Khim.*, 44, 1361.

[12] Artamonov, A. A., Oksenoid, K.G., and Ramendik, G.I., *Zh. Anal. Khim.,* 46, 1881.

[13] Kryuchkova, O. I., Ramendik, G.I., Khromov, A.Yu., et al., 1988, *Zh. Anal. Khim.*, 43, 1397.

[14] Bykovskii, Yu. A., Timoshin, V.T., Laptev, I.D., et al., 1988, *Vysokochist. Veshchestva,* 323.

[15] Ed. Brouillard, F,, 1986, *Physics of Ion-Ion and Elec- tron-Ion Collisions* (New York: Plenum Press).

[16] Zel'dovich, Ya.B. and Raizer, Yu.P, 1966, *Physics of Shock Waves and High-Temperature Hydrodynamic Phenomena* (Moscow: Nauka) (in Russian).

[17] Afanas'ev, Yu. V., Basov, N.G., et al., 1978, *Interaction of High-Power Laser Radiation with Plasma* (Moscow: Nauka) (in Russian).

[18] Bulanov, S. V., Inovenkov, I.N., Naumova, N.M., et al., 1990, *Fizika Plasmy,* 16, 764.

[19] Gurevich, A. V. and Meshcherkin, A.P., 1983, *Fizika Plasmy, 9,* 955.

[20] Mueller, A. and Shevelko, V.P., 1990, *Zh. Tekh. Fiz*., 50, 985.
[21] Aref'ev I.M., Benyaev N.E., Komleva A.A., et al. *Zh. Analit. Khim*. 61, No. 1, 50–54 (1986).
[22] Ramendik G. I., Tyurin D.A., Kryuchkova O.I., Chernoglazova G.I. *Zh. Analit. Khim*., 40, No. 7, 1210 (1985) and 41. № 1.p. 50.
[23] Ramendik G. I., Manson B. M., Tjurin D. A. et al.//*Talanta*. 1987. V. 34. I.P.61
[24] Ramendik GI, BM Manzon, Tyurin DA / / *Zh. analyte. chemistry*. 1989. T. 44. № 6. S. 996.
[25] Bykovskii, Yu. A.., Basova, T., Belousov VI and others / *Zh. tehn. physics*. 1976.T. 46. Number 6. S. 1338.
[26] Ramendik GI, Tyurin DA, Chernoglazova GI / / *Zh. analyte. chemistry*. 1989.T. 44. Number 12. S. 2157.
[27] Beusen J. M., Surkvn P., Gijbels R., Adams F. // *Spectrochim. acta*. 1983. V. 38B.№ 5-6. P. 843.
[28] Sysoev AA, Samsonov, G., "*Theory and Design of static mass - analysistors*" - Preprint MiFi "Vol.1, 2., 1972.
[29] AN Oblizin The physical features of ion emission PSLs for laser mass spectrometry, dis. by the competitor. PhD, MA, 1990 Christie W. H., Smith D. H., Eby R. E., Carter J. A. *Intern. Laboratory*. 1978.
[30] Ramendik GI C Zh. analyte. chemistry. 1990. T. 45. Number 10. S. 2044.V.T.Timoshin, S.M.Silnov and other studies of hydrocarbon background in laser mass spectrometry, BB, № 6, page 117.1989 year. GN Flerov, Ter-Akopyan G. M. *Pure and Appl. Chem.,* 1981,53, r.909.

Chapter 15

3.6. LASER REMOVAL OF SPACE DEBRIS

3.6.1. ORIGIN OF SPACE DEBRIS

Space debris are man-made orbital objects, which no longer serve any useful purpose, and objects of natural origin, such as meteorites, asteroids. Based on their origin, all space debris can be divided into several categories: exhaust upper stages and upper stage launch vehicles; broken-down satellites; debris, formed by the collisions or after the testing of anti-satellite defense; and objects of natural origin.

Many small-sized (less than 10 cm in diameter) particles constitute the largest part of space debris. The most significant contribution to the 'littering' of the near-Earth orbits have been made by anti-satellite missiles (during the bombing of worn-out satellites), which led to the formation of new fragments ranging from centimeters to several meters in size.

The problem of space pollution, which arose with the launch of Sputnik in 1957, received the official status at the international level only at the end of the 1980s. The space debris problem is of international and global nature. The UN General Assembly endorsed the space debris mitigation guidelines, including development of 'clean' (non-waste) technological processes in space, removal of defunct spacecrafts from the orbit and the rejection of anti-satellite defense test. Such measures are necessary to avoid the so-called Kessler effect, a scenario in which the density of objects in Low Earth orbit (LEO) is high enough that collisions between objects could cause a cascade – each collision generating space debris which increases the likelihood of further collisions. This development of events will inevitably stop for many decades the further outer space exploration. According to the forecasts of many experts, for the most littered LEO [sun-synchronous, geostationary Earth orbit (GEO)], the critical density of the space debris can have been reached by the 2050s.

The number of space debris annually increases exponentially not only because the orbits are more and more full of non-working spacecrafts, but also because any collision of the two fragments results in tens, hundreds and thousands of fragments of smaller diameter. For example, in 1996 a French satellite collided with a fragment of the third stage of the French rocket Ariane. In 2009, the US commercial Iridium communications satellite hit a defunct Russian military Kosmos-2251 communications satellite, launched in 1993 and decommissioned in 1995.

One of the first victims of the space debris was the crew of the shuttle Challenger in 1983, when the spacecraft collided with a small-size particle (less than 1 mm in diameter); as a result, there appeared a crack in the porthole. Later, the experts concluded that it was a microscopic particle of paint detached from some spacecraft. The Soviet space station 'Salyut-7', which was also hit by microscopic particles, continues the list of victims of space debris. The Russian space station Mir whose solar battery was punched by a piece of space debris, which resulted in a 'hole' of more than 10 cm in diameter, was no exception.

Now the ISS can maneuver to avoid possible collisions with space debris (this year it has been done seven times), but the safety of the station is still the main concern to the experts. Thus, in 1999 the ISS nearly collided with a fragment of the upper stage of the rocket, long wandering in space. In 2001 the station had a chance to be hit by a 7-kg instrument lost by U.S. astronauts during a space mission.

Besides the fact that space debris is a danger to space explorers, it is also unsafe for the inhabitants of the Earth, because it can fall down on heads in the literal sense of the word. Suffice it to recall the fall of the Skylab satellite in Australia. Fortunately, no one was killed except a cow in this accident. In 1991 the Soviet station 'Salyut-7', which had already experienced much during its operation life, got broken up into fragments over Argentina. Of particular concern is the space debris, which contains radioactive materials. The worst accident involving a Soviet or Russian nuclear space system was the fall from orbit in 1978 of the Cosmos 954 satellite powered by a nuclear reactor. It also broke up in the atmosphere spreading radioactive debris which scattered over 77,000 square miles of the Northwest Territories of Canada.

In addition, there were cases when fragments of space debris, which did not burn up during the atmosphere re-entry, injured people. For example, in 1997 a fragment of the second stage of the Delta launch vehicle injured a woman's shoulder.

Gradually it becomes clear that in the near future the problem of collision with space debris fragments will have to be taken into account in ballistic designing any space mission (nowadays, space debris are considered in calculating the orbit trajectories of space stations and large satellites).

The increasing trend towards miniaturization of spacecrafts and the use of a group of small spacecrafts rather than one large spacecraft only exacerbates the problem by increasing the total number of objects in LOW and GEO.

To date, the problem of protection of actively functioning spacecrafts against debris and objects of natural origin has become of fundamental importance due to a significantly increased density of space debris in LEO and GEO, especially near the regular orbits of spacecrafts.

Virtually the entire man-made space debris are metal fragments of former spacecrafts (steel, titanium and other metals), which move in elliptical orbits around the Earth. In size, they are divided into 4 categories: small-size particles, less than 1 to 10 mm; medium-size particles, from 1 to 10 cm; large-size particles, more than 10 cm for low orbits and more than 1m for geostationary orbits; and micro-size particles, less than 1 mm.

The size of particles, the damage they cause and the possible solutions are presented in Table 1.

Large fragments of space debris can be observed by using ground-based optical and radio-tracking means, and many of them are followed by terrestrial radars and catalogued. Due to the presence of different space debris categories and methods of their removal with different laser systems, one can single out the following separate independent tasks which differ in formulation, criteria for space debris irradiation and, correspondingly, in the average power and the output pulse periodic radiation parameters:

1. Early removal of threats caused by hitting a spacecraft by space debris.
2. Spacecraft protection from the threat of collision with approaching space debris.
3. Removal of space debris from LEO and GEO.

The first two problems are directly related to the protection of a particular spacecraft from space debris, whereas the third one refers to the task of global LEO and GEO cleanup.

Table 1. The size of particles, the damage they cause and the possible solutions

Size	> 10 cm, LEO > 1m, GEO	1 to 10 cm	0.1 to 1 cm	< 0.1 cm
Number	14 thousand fragments, under supervision	300 thousand fragments	70 to 80 million fragments	10 to 100 trillion fragments
Effect	Loss of a spacecraft	Destruction of a spacecraft's elements	Significant damage of a spacecraft	Spacecraft surface erosion
Solution	Mechanical removal, maneuvering, laser removal	A ~500-kW laser	Architectural protection, a spacecrafts' surface hardening	Unnecessary

3.6.2. Laser Methods of Space Debris Removal

As an effective means of addressing these problems, it is proposed to use high-frequency, high-power pulse-periodic laser systems. The principle of debris removal is very simple: Radiation of a laser system rapidly heats the debris surface and removes part of the material owing to evaporation. As a result of exposure, depending on the absorbed energy and the exposure time, the space debris can break down into smaller fragments which do not threaten the spacecraft, or can change the flight trajectory due to a recoil momentum, and prevent a collision with the spacecraft. Complete evaporation of small fragments of space debris is also possible.

The most powerful at the moment are DF chemical lasers. When the laser beam passes through the atmosphere, DF lasers with a wavelength $\lambda_{DF} \sim 3.8 \times 10^{-4}$ cm. When use is made of a spaceborne laser station, advantage can be taken of a solid-state diode-pumped laser (Nd:YAG ~ $1.06 \cdot 10^{-4}$ cm) High quality of output radiation of chemical lasers provides the beam divergence, which is close to the diffraction limit. To remove efficiently the space debris, it is proposed to use a DF chemical laser and a solid-state laser in the *Q*-switched or the gain medium modulation regime, i.e., in the pulse-periodic regime. In this case, the peak values of the intensity of the radiation incident on the space debris increase by orders of magnitude compared to the cw regime. The time between pulses is determined by the change of the active medium in the chemical laser zone or by restoring the population inversion of a medium of a solid-state laser. Experiments performed at the A.M. Prokhorov General Physics Institute of the Russian Academy of Sciences showed that the optimal (in terms of the maximum energy output) modulation frequency of a DF laser is ~ 150 kHz, and the excess of the peak power over the average output power is 2-3 orders of magnitude. In the case of solid-state laser the optimum frequency is about 100 kHz, the pulse duration lying in the range of $10^{-7} - 10^{-8}$ s.

To solve the above-mentioned problems, use can be made of various laser systems, both with a ground-based and a space-based power plant:

- a stationary ground-based laser system (GBLS) ensuring beam focusing and pointing on a space debris fragment;
- a standalone space-based laser system (SBLS) also ensuring beam focusing and pointing on a space debris fragment.

3.6.3. DESIGN PARAMETERS OF LASER DEVICES

Let us estimate the solutions of the above-mentioned problems relying on the use of different laser systems. To determine the ratio of the chemical laser power in the CW and pulse-periodic regimes, use was made of the models based on the Navier–Stokes equations, with all the essential chemical and vibrational kinetics processes taken into account. Radiation on various vibrational–rotational transitions in the pulse-periodic regime was calculated using the equations for a free-running and *Q*-switched laser resonator. At small pulse durations (<100 ns) non-equilibrium distribution of emitting molecules in the rotational levels of the generation process was not taken into account.

A considerable part of space debris fragments is controlled by means of tracking, their trajectories are known, and the largest space debris fragments are catalogued. The space debris fragments that represent a threat to the spacecraft and the time of a possible future collision are well known. To remove the threat of a collision, it is needed during the interaction to change the trajectory of the space debris so that the spacecraft is not hit by a fragment, or the threat of the next collision with the fragment is postponed indefinitely. Due to collision threat warning technology, there is an opportunity to remove a fragment in advance.

This problem can be solved by using a ground-based DF laser, which directly shots the fragment from the Earth (from a mountain of height 2.5–3.5 km). More preferable are mobile airborne laser systems (ABLSs), which allow their transport at the right moment. The most preferred mobile system is an ABLS at a height of ~ 20 km, which allows one to not distort radiation during its passage in the upper atmosphere and to transport the installation to any part of the Earth. The advantage of mobile systems in case of low repetition rate lasers is the fact that in case of failure when the fragment is not destroyed by a train of pulses or in the case of the system failure, the possibility of rapid transport of the system to the nearest point of the space debris fragment passage allows the laser firing to be repeated after a short period of time (less than one day).

A stationary GBLS is used to destroy space debris fragments moving at a low orbit over the laser. The beam impact is effective in the sector with a full opening angle of ~30° with respect to the vertical, and the region of interaction at a height of 300 km is a circle with a diameter of ~160 km. When flying over the Earth the space debris leaves a spiral 'footprint' with a width of 160 km, making 16 orbits a day, which are spaced at a distance of about 2.5 thousand km from each other. Thus, ~ 5200 km of the circumference of the Earth is covered each day, the circumference being approximately 40000 km (2.6 thousand km = 8 days). Therefore, stationary lasers are not always applicable to the task of removing space debris from LEO and GEO; they can be used only when the global early warning system tracks a fragment and the laser itself is mounted in the right place. However, with a significant increase in the laser pulse repetition rate (high-frequency pulse-periodic regime), there can arise a situation when the space debris fragment is destroyed during its orbiting over the stationary laser system.

As a result of the beam action, the debris fragment and the spacecraft become spatially separated by the size of the interaction region. In the case of non-coplanar orbits, this region is of the order of the spacecraft size. In the case of close-to-coplanar orbits the interaction region is much larger with the borders of this region being within the distance of the spacecraft size. For estimates we consider the following example. Let a spacecraft with characteristic dimensions $L_{SC} \sim 100$ m move along a circular orbit at an height of $H = 300$ km at a speed of $\mathrm{v} = \sqrt{\dfrac{\gamma M_з}{R_з + H}} \sim 7.73 \cdot 10^5$ cm/s and an orbital period of $T = \dfrac{2\pi(R_з + H)}{\mathrm{v}} \sim 5.42 \cdot 10^3$ s ≈ 90 min.

Let the fragment move along the same trajectory. We shall consider the impact of the laser pulse at the fragment instantaneous, which increases its speed by $\Delta \mathrm{v}$. Then, it is easy to obtain that orbiting period will change by $\Delta T_\kappa = 3T\Delta \mathrm{v}/\mathrm{v}$. During one orbiting of the fragment the traveled distance will change by $\Delta S = 3T\Delta \mathrm{v}$. To reach $\Delta S > L_{sc}$ it is needed to change speed of the fragment as a result of irradiation by $\Delta \mathrm{v} > L_{sc}/3T = 6$ mm/s. If this irradiation has been performed during N orbits before the collision, the necessary change in the speed is reduced by N times. For example, with the fragment irradiated during 100 orbits, the debris and the spacecraft will be spatially separated by a distance of 100 meters and the fragment speed will be equal to $\Delta \mathrm{v} \sim 0.06$ mm/s. Therefore, one must first choose the number of orbits before the first irradiation, the recoil momentum and the number of pulses needed to remove the space debris fragment from the spacecraft trajectory.

Consider the mechanism of irradiation of a metal space debris fragment. Metals are good absorbers of radiation in the mid-IR range. It is known that radiation in metals is absorbed in a layer of thickness $\sim\lambda$. During radiation absorption at a depth of 0.05–0.1 μm, the melting and evaporation zone is localized. Because of this, a part of the melt is removed in the form of droplets with a sharp increase in the interaction zone. In the absence of gravity, this effect is most pronounced: The melt cannot be kept on the surface. In weightlessness, liquid droplets are quickly destroyed by heating, because the internal pressure overcomes the surface tension before the process of evaporation. Thus, the task is to ensure the conditions of intensive evaporation of the space debris material. In this case, we should 'pump' by a pulse train the required energy into a thin surface volume for the time when the material have no time for the heat spreading to occur throughout the entire volume, which ensures the destruction of the fragment surface and the creation of a mechanical recoil momentum.

When $W > 10^5$ W/cm^2 and the pulse repetition rate is $> 10^3$ Hz, the debris material is fairly intensively evaporated. For fusible materials the threshold intensity(Wt) is lower. The impact of such a field on the debris fragment can be called 'soft.' Since the recoil momentum of each pulse is small enough, the addition of pulses at a high frequency of the pulse-periodic regime leads to significant destruction in the debris material. We carried out assessments on the basis of a DF laser (Table 2) and an Nd:YAG laser (Table 3)

Table 2. LEO / MEO ESD removal data for DF laser

λ	τ	D_b	W	f	<P>	d_s	Z	I
3.8 μm	10 ns	30 m	150 J	10 KHz	1,5 MW	18 cm 2Dif	300 km	0,6 J/cm

Table 3. LEO / MEO ESD removal data for Nd YAG laser

λ	τ	D_b	W	f	<P>	d_s	Z	I
1.06 μm	10 ns	30 m	60 J	3000 Hz	360 kW (0,5)	5,2 cm 2Dif	300 km	3,0 J/cm 2

Consider their content in more detail. When the diameter of the telescope is D = 3 m, Wt = 100 KW/cm^2, d = 2.8 m, the DF-laser output powers at the corresponding pulse durations and frequencies are as follows:

100 ns – 100 kHz – 120 MW
100 ns – 10 kHz – 25 MW
10 ns – 100 kHz – 30 MW
10 ns – 10 kHz – 6 MW

Let the output mirror of the telescope have a diameter D = 3 m. With the necessary power level, the beam divergence equal to three diffraction limits at the laser output $\Delta\theta \sim 3\theta_d$ can be considered a good achievement. In addition, the propagation in the atmosphere introduces an additional distortion of the beam accompanied by the absorption loss. When the light spot is circular, the divergence (full angle) will be $\Delta\theta \sim 3\theta_d \approx 2.44 \cdot \lambda_{DF}/D \approx 9 \cdot 10^{-6}$rad. When focusing laser light on a target at a height of H = 300 km, the spot diameter will be equal to $d = H \cdot \Delta\theta \cong 2.8$ m. Therefore, to ensure the emission intensity $Wpeak$=10^5 W/cm^2 on the target, an average peak pulsed power must be equal to $Ppeak=10^5 \pi d^2/4 = 6 \cdot 10^9$ W at the output telescope mirror (excluding losses during the beam propagation).

Let us now estimate the expected time-averaged power of a DF laser with a pulse duration of ~ 100 ns. Because the characteristic time of energy storage in the active medium of the chemical laser is about 10 μs, then at a characteristic pulse duration of ~ 100 ns (duty cycle is equal to 100), the average power of the DF laser in the pulse-periodic regime should be about Pa^{pp}=0.01 $Ppeak$ ~ 60 MW. However, due to strong relaxation in the medium at the moments when lasing is absent, transition from continuous operation to the pulse-periodic regime is accompanied by a substantial energy loss. What we can expect is ~ 50% of the average output power : $Pa^{pp} \sim 0.5\ Pa^o$. Therefore, at a pulse duration of ~ 100 ns for three diffraction divergences an average power of ~ 120 MW is required.

At a pulse duration of ~ 10 ns and a repetition rate of 100 kHz the laser pulse peak power can exceed the average power of the chemical laser by virtually 1000 times: Pa^{pp} =0.001 *Ppeak*. However, in the absence of rotational equilibrium of emitting molecules the power falls further relative to the equilibrium case (at ti> 100 ns). The necessary level of the average laser power in the pulse-periodic regime decreases with the rotational equilibrium taken into account by approximately 3-4 times as compared to the cw laser Pa^{pp} ~0.25 Pa^{o}: at three diffraction divergences Pa^{pp} ~25 MW.

For comparison, the table shows the results of estimates of the average laser power at a pulse repetition rate of t~ 10 kHz and pulse durations of 100 and 10 ns. However, the values of the average laser power cannot be obtained even at the present level of technology.

In principle, one could try to further reduce the laser pulse duration down to 1 ns, which, other things being equal, would enable us to decrease the average power of the DF laser by one more level: At the divergence equal to three diffraction angles the power will be equal to Pa^{o} ~ 2.5 MW. However, we should keep in mind that it is virtually impossible to obtain at these ultrashort pulses a beam divergence, which is close to the diffraction limit (for such a short time the mode structure of the radiation field in a cavity does not have time to form – the round-trip time for light in the cavity is about 10 s at the time of the pulse existence ~ 10^{-9} s). A solution to the problem by shortening the master oscillator is limited by a threshold condition for lasing. Moreover, the weak power of the master oscillator requires a significant lengthening of the active medium of the amplifier, which further complicates the solution of problems related to the amplified spontaneous emission confinement.

The above estimates are consistent with a minimum achievable recoil momentum in the laser radiation – refractory target surface interaction. It is assumed that this is sufficient when the onset of irradiation is appropriately chosen in order to remove a space debris fragment from the interaction region of the spacecraft's and debris fragment's orbits. For more fusible materials requirements to the average laser power are lessened. It is clear that the increase in the average power of a laser source simplifies the solution to this problem, thereby making it possible to destroy or remove a space debris fragment by using fewer pulses.

3.6.4. Protecting Spacecrafts from Collisions with Space Debris

Protecting a spacecraft from collisions with space debris or objects of natural origin is needed when a spacecraft is subjected to the possibility of being hit by approaching uncontrolled space debris fragments or meteoroid particles, or by controlled fragments when their trajectories coincide with those of a spacecraft at some orbit. In this case, the fragment can be destroyed or removed from its orbit using a standalone space-based solid-state diode-pumped laser system, mounted directly on the protected spacecraft, or running along the same orbit near the spacecraft. The laser must have sufficient electrical power to operate continuously in orbit. Effective irradiation of the space debris target becomes possible if the time of its detection does not exceed much the total time of target tracking, preparation of the device to shoot, power beam pointing, and duration of a pulse train.

To remove the threat of a collision requires endowing the space debris fragment with such an amount of energy that the fragment will start braking and lag behind the spacecraft or its trajectory will change so much that it will miss the spacecraft. It is clear that the net effect on the target, compared to the stationary problem, should be much more powerful. This is accompanied by intense evaporation of the space debris material; therefore, ionization of molecules vaporized and plasma formation are also possible. To evaluate the braking effect, there exists a semi-empirical formula, which shows that under the action of the laser pulse, the fragment velocity Δv, associated with the absorbed pulse energy, changes due to the expansion of the plasma produced

$$\Delta \mathrm{v} = C_m \cdot E / m$$

where E is the absorbed energy, m is the target mass, and C_m is a coefficient determining the energy efficiency and to a large extent depending on the type of target material, the radiation intensity and pulse duration. The table 4 below shows the experimental data on the optimal (with a maximum C_m) peak intensity I_{peak} as a function of the pulse length τ:

Table 4. The experimental data on the optimal (with a maximum C_m) peak intensity I_{peak} as a function of the pulse length τ

τ, ns	10^{-7}	10^{-8}
I_{peak}, W/cm^2,	$2\text{x}10^8$	$6\text{x}10^8$

It is seen that $I_{peak} \sim \tau^{-0.5}$ the conditions being far from optimal, the laser pulse action decreases rapidly; therefore, we assume that the peak intensity is optimal. Note that the power density of ~ 10 W/cm is widely used for laser drilling and cutting of materials under sublimation or ablation. This irradiation regime compared to that previously discussed can be called 'hard.' In the interval between two successive pulses at a repetition rate of ~ 100 kHz the plasma expanding into vacuum has time to travel a distance of about 1 to 2 cm, which is enough to get rid of the plasma screening effect on a target.

It is easy to show that the average laser power is associated with a mirror diameter D, distance to the target L and beam divergence $\Delta\theta/\theta_d$ by the ratio $P_a^o \approx \left(\frac{\Delta\theta\lambda D}{\theta_d L} \right)^2$.

Assessment of the impact of molecular absorption on the efficiency of the laser radiation passing through the atmosphere shows the advantage of the DF-laser radiation over the Nd:YAG-laser radiation. The integral attenuation coefficient in the case of the DF laser is two times lower in comparison with that of the Nd:YAG laser. The tables 5a,b below show the average power of a pulse-periodic solid-state laser with a mirror 3 m in diameter (medium) operating at a pulse repetition rate of 100 and 10 kHz in the 'hard'(5a) and 'soft'(5b) regimes as a function of the distance L and divergence $\Delta\theta/\theta_d$ for pulse durations of $\tau=10^{-7}$ and $\tau=10^{-8}$s.

Obviously, such lasers are very expensive and technologically imperfect. To date, the output power of these lasers in our country has reached 3–5 kW. Thus, space-based diffraction-limited pulse-periodic solid-state diode-pumped lasers with an output power of 100−1000 kW and a pulse duration of 10−100 ns will provide in the 'soft' regime the necessary impact on a space debris fragment at an aiming distance of 100 km. The light spot on the fragment at this distance will be within 8 cm.

Table 5 (a, b). The average power of a pulse-periodic solid-state laser with a mirror 3 m in diameter (medium) operating at a pulse repetition rate of 100 and 10 kHz in the 'hard'(5a) and 'soft'(5b) regimes as a function of the distance L and divergence $\Delta\theta/\theta_d$ for pulse durations of $\tau=10^{-7}$ and $\tau=10^{-8}$s

In the 'hard' regime
a) $D = 3$ m, Wpp = 1 GW/cm^2, $d = 0.08$ m, Nd:YAG laser
100 ns – 100 kHz – 1200 MW
100 ns – 10 kHz – 120 MW
10 ns – 100 kHz – 120 MW
10 ns – 10 kHz – 12 MW
In the 'soft' regime
b) $D = 3$m, Wpp = 10 MW/cm^2, $d = 0.08$m, Nd:YAG laser
100 ns – 100 kHz – 12 MW
100 ns – 10 kHz – 1.2 MW
10 ns – 100 kHz – 1.2 MW
10 ns – 10 kHz – 120 kW

It is also possible to make use of ground-based DF chemical lasers (both mobile and stationary) with a receiving mirror installed on the spacecraft. This variant is suitable only when the point of collision of a spacecraft with a space debris fragment lies in the region of the GBLS action, which is a highly unlikely event. However, we note that in this case the receiving glass mounted on the spacecraft needs to be large enough to intercept the entire radiation beam. Estimates show that the required diameter of the receiving mirror, depending on the beam divergence, must be several meters. The use of a fully terrestrial laser system for 'hard' irradiation of the space debris fragment by a DF laser directly from the Earth would require a laser system with high average powers (presented in Tables 6a and b):

At the size of the telescope D = 30m, W_a = 10 MW/cm^2, d = 0.28 m (for a DF laser at three angles of diffraction beam divergence).

Table 6 (a, b). The use of a fully terrestrial laser system for 'hard' irradiation of the space debris fragment by a DF laser directly from the Earth would require a laser system with high average powers

a) 100 ns – 100 kHz – 60 MW
100ns – 10 kHz – 24 MW
10 ns – 100 kHz – 6 MW
10 ns – 10 kHz – 2.4 MW

In the 'soft' mode, when the size of the telescope is D = 30m, W_a = 100 KW/cm^2, d = 0,28 m (for a DF laser at three angles of diffraction beam divergence).

b) 100 ns – 100 kHz – 600 kW
100 ns – 10 kHz – 250 kW
10 ns – 100 kHz – 60 kW
10 ns – 10 kHz – 25 kW

The effective action is achievable because the desired power level of the chemical laser has already been reached!

3.6.5. Clearing Near-Earth Space Debris

Clearing near-Earth space debris with lasers is not subject to a detailed review and quantitative estimates in this paper. This problem has been considered at one time or another by many authors in Russia and abroad. However, we note that the solution of this problem is also possible through the use of a standalone space-based HF laser with a focusing and a beam pointing systems. To speed up the fall of a space debris fragment to the atmosphere requires the fragment braking and transfer to a lower altitude with a shorter 'lifetime.' It is known that the dwell time ('lifetime') of a space debris fragment in orbit is highly dependent on the height of the orbit. According to the data available in the literature the lifetime of a fragment at an altitude of 1000 km is about 2000 years, at an altitude of 600 km – ~ 25–30 years, at an altitude of about 200 km – about a week. In the altitude range of 100–1000 km the dependence of the space debris fragment lifetime on the height above the Earth can be approximated (in accordance with the above data) as $T \sim H^7$. With such a strong height dependence even a small deceleration and a decrease orbit leads to a significant reduction in lifetime. Thus, the lifetime decreases from 120 to 6 days with decreasing the orbit from 300 to 200 km.

The estimates require special modeling and calculations for each particular orbit and space debris fragment. Theoretically, in the entire range of laser irradiations (from 'soft' to 'hard'), the fall of space junk on the Earth is accelerated. As follows from the estimates, in the 'hard' regime a train of high-power laser pulses (when a fragment flies over the stationary laser) can decrease the orbit of a space debris fragment to the required level. If this fragment enters the upper atmosphere ($H \sim$ 100 km), it slows down and burns out during 1-2 orbits; thus, the problem is solved. However, this formulation seems excessive, because it is quite often sufficient to lower the fragment's orbit, so that its orbit is below that of the spacecraft.

Based on the available literature, we can assert that the space-based laser power of about several tens of kilowatts is sufficient to significantly reduce the lifetime of a small fragment. Naturally, the further increase in the laser power further reduces the lifetime, i.e., increases the effectiveness of the fragment entry to the atmosphere; therefore, near-Earth space debris can be rather rapidly cleared using a standalone SBLS with an output power of several hundred kilowatts. The number of laser shots will be determined by the accuracy of direct hits and the time needed for a space-based laser to become operational again.

When use is made of a ground-based laser system with a receiver mirror installed on a spacecraft, the energy loss during the passage of radiation through the atmosphere and the mirror loss will require a much more powerful installation (apparently, the level of several MW). The number of laser shots is determined by the number of passes of the mirror over the ground-based laser (on average no more than once every 8 days) and the probability of the fragment location within the laser hitting range in these periods. In this case, less laser shots are needed compared to the case when use is made of a standalone SBLS.

Finally, we note that a more detailed study of the problem presented, in particular, the effect of the pulse repetition rate and duration, relative velocities of the space junk, laser power, etc. on the space debris destruction should rely on a more accurate mathematical modeling of pulse-periodic laser processes, radiation propagation, effectiveness of its action on the target, etc.

3.6.6. Space Clearing With Lasers

The need for a pulse-periodic chemical laser operating at a pulse repetition rate of ~ 10 – 100 kHz and duration of 10 – 100 ns leads to the fact that the laser system should rely on the use of an optical master oscillator – power amplifier scheme with two small- and medium-scale active medium generators, an average output power level in the master oscillator not exceeding 5 – 10 kW. First, this will allow one, for the pulse-periodic regime implementation, to use well-proven methods, based either on the electro-optic effect, or on the method of a passively Q-switched optical resonator in the master oscillator. Second, it is well known that at a relatively low power level near-diffraction divergence of master oscillator radiation is easy to obtain.

A solid-state diode-pumped laser should be compact and lightweight to transport it in outer space. A prototype of such a laser with a power of over 100 kW has been recently designed in the USA. Soon, we can expect this laser to be scaled up to a level of 500 – 1000 kW. The pulse-periodic regime in the case of such technological solutions is also possible.

Fabrication of solid-state lasers with a large cross section of the output beam (wide-aperture lasers) is a complex scientific and technical problem of modern laser physics. One of the possible solutions is based on a multi-channel laser principle. In accordance with this principle, the laser represents an array of identical channels, each of which ensures lasing. Such an array of lasers forms an optical source with a composite aperture, whose output radiation is a collection of individual laser beams. The small cross section of the channels allows efficient and easy energy pumping into the active medium and the heat removal from its volume.

Despite large structural advantages of multichannel lasers, high directivity of their radiation is a challenging physical problem. In a multichannel laser each channel emits a wave, the phase of which is independent of the phase of the waves from other channels. The channel fields are incoherent, and their intensities are added. In the incoherent regime the resulting intensity I_{inc} of a multichannel laser is

$$I_{inc} = N.\ Io$$

where I_0 is the intensity produced by a single channel, and N is the number of channels.

In the multichannel laser beam, when collective lasing is in phase, the phases of some channels coincide. The output beam in this case is formed by the interference of waves whose phases on its axis are equal. The resulting intensity I_c on the axis of the beam is N times greater than the intensity of one channel:

$$I^2_c = N.\ I^2o.$$

The angular divergence of the beam in the axial region in the case of in-phase oscillation is close to the divergence of the laser with a wide-aperture of diameter *D*. Therefore, the key challenge for a multichannel laser is the realization of in-phase collective oscillation.

The system of detection, beam formation and pointing – the information-sighting system (ISS) – should be directly mounted on a spacecraft. The issues concerning the system elements' arrangement in the spacecraft modules (together with the elements of the system of radiation generation) are not discussed here.

Note that the ISS makes use of two pointing channels (radar and high power), built using a CCD-array of the corresponding channels. In principle, more complex ISS schemes relying on the use of two, three or more channels are also possible.

The ISS telescope, forming the output radiation of a chemical laser, is placed on a rotating platform. The ISS must provide the following basic functions:

1. to receive target designation from external systems;
2. to point the shaping telescope in the direction of the target;
3. to determine target angle;
4. to make the laser radiation power axis coincident with the target angle by controlling the angular position of one of the ISS mirrors;
5. to carry out real-time tracking of the target with the required accuracy;
6. to provide automatic readjustment of the cavity mirrors and power channel as a whole; and
7. to record the radiation arrival at the target.

A wide range of functions performed by the ISS suggests a rather complex structure, which should be developed in line with the objectives imposed on a laser system. We only note that at an average power level of 500 kW the specific consumption of fuel components is estimated to be 3.5 kg per second. Despite the pulse-periodic irradiation of a space debris fragment, working components should be supplied continuously. With the onset and termination of laser operation taken into account, the typical irradiation duration of a space debris target will take at least 2 – 3 s. It is clear that a sufficiently effective use of the laser implies that the fuel supply should be sufficient to destroy or remove quite a large number of targets (~ 100), which means that the desired amount of fuel before the next refueling should to be at least 600 kg.

As noted in the previous section, due to the significant absorption of radiation in the atmosphere, in the case of a GBLS use should be made of a DF chemical laser, which is placed at a sufficient height above sea level (~ 2.5 – 3.5 km). Besides, the DF laser should operate in the pulse-periodic regime with a repetition rate at 10 – 100 kHz, the optical quality of output radiation being better than three diffraction limits (the latter can be achieved, for example, by the method of adaptive optics). The optical schematic of radiation generation should be based on a master oscillator – power amplifier scheme. With a shaping telescope having a main mirror of diameter $D = 30$ m the DF-laser radiation is directly transmitted to a space debris fragment or a receiving mirror mounted on the spacecraft. Apart from a focusing telescope, the spacecraft should have the systems installed which allow detection, tracking and pointing of power radiation at a space debris fragment.

The estimates show that the current level of chemical laser technology used in civil and military sectors of economy makes it possible to construct a stationary system for efficient removal of space debris fragments from orbits. International cooperation is possible and, which is more important, desirable in this field. The conference in Ludwigsburg (Germany, 2011) confirmed this. The space of the Universe is vast. However, here in our solar system, as at home, we need to put things in order, to throw away old and buy new stuff, and ensure that the 'space casket' is not overfilled one day. To date, the problem of space debris removal is one of the challenging and cannot be solved by any single country separately. However, both politicians and scientists in space industry are united in one thing: Space debris must be controlled to ensure space activity in the future. Members of all space-faring nations must sit down at a round table and discuss the problem, as it is very important for all of us today.

3.6.7. CONCLUSION

1. Pulse-periodic laser systems with a high repetition rate are an important tool in solving the problem of destruction of space debris and objects of natural origin or their removal to a safer orbit.
2. The problem of advance protection from space debris can be solved with the help of ground-based high-frequency pulse-periodic DF lasers directly irradiating the debris from the Earth in the 'hard' regime, when the size of the telescope mirror is 30m. At a pulse duration of ~ 10 ns and repetition rate of 10 kHz the laser system should generate an output no less than ~ 2 – 2.5 MW. In the 'soft' regime, the average output power of the laser system is lowered to ~ 250 – 600 kW.
3. The problem of protecting the spacecraft from space debris or meteor particles approaching the spacecraft can be most effectively solved with the help of a standalone space-based high-frequency pulse-periodic solid-state laser system (mounted on a spacecraft, or near it). At a power of ~ 12 MW and pulse duration of 10 ns, in the 'hard' regime the spacecraft can be reliably protected from typical medium-sized refractory materials (~ 10 cm) from a distance of about 100 km. In the 'soft' regime, when the size of the telescope mirror is about 3 m the average output power of a solid-state laser system is lowered to <1 MW, depending on the pulse repetition rate.

Chapter 16

3.7. Oil Films Elimination from the Water Surface

3.7.1. Introduction

The contamination of large water areas (oceans, seas, lakes, and rivers) with petroleum products as a result of accidents and industrial activity of the mankind is one of major problems of protection of environment. Any other contaminant, how dangerous is it, can not be compared with petroleum on a basis of universal utilization, number of contamination sources and degree of effect on all components of environment. Through penetration of petroleum products into water, there are deep, frequently irreversible changes of its chemical, physical and microbiological properties.

The following methods of disposal of petroleum contaminations of water: mechanical, physical-chemical, chemical, biochemical are used nowadays. The mechanical methods are as such: collecting of petroleum from a surface manually or with different installations. All these methods are effective during limited time (from several hours to several days) only – time period during which the thickness of a petroleum film is great enough.

As for physical-chemical methods it is necessary to mention, first of all, application of various adsorbing materials (polyurethane foam, coal dust, sawdust etc.); however, all these methods are labor-consuming and low efficient. Besides they require secondary reprocessing of adsorbents.

Chemical methods – removal of petroleum with the help of chemical substances. The basic disadvantages of this method are high price and the fact, that detergents frequently are more toxic for water microorganisms, than petroleum.

In the case of microbiological decomposition of petroleum, the bacteria will be utilized, for which the petroleum is a nutritious medium. However natural velocity of microbiological decomposition of petroleum is too small, and this process requires long time. Besides, the biological activity of microorganisms strongly depends on the temperature of water.

The laser method of cleaning of water surface from a thin petroleum film is one of physical-chemical methods. Some time later after penetration into water, the petroleum spreads on a surface of a water and forms very thin film (thickness several microns). This film cannot be forced to burn, since because of a good thermal contact to a surface of water the film cannot be heated to a temperature ensuring steady combustion.

3.7.2. THE LASER METHOD OF CLEANING OF WATER SURFACE

The principle idea of a laser method consists in following. The laser beam passes through a petroleum film and then is absorbed in thin layer of water. The water heats very fast up to the boiling temperature, and the forming vapor destroys a film, bursting it in small-sized fragments, which, mixing up with the hot vapor, are decomposed quickly with formation of simple non-toxic substances.

The main advantages of a laser method consist in following.

- This method is 'fast-response' one, since it does not require any special preparation; in emergency situations the time from the moment of obtaining of the alarm signal to the beginning the laser installation operation is determined only by the time necessary for arrival of the ship or the helicopter with the installation on board in a given area.
- The method is contactless, i.e. does not require realization of preparatory or other activities in the oil spillage.

For realization of the proposed technology, it is possible to utilize different types of lasers, as continuous (power from 20 up to 100 kW), and pulse (from 1 mW, 1 J and higher in a microsecond pulse with a pulse repetition rate from 1 Hz and above). The laser radiation density should be about 50 - 100 W / cm^2 at pulse duration no more than 1 millisecond.

Theoretical estimations and the experiments have shown, that as a result of a laser beam action on a surface of a water and land, covered with film of hydrocarbon contamination, following effects can be fixed:

Evaporation and burning, and in a continuous mode the consumption of energy for 1 gram of vaporized liquid on the order of value surpasses energy necessary for heating up to boiling temperature and evaporation of a film, that is explained by a heat consumption for heating of water. In a pulse mode, the consumption of energy it is per unit of vaporized mass of a liquid (film) approximately in 5 times less, thus the process of film ignition start more easily.

Knocking out the particles of polluting substance above a surface of water under action of pulse or powerful scanning laser radiation. The physical mechanism of this effect is explained by sharp evaporation (boiling-up) of a thin layer of water under a layer of polluting substance. This process takes place whereas hydrocarbon polluting thin film have an absorption coefficient less, than water. Knocked out particles of petroleum at an height up to 30 cm is possible. For this case, the energy consumption per unit mass of a raised liquid is sufficiently less than energy necessary for evaporation and ignition (mode number one) or for sucking up and saving for future efficient usage (mode number two).

3.7.3. THE LASERS OF CLEANING

The character of a task to be solved superimposes certain conditions on operating characteristics of the laser device concerning both parameters of laser source, and concerning auxiliary systems.

Technological lasers of 10 - 15 kW power range, that widely are used in production, will have the output completely insufficient for liquidation of large scale contaminations presenting the greatest danger according to the above-stated estimations. Such lasers can be utilized, at the best, for improvement of a process engineering of cleaning under modeling conditions.

Besides, the character of this task dictates impossibility or extreme undesirability of use of the stationary civil engineering service line (electric power network, waterpipe, main gas line etc.), in view of the requirement of a high self-sufficiency and mobility. The power unit should not limit mobility and thus provide totally energy needed of all installation, the necessary reserve of expendables and fuel onboard a complex. A capability of a fast redeployment from one type of a vehicle on another is also desirable.

In the present technical proposal, the basic design concept of the mobile laser installation on basis of gas dynamic CO_2-laser is developed. The activities were executed in accordance with the working plan of OOO "Energomashtechnika"

In 3.1 the possibility of using of various types of lasers for solution of required tasks is considered and the selection of gas dynamic laser is justified.

3.2 is devoted to selection and substantiation of basic performances of the laser installation and principal schematics of GDL.

In 3.3 the possible schemes of organization of air supply in the laser installation are considered. The selection of an aero-engine for use in quality power unit of the laser installation is justified. The necessary volume of adaptation/modification of this engine is presented.

3.4 of the technical proposal is devoted to the description of the design concept of gas dynamic laser, and also laser installation as a whole. The pneumo-hydraulic schematics is described, the structure of the equipment, fuel tanks, operation control units are determined. The description of operation and cycle sequence of start and shut down of the laser installation are also presented.

3.1. The substanntion of selection of laser type according to specific parameters, operation autonomy, and mobiliti

Presented the activities during the first phase, the modern high power continuous operation gas lasers: electro discharge (EDL), gas dynamic (GDL), continuous chemical (CCL) and oxygen - iodine (OIL) lasers were analysed.

The main results of this activity are reduced to following:

3.1.1. Analysis of applicability of various types of lasers to the task of the present project.

The laser with output power of 100 kW of continuous operation during 0,5 hours (minimum requirement) is required for realization of the task. besides, according to physics of process of destruction of a petroleum film, explained in Introduction, laser radiation should be weakly absorbed by petroleum film and it should to be absorbed effectivelly by water.

In the table 1, the computational data for depth of radiation penetration for 4 types of lasers in petroleum and water are presented.

The range of values of depth of penetration for petroleum is the sequence of the fact, that absorption coefficient of various grades of petroleum and water (for the CCL case) in relation to wavelength in CCL radiation spectral band were considered.

Table 1. Effects of petroleum and water on depth of penetration

The laser type	Depth of penetration, micron	
	Petroleum	Water
GDL, EDL	100 – 260	10
CCL	26 – 200	0,8 – 40
COIL	50	2000

From the table 1, it is obvious, that the COIL radiation is rather weakly absorbed by water and is rather strongly absorbed by petroleum. However to heat a petroleum film by thickness ~100 microns through absorption of COIL radiation practically is impossible because of strong heat transfer from a film into water through heat conductivity. The conclusion is, that use COIL for the solution of this task is impossible. As the lengths of waves of radiation GDL and EDL are identical, the comparison of these two types of lasers is carried out on the basis of mass - dimension characteristics and other parameters that are listed in table 2. The comparison is carried out for two real installations, which were developed and tested in the past to sufficient degree, namely - a 100 kW GDL, and a 20 kW EDL.

Table 2. Comparison of GDL and EDL

№	The name	GDL	EDL
1.	Power laser (max), kW	100	20
2.	Overall dimensions of the laser installation, m	9x2.4x2.4	2.5x3x3.5 + 2 x 0,8 x 0,8
3.	"Dry weight " (free of fuel), metric ton	6	18.6
4.	Specific, 1 kW of laser power dry weight, t/kW	0,06	0,93
5.	Specific, 1 kW of laser power, dry volume, cubic m/kW	0,5	1,5
6	Specific volume fuel consumption for 1 kW of laser power, cubical meter/ hour / kW	0,03	0,01

The analysis of the data, presented in Table 2 shows, that the GDL has decisive advantages in comparison with EDL. It is necessary to pay attention, that EDL specific fuel consumption is 3 times less, than that of the GDL. However influence of this factor on the total complex weight will have negative influence only for large operation time (5 hours or more). At the arrangement of a laser complex on the board of helicopter, the operating time will not exceed 0.5 - 1 hour, therefore in this case smaller EDL specific fuel consumption will not have decisive value. Thus, the GDL and CCL remain only as competing systems. The comparative analysis of these two types of lasers with reference to a considered task is stated below.

3.1.2. Specific power of laser generation

The GDL specific power, Wsp is 10 J/gram; this value for CCL is much higher and reaches 150 J/gram. From this point of view, CCL has conclusive advantage in comparison with the GDL.

3.1.3. Wavelength of radiation

The GDL radiation wavelength, λ = 10,6 microns and CCL wavelength, λ= 2,7 - 3,3 microns. From this point of view, CCL has one more potential advantage over the GDL, as the diffraction limited angle of divergence of CCL beam is 3 - 4 times less in comparison with the GDL (at identical aperture). However this problem requires the further arguing.

- Radiation extraction from the resonator.

In high-power lasers, extraction of the laser beam from the resonator, where the pressure is much lower than atmospheric pressure, in external space through a rigid window is practically impossible for the reason of thermal destruction such window. For a extraction of such beam out of installation, the aerodynamic window (AW) usually is utilized. The supersonic gas stream cross-sectional to beam fulfills the task of transparent boundary. As there are areas of various density gas sheaths limited by curvilinear surfaces in such flow, the beam transmission through this gas non-uniformity introduces the distortions of a wave front resulting in increase of an angular divergence of a beam. These distortions do not depend on a wavelength, as the dispersion of an index of refraction of gas is extremely small and therefore these distortions have the same absolute value for the GDL and CCL. However relative values of these distortions of a wave front set (i.e. reduced to wavelength) are for CCL 3-4 times higher, than for the GDL, therefore from the point of view of an angular divergence of radiation the advantage CCL in comparison with the GDL can be not so essential.

- Use of an external optical system.

The optical GDL and CCL systems for delivery beam energy to remote objects include necessarily external (in relation to the resonator) optical elements and systems, intended for expansion of a beam with the purpose of reduction of its divergence. A principal component of a telescope is the main mirror, which diameter can reach several meters (depending on required power density on the target). The one piece glass or metal mirror will be too heavy. For this reason the main mirror is usually considered as consisting from large number of small mirrors (facets), assembled in a unit. Every facet has the own adjusting device. The accuracy of relative positioning these facets should be not worse $\lambda/4$, otherwise distortions of a wave front after reflection of a laser beam from this mirror will cause the essential increase of an angular divergence in comparison with it diffraction limit. This is $\lambda/4 \cong 2{,}5$ microns for the GDL, $\lambda/4 \cong 0{,}75$ microns for CCL. Therefore in practice required accuracy is much more difficult to ensure for CCL.

This brief discussion shows, that the efficiency of CCL external optical system can be much below, than that of the GDL and, hence, reduces advantage of shorter wavelength of CCL radiation. Besides at a small length of beam paths from the laser down to a water surface, the angular divergence has no importance, since density of power on a water surface appears sufficiently high without application of an external optical system.

3.1.4. Length of a zone of radiation generation

In the GDL working section (the resonator region) the transfer of oscillation energy from N_2 to CO_2 has taken place. The characteristic time of this process is relatively large. Besides the relaxation time of oscillatory - exited molecules CO_2 at their collisions with other particles is also large. Therefore for full extraction of N_2 accumulated oscillatory energy, and transformation it in laser radiation, length of a working zone in flow down direction should be 10 - 15 cm. Such large length provides certain freedom of selection of the resonator scheme (one pass or multipass, symmetrical or asymmetrical, etc.). Thus the mirror operating surfaces are large enough, that results in a reduction of its thermal loading and distortion of a surface due to inhomogeneous heating.

In CCL, the length of beam generation section is small, less then 2 cm. This circumstance causes difficulties in optimization of the laser resonator and leads to increase of thermal loading of mirrors that results to significant thermal distortion of mirrors and distortion of a wave front, i.e. in increase of an angular divergence of radiation.

3.1.5. Direct exhaust of used gas in atmosphere

If the GDL or CCL is placed on ground or on board of flight vehicle (plane or helicopter), there is a problem of exhaust of spent gas in atmosphere. In the case of the GDL this problem is decided rather simply by use of the diffuser, as the operating pressure in a zone of the resonator is rather high. For reaching a maximum degree of pressure recovery in the diffuser it should have special geometry.

In the case of CCL, the operating pressure in the resonator is much lower in comparison with the GDL; therefore, the direct exhaust in atmosphere with the help of the diffuser is impossible, it is necessary to utilize additionally gas or water vapor ejector to ensure necessary pressure in output section of the diffuser. Such ejector requires large additional gas flow rate, that results in fundamental complexity of the laser installation design and significantly decrease specific power of generation, i.e. in this case one of main CCL advantages is lost in comparison with the GDL.

3.1.6. Toxic characteristics

The degree of CCL danger from the point of view of toxic characteristics of working components and exhaust gases is much higher in comparison with the GDL. In CCL, for creation of a working mixture the extremely toxic substances containing fluorine will be utilized. Besides the exhaust CCL contains a significant quantity of fluorine hydride HF, which is supertoxicant. In this connection with use CCL in a system of laser cleaning of water areas it will be necessary to ensure special safety measures, that, naturally, will cause the complication of the installation design and rise of its price.

In the case of the GDL the exhaust gas is ecologically clean ($N_2 + CO_2 + H_2O$). However, CO, carbon monoxide which is a toxic gas can be used as a fuel for the GDL. Hence, system of storage and supply of CO should ensure absence of its leakage. On the other hand, CO is lighter-than-air, therefore it floats in atmosphere and is fast blown away by wind.

If liquid hydrocarbon fuel (benzole, toluene) will be used as fuel in the GDL, the system of its storage and supply does not require special measures to secure the ecological safety.

3.1.7. Absorption of laser radiation in atmosphere

The wavelengths of the GDL and CCL radiation are within of windows of atmosphere transparency. These windows are usually determined for vertical transmission of beam through atmosphere, and for this case power dissipation on track is ~20 % for the GDL and

25 % for CCL. For vertical tracks of length 50 - 150 m, which are of interest with reference to a considered problem, the radiation power dissipation can be neglected.

3.1.8. Substantiation of selection of a gas dynamic CO_2-laser

Proceeding from the comparison of various type lasers, explained in the previous chapter, it is possible to make a conclusion, that the most suitable laser for development of the installation for disposal of petroleum films is the GDL for following reasons:

1. Simplicity of design.

From design point of view, the GDL components are most simple of the considered lasers. The combustion chamber operates on usual components, for example, kerosene + air. The designs of such chambers are virtually perfect; there is a wide experience of their operation in various technical devices. The nozzle unit is the GDL made from heat resisting steel and the long time without forced cooling can work. The long zone of population inversion behind the nozzle unit (10-15 cm) stipulates simplicity of selection of the scheme of the resonator for obtaining laser generation with maximum efficiency. Thus the characteristic size of a mirror of the resonator is approximately equal to length of a zone of generation, i.e. 10-15 cm, therefore manufacturing of these mirrors does not call serious technological difficulties.

The capability of the GDL activity at high pressure of gas in the resonator in front of the nozzle unit (up to 3 MPa and more) provides with a straight line an exhaust of spent gas in atmosphere with the help of the supersonic diffuser of a special design. Such capability all remaining considered lasers are deprived. The aft ejector is necessary for maintenance of a direct exhaust of these lasers which requires additional gas flow rate that complicates the installation and reduces specific energy of radiation.

It is necessary to note also, that the GDL output power rather weak varies at change over a wide range of parameters of gas (pressure, temperature, chemistry) in the combustion chamber, i.e. the GDL is not critical to accuracy of a task and maintenance of an operational mode.

One more virtue of the GDL is the small time (some seconds), necessary for start.

Thus, the GDL is the rather simple, reliable and flexible tool ensuring high output power of laser radiation.

2. Mobility.

From all considered types of lasers the GDL for today has, apparently, least “weight” and “volume” of 1 kW of laser power. It allows rather simple to place the GDL on any vehicle, whether it will be airplane or helicopter, ship or railway platform, etc. Such GDL mobility allows one with its help to solve many tasks, including connected with disposal of petroleum films.

3. Selection of propellant components.

One of specific singularities of working process of the CO_2-laser is, that the working mixture of gases should contain in a receiver rather small quantity of water vapor (no more than 5-6 volume %). On the other hand, the operation temperature of gas in a receiver should be enough high (1300-1600^{0}K). It superimposes limitations on selection of propellant components (fuel and oxidizer). From the point of view of maximum specific energy of radiation (J/kg) optimum components are gaseous at standard conditions with damp (CO) and liquid nitrous oxide (N_2O), thus additional nitrogen (or air), necessary for working process, moves in a receiver from a separate source, and the water vapor in quantity 1,5-2 % is formed by burning small quantity of hydrogen or alcohol in air.

4. Wavelength of radiation

The analysis of the data presented in tab. 1 show, that for the solution of a problem of disposal of petroleum films, the optimal wavelength of radiation is that of the CO_2-laser, as this radiation is weakly absorbed by the film and is strongly absorbed by water. From the point of view of a diffraction limited angular divergence of radiation, the CO_2-lasers (EDL and GDL) lose for CIL and CCL. However with reference to a considered problem, the value of a divergence is not critically important. The estimations show, that at length of a beam from the laser source down to a surface of water ~ 50-100 m and power of radiation ~100kW, an precision focussing of a beam is not required, for the reason of obtaining required density of power in a spot on water surface. It is possible to utilize nonfocussed or partially focussed beam. Therefore with reference to a considered problem, the relatively large wavelength of the GDL radiation is not an insufficiency factor.

The explained above reasons allow to make a unequivocal conclusion for the selection of the GDL as the generator of laser radiation for the mobile installation intended for disposal of a petroleum film on a surface of water. At the estimated power of radiation, the GDL is preferable in weight and volume factors in comparison with EDL, HCL and COIL: it does not require the electric power, will utilize low toxic fuels, provides an direct exhaust of utilized gas in atmosphere. The GDL simple in control, it is not critical to changes of working parameters in sufficiently broad range, and it is convenient in operation. The important circumstance is that there exists a reliably operable GDL prototype with power range of 100 kW, on which the extensive researches are carried out. In particular, the design of an aerodynamic window permitting to extract a laser beam from a zone of the resonator in atmosphere without application rigid transparent for a working wavelength of materials was developed.

3.2. Selection and substantiation of basic performances of the laser installation and GDL principal diagram

In the previous chapter the comparison of various types of high power lasers was carried out from the point of view of their application for the solution of this task, where was shown, that the most appropriate type of the laser is CO_2 - GDL. However there are different types of the GDL, therefore it is necessary to choose among them optimal version, and also to develop the general concept of the power installation for optimal supply of working components in the GDL.

In the present chapter the analysis of the various GDL schemes, their advantages and disadvantages are given from the point of view of application in the considered installation, the selection of propellant components for the GDL is justified from the point of view of their power efficiency, production and toxic characteristics, and ecological safety.

Last portion of the chapter is devoted to the substantiation of the GDL characteristics working on the chosen components. For this purpose the calculation results permitting to define a specific power of laser radiation at the given initial parameters, such as fuel chemistry, temperature and pressure in front of the nozzle block, expansion ratio of the nozzle, etc. are used. The calculation results and their analysis are shown.

For definition of the main dimensional characteristics of the laser and its flow organizing parts it is necessary to define an basic level of power of laser radiation, which, in turn determines the total fuel consumption and remaining geometrical characteristics of the laser.

3.2.1. Selection of optimal radiation power of gas dymamic laser

3.2.1.1. Theoretical substantiation of power level

The basic capability of disposal of petroleum contaminations from water surface by means of irradiation of a petroleum film by a high flux of laser radiation is proved experimentally in our publications. However, the published data give only the general representation about basic possibility of using of such method. It is obvious, that the combustion is complex multiparameter process and, if we talk about operational use of the mentioned principles in the particular technical device intended for practical use, the optimization of parameters of the installation is necessary. The important value in this case is the efficiency, which it is possible to understand as the area of cleaned water surface divided to mean energy, used on clearing, (m^2/kW). For optimization of parameters of the installation it is necessary to conduct a cycle of research works on study of evaporation, ignition and combustion processes of petroleum used for the above mentioned method. The experimental research of the given problem even without qualitative understanding of involving processes represents a very difficult task in connection with vast quantity of optimization parameters and requires heavy material costs and time. In this connection the preliminary development of theoretical model representing process of disposal of petroleum film is desirable.

The second phase presented the various interaction mechanisms of a laser beam with a petroleum film on a water surface are theoretically considered. The following mechanisms were included:

- laser heating active absorbing, translucent and transparent films;
- an explosive boiling-up and conditions of process efficiency;
- evaporation and vaporization.

The basic conclusions of the technological report are the following.

1. For reaching temperatures of evaporation of a film the necessary time of radiation action is about one milliseconds at a laser power of 50 kW and diameter of an irradiated zone 10 cm. Temperature of water layer under the surface of film can essentially exceed boiling point. Such situation arises due to superior velocities of laser heating (up to several millions degrees of Celcius per one second) and inertial character of heat transfer process. At such temperatures the water passes in a metastable state with active outgassing. Under a surface of a film there will be

microexplosions of air - steam bubbles. As result, there will be a separation of a film from a surface of a water, its breaking and outflow of drops of petroleum in air.

2. The scanning velocity at power of the laser of 50 kW and diameter of a spot 10 cm ought to be in range of 10-12 m/s. Thus the velocity of cleaning of a surface 6000-8000 m^2/hour is provided.

It is necessary to note, that all calculations are fulfilled for thickness of a petroleum film 100 microns, that is rather large value and is observed only in the initial moments of formation of oil spillages. At thinner films, the velocity of cleaning will increase proportionally.

3.2.1.2. Experimental researches

For experimental confirmation of basic physical principles included in the basis of the Project, the participants carried out preliminary experimental research of interaction of laser radiation with films of various petroleum types on a surface of water. The research works were conducted within the framework of financing of activities. During research works gas dynamic laser developed by JSC "Energomashtechnika", and established on the bench was used. The optical scheme of realization of experiments is presented in Figure 7.1.

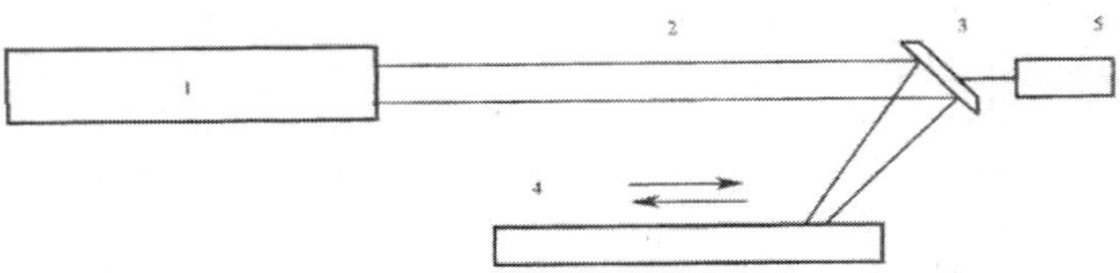

Figure 7.1. The optical scheme of realization of experiments.

The gas dynamic laser 1 generates radiation 2 with a wavelength 10.6 µm. The radiation is directed on a concave focusing mirror 3 and, being reflected, impacts on a surface of test heterogeneous structure in a cuveta 4. The mirror realizes periodic oscillations with the help of the mechanical device 5, thus the laser beam scans on a surface of a dish.

Diameter of laser beam cross section on a surface of a dish changed from 5 up to 10 cm. The laser radiation power was within the limits of 50 - 100 kW.

In Figures 7.2, 7.3 and 7.4 the results of experimental research works are presented, on which it is possible to make following preliminary conclusions.

1. The carried out experiments testify the possibility of using of the laser installations for cleaning of an aqueous medium from contaminations of petroleum and of it processing products.
2. For definition of optimal parameters of the laser installation and operation modes in practical conditions it is necessary to continue experimental research works under the special program approved by a potential customer.
3. For necessary reserve, the radiation power of gas dynamic laser of 100 kW for the installation to be developed according to present technical proposal is chosen

Make following preliminary conclusions

Figure 7.2. Intensive combustion of petroleum contamination film under laser action.

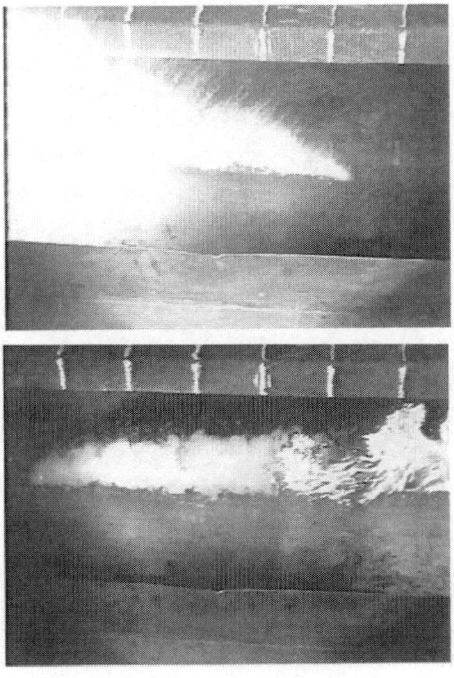

Figure 7.3. Combustion of a film of kerosene at laser action.

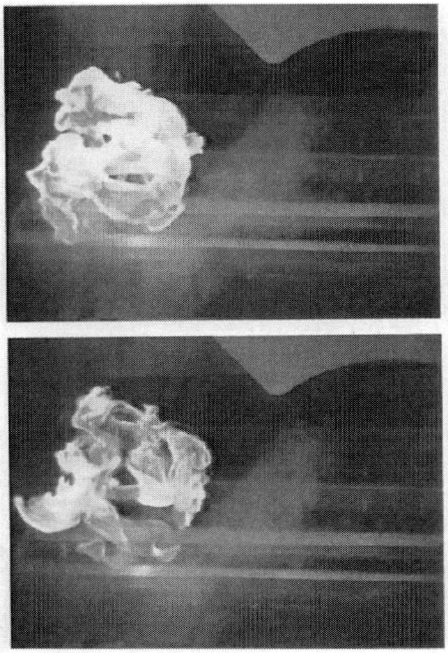

Figure 7.4. Fire of used engine oil vapors above water surface at laser action.

3.2.2. Selection of the GDL scheme and ITS substantiation

3.2.2.1. Conventional scheme of the GDL

The principle of the GDL operation is based on non-equilibrium fast expansion of the heated up working gas in the supersonic nozzle, at which there is a partial freezing population of oscillatory levels of molecules CO_2 and N_2 and formation of population inversion of upper and low laser levels. The heating of gas to an operation temperature 1300 –1700^{0}K at pressure 2-3 MPa is conducted in the combustion chamber through of combustion of chosen fuel – oxidizer components. The transformation of the accumulated oscillatory energy of gas to laser radiation has taken place in the optical resonator located down the nozzle. After the resonator the working gas passes through the supersonic diffuser and will be exhausted in atmosphere.

The most fast gas expansion is near to critical cross-section of the nozzle on length equal on an order of magnitude to the size of critical cross-section. Therefore for fast freezing of oscillatory energy it is necessary to use the nozzles with a small size of critical cross-section (< 1 mm). On the other hand, the pressure of gas after the nozzle (i.e. in the resonator region) should be small enough (< 100 mm Hg) to reduce to a minimum the losses of oscillatory energy due to collision relaxation. It is solved with use of nozzles with large expansion ratio (area ratio of output to critical cross-sections $A / A^* \geq 20$, that corresponds to value of a Mach number M 4).

For effective transformation of the accumulated oscillatory energy to laser radiation the amplification of a small signal during the pass the length of an active zone should be large enough to compensate optical losses (removal of radiation, absorption in mirrors, dissipation in medium) and to ensure a high enough radiation power in the resonator. In the GDL of very low powers (tens W) is possible to use the so-called mononozzle, which represents the flat (rectangular) nozzle with critical cross-section as a narrow slot. However for high-power GDL implementation of the mononozzle is practically impossible because of its large length (1 m and more); arising thermal deformations result in essential to nonuniform width of critical cross-section, h*, down to full blocking of critical cross-section in some zones. Besides the mononozzle does not allow to receive enough large transversal size of an active zone, that causes certain problems at resonator selection. Therefore in a high-power GDL the so-called nozzle unit representing a set of a many relatively short (~100 mm) slot-type nozzles is used. These nozzles are formed by contoured nozzle blades made from thermal resisting metal. Such design of the nozzle unit allows to vary over a wide range width and height of an active zone. Besides the nozzle unit provides a uniform enough density field of gas in an active zone, that allows to assure a small angular divergence of GDL radiation.

In view of such advantages of nozzle blocks, they are now widely used in the GDL, and therefore it is possible to name such a GDL a "GDL of traditional scheme". The radiation power of such a GDL is up to 100 kW and more in a continuous operation radiation mode obtained.

3.2.2.2. GDL with a set of axisymmetrical nozzles

The nozzle unit of the conventional scheme has the disadvantage that it is complex in manufacturing, and also is subject to influence of thermo-deformations. There is an alternative approach to development of nozzle units, which consists in following.

Instead of two-dimensional nozzles the axisymmetrical nozzles with small diameter of the critical cross-section are used. The supersonic part of such nozzle can be made conical or contoured. In this case the nozzle unit looks like a plate, in which the numerous small-sized nozzles located close as possible to each other are made. The manufacturing of such nozzles is possible, they are not subject thermo-deformations, by the appropriate arrangement of these nozzles on a plate (quantity of rows and quantity of nozzles in a row) it is possible to give output cross-section of the nozzle unit the required configuration.

However such nozzle unit has at least two disadvantages caused by presence of "empty" zones between nozzles (since complete filling of output cross-section by round nozzles basically is impossible).

The first disadvantage consists that the empty zones are sources of shock waves in a supersonic flow. These waves result in density inhomogeneous structure of gas in a zone of the resonator and, hence, to increase of an angular divergence of laser radiation.

The second disadvantage consists that in empty zones the viscous vortices flows are formed, that results in reduction of total pressure of a main flow. In turn, the reduction of total pressure has negative consequences on the supersonic diffuser operation, that can be appreciated as reduction a pressure recovery coefficient. In this case for realization of direct exhaust of working gas in atmosphere it is necessary to raise combustion-chamber pressure, that results in a decrease of output power of laser radiation.

For these two principles the GDL with the nozzle unit as a set of axisymmetrical nozzles is not used at noticeable production scale.

3.2.2.3. GDL with mixing of working gases in a supersonic flow

In conventional GDLs a simultaneous heating of all components of a working mixture - N_2, CO_2, H_2O takes place in the combustion chamber. However initial reserve of laser energy contains, primarily in oscillatory - exited molecules N_2, which part in gas mixture makes ~ 90 %. The role CO_2 and H_2O consists only of transforming accumulated in N_2 oscillatory energy to laser radiation.

The pure nitrogen has extremely large time of an oscillatory relaxation stipulated by collisions of molecules N_2 among themselves. As the result, the oscillatory levels N_2 are easily frozen at fast expansion of the working mixture in the nozzle. However other components of a mixture (CO_2 and H_2O) considerably accelerate a collision relaxation N_2 and by this reduce a reserve of oscillatory energy.

For increase of a reserve of oscillatory energy it is desirable to raise temperature in the combustion chamber. However in conventional GDLs temperature is limited to value about 2300^0K; the dissociation CO_2 begins at higher temperatures, and the efficiency of conversion of the accumulated oscillatory energy in laser radiation decreases. In contrast to CO_2 the molecular dissociation N_2 begins at temperature ~ 4000^0 K.

The presented ideas have resulted in the new GDL scheme, which concept consists in following: the pure nitrogen, heated up in a prechamber, expands in the supersonic nozzle, and at the exit of the nozzle to it is mixed with cold CO_2 and H_2O. Such a scheme has three basic advantages over the conventional scheme of the GDL: a) temperature of nitrogen in a prechamber can be essentially increased, b) more effective freezing of oscillatory energy of nitrogen is reached, c) energy for CO_2 and H_2O heating is not necessary.

In view of the indicated advantages mixing GDLs can basically ensure an increase in the specific power of radiation by many times in comparison with the conventional GDL scheme.

A main problem arising in realization the mixing GDL is the necessity to ensure fast intermixing of hot N_2 with cold CO_2, but the disturbances of a supersonic flow of N_2 should be kept at minimum.

Some solutions of this problem are possible. Technically, the most simple solution is to organize mixing of CO_2 and N_2 in wakes. In this case the flow of CO_2 moves in parallel to the flow of N_2; in this case, two sets of supersonic nozzles - one set for N_2 and the other for CO_2 - are used. An advantage of such a supply scheme is that wake disturbances arising upon interaction are minimum in comparison with other schemes of intermixing.

The basic disadvantage of this scheme is that the mixing of wakes is rather slow; therefore, for effective operation of the laser the working section (from nozzle exit down to the diffuser inlet) should have a sufficient length, that results in large relaxation losses and reduction of a pressure recovery coefficient of the diffuser. For these reasons the GDL scheme with intermixing of supersonic wakes is, apparently, not effective.

The theoretical analysis and the experimental data show, that the GDL schemes are more effective, in which the injection of cold CO_2 is made near to critical cross section of the nozzle in its transonic part. In experiments with heating of N_2 in the electric arc plasma generating device and supply of CO_2 was received a record-high value with a gain equal 3 % cm^{-1}, that exceeds the maximum value of the gain for conventional GDLs by almost an order of magnitude. However, the specific output power was insignificant, which is explain by a small size of the installation.

Despite basic promising characteristics of a mixing GDL, there are a lot of problems connected with their practical realization.

One of the most important problems is obtaining hot nitrogen at temperature 3000-4000^0K. Application of the electric arc plasma generating device for a high-power mixing GDL is unreal, since under consumption of nitrogen at 10 kg/sec and the electrical power exceeds 100 MW. It is possible if use is made of special hot nitrogen fuels; however, all of them are strong toxicants.

As a source of hot nitrogen, the devices can also be considered, in which combustion of metal in air takes place. As a result of this strong exothermal reaction hot nitrogen and oxide of metal as particles are formed, which are extracted from the formed flow with the help of a cyclone separator. However necessity of a very high degree of cleaning of a nitrogen stream from particles makes this way problematic.

Other important technical problem is the development of the nozzle (selection of a material, development of a robust design, which would maintain without destruction large heat flows in the region of critical cross-section.

The supersonic diffuser causes one more important problem. For an effective work of the laser translation temperature of working gas in an active zone should be sufficiently low, whereas temperature in a prechamber should be relatively high. For the coordination of these two conditions it is necessary to use supersonic nozzle with large Mach numbers (M=5 and more). However is known, the higher Mach number is, the lower is pressure recovery coefficient for the diffuser. Therefore for a direct exhaust of gas in atmosphere it is necessary to develop special diffusers, thus can appear, that such task basically is impracticable without use of a fodder ejector. However consumption of inducing gas in some times exceeds the consumption of induced gas, and in this case specific laser power designed on summarized gas flow rate, considerably decreases.

From explained above it is possible to make a unequivocal conclusion: taking into account problems connected to development of the high power mixing laser, and also absence of the prototypes of such lasers, for the solution of a task of disposal of petroleum film on water surface it is necessary to choose GDL of conventional scheme.

3.2.3. Selection of fuel for mobile GDL

In the present part the analysis of possibility of using various fuels for support of operation of the mobile gas dynamic CO_2-laser (CO_2-GDL), intended for disposal of petroleum film on a surface of water is carried out.

CO_2-GDL provides obtaining high power laser radiation generation in a continuous mode with a wavelength of radiation 10.6 microns. A basic physical principle of operation gas dynamic of the CO_2-laser is the fast expansion in the supersonic nozzle of a mixture of gases (CO_2, N_2, H_2O) preheated up to temperature 1500- 2000^0K. In a supersonic flow with Mach number M = 4 - 5 temperature of gases down to values 300 – 350^0K, necessary for obtaining, population inversion of laser levels of a molecule CO_2. For nozzle operation at direct exhaust in atmosophere after pressure recovery in the diffuser (the static pressure in a zone of generation is equal 5 10^3-1 10^4 Pa) the nozzle inlet pressure (stagnation pressure of a flow), equal 2-3 MPa is necessary. At small transversal and longitudinal sizes of the supersonic nozzle (the height of critical cross-section is equal 0.3 - 0.5 mm, and length of the nozzle 20 - 40 mm) characteristic time of the gas flowing through the nozzle appears small and comparable with time of oscillating relaxation of molecules of a mixture of gases, i.e. oscillating - oscillating exchange (*V-V* process) and oscillating - translation relaxation (*V-T* process) has taken place in nonequilibrium mode. As a result of nonequilibrium processes in oscillating exited molecules at certain gas mixture ratio and flow parameters in the supersonic nozzle the inversion population of the upper laser level of molecules CO_2 (0001) is formed. The molecules CO_2, are light generating molecules (laser radiation is realized on transition

00^01 -10^00), N_2 molecules are donors molecules transmitting oscillating energy to the upper laser level of molecules CO_2 (00^01), the molecules H_2O are intended for population of the lower laser level of molecules CO_2, (10^00). The optimal ratio of mixture of gases for an effective work of gas dynamic CO_2-laser should contain CO_2 = 0.1, N_2 = 0.89, H_2O = 0.01 (volumetric fraction). It is necessary to note, that the increase of a volumetric vapor fraction of (H_2O) results in decrease of the laser radiation power with other things being equal.

For mobile GDL, laser radiation, ensuring high power in a continuous mode, a basic method of thermal excitation is the combustion of such fuel components, which as final products give necessary components of a laser mixture at relatively high temperature. The optimal mixture ratio of gases at high temperature of mixture can be obtained, for example, at burning of carbon oxide (CO) and hydrogen (H_2) in air or other oxidizer (N_2O, N_2O_4) with subsequent use of ballast nitrogen.

The specific requirements to the analyzed mobile installation, impose some limitations on use this or that fuel, oxidizer and ballast gas. The main difference of this installation from existing lasers is the requirement of large duration of continuous operation.

For processing of the greatest possible area of the polluted basin, the laser installation should have an operating time appropriate to operation capabilities of a vehicle, on which it is secured. For the helicopter this time makes $t \sim 30$ min. The carried estimations of interaction of a beam with a water surface covered with thin petroleum film, have shown, that the power losses by heating water can be reduces through fast scanning of a surface by a beam of power $W \sim 100$ kW. These requirements allow to estimate total of fuel and ballast gas necessary for mobile GDL operation. The specific power *(W_{out})* existing homogeneous CO_2-GDL, as a rule, does not exceed 10 kJ/kg. Thus weight of fuel and ballast gas for such GDL operation should be not less than $m = Wt\ /\ W_{out} = 18000$ kg. This value essentially will increase at the account of weight of the equipment for storage of working components. Apparently, that such large weight is unacceptable for laser arrangement on a flight vehicle.

At the same time, the main part (80 - 95 %) of weight of design components used for creation of a propulsive mass GDL, is the weight of an oxidizer and ballast gas. Thus, use of free air as an oxidizer and a ballast gas is a unique opportunity to guaranty this installation operation.

As a CO_2-GDL fuel, carbon monoxide is usually used with the hydrogen addition. Carbon monoxide allows one to receive high values of specific output power in comparison with liquid fuels, such as toluene, benzole, kerosene, but the weight of this fuel together with a storage system is rather high. Weight of CO is 13 - 15 % of the total mass of spent components, i.e. in our case, 2000 - 3000 kg. Weight of a storage system of gaseous carbon monoxide is ≈7000 kg, and summarized weight of the fuel and storage system is $m_\Sigma = 9000$ - 10000 kg. When advantage is taken over the cryogenic state of carbon monoxide the summarized weight m_Σ can be reduced approximately twofold, but nevertheless it considerably surpasses appropriate weight for liquid hydrocarbon. Weight of liquid hydrocarbon is 4 - 5 % of the total mass of spent components, and the tanks for storage are much lighter than for carbon monoxide, as the hydrocarbon is stored at atmospheric pressure. In our case the weight of liquid hydrocarbon is 700 - 900 kg, and summarized weight of fuel and storage system is $m_\Sigma = 1000$ - 1500 kg.

Taking into account the above-mentioned reasons, in the present operation as possible fuel are considered benzole, toluene and kerosene, and as an oxidizer and a ballast gas (instead of nitrogen) – free air compressed by the compressor.

As it was mentioned above, for normal operation of supersonic nozzles of the laser it is necessary to ensure a high total pressure of a stream (P_0 = 2 - 3 MPa). The aviation compressors can ensure a compression ratio up to P_0 / Pa = 20. Here P_0 is the stagnation pressure of a working mixture in a GDL channel, Pa is the atmospheric pressure. In all further calculations the value of stagnation pressure of a flow P_0 = 2 MPa and maximum values, appropriate Mach number of the nozzle M = 4.7 with the area ratio of an exit of the nozzle to the critical cross-section $F/F^* \approx 25$ was accepted.

At comparison various fuels, basic parameter is the power efficiency, namely specific power of laser radiation W_{out} (power of laser radiation referred to summarized mass flow of laser components), achievable at use of the given fuel. In this connection the analytical-theoretical research of influence of various parameters on value W_{out} was carried out.

Temperature and chemistry of a working mixture formed upon combustion of various fuels in air, were determined by thermodynamic calculation. For the reason of uncertainty of efficiency of the compressor (approximately it is 0.75-0.83) in a main part of executed calculations temperature of inlet air in the combustion chamber was assumed to equal appropriate temperature at adiabatic compression up to value P_0 / Pa = 20. This temperature calculated according to the following formula

$$T = T_a \cdot \left(\frac{P_0}{P_a} \right)^{(\gamma - 1)/\gamma},$$

where T_a is the temperature of free air, γ is the isentropic exponent. At adopted values T_a = 298^{0}K, γ= 1.4 temperatures of inlet air in the combustion chamber is T ~700^{0}K. The enthalpy, appropriate to this temperature of air is equal I = 400 kJ/kg.

At thermodynamic calculation the thermal losses which depend on a design of the combustion chamber and beforehand are not known were not taken into account. In this connection in the subsequent calculations of flow of a working mixture in the nozzle block inlet temperature (^0K) was taken equal to 0.95 of temperature obtained in thermodynamic calculation.

The thermodynamic calculations have shown, that the working mixture contains in the investigated temperature range (or oxidizer-to-fuel ratios) only necessary for operation CO_2 - GDL components (CO_2, N_2, H_2O) and oxygen (O_2), as the content of remaining components is insignificant. Therefore in the further researches it was supposed, that the working mixture consists only of carbon dioxide, nitrogen, water and oxygen.

In the calculations of the non-equilibrium flow of the CO_2 - N_2 - H_2O - O_2 mixture in the supersonic nozzle, use was made of the kinetic model describing population change of separate oscillating levels of molecules CO_2 and N_2 at processes of *VT*- and *VV*- of exchange. All oscillating levels of a molecule CO_2 with characteristic temperatures $\theta < 5000^0$K were taken into account. It was supposed also, that the velocity of exchange of oscillating energy to a Fermi-resonance considerably surpasses a velocity of all remaining *VT*- and *VV*- processes.

Thereof, the modeling system of levels was introduced, some of which have a large degree of degeneration and correspond to several levels in a full system of levels of a molecule CO_2.

As a result of calculations the population inversion and gain coefficient of the nozzle unit outlet was determined.

The calculations of specific output power of laser radiation were conducted at following input data.

1. The working mixture of gas dynamic laser is formed at combustion hydrocarbon fuel in air. The stagnation pressure of a flow is equal 2 MPa, total temperature is equal 1600^0K.
2. The nozzle unit of the installation consists of flat contoured nozzles with the height of critical cross-section 0.8 mm and expansion ratio 25.
3. The sizes of exit cross-section of the nozzle unit: length - 1600 mm, height - 130 mm.
4. The size of a cavity of the resonator along the flow direction - 200 mm.
5. The three pass unstable resonator with a multiplication factor

$M = 1.4$. A reflection coefficient of mirrors $r = 0.98$, the losses on dissipation make $\beta = 0.03\ m^{-1}$.

The contour of a supersonic part of the nozzle was obtained as follows. The flow of gas with an isentropic exponent equal 1.35 in the contoured nozzle with an angular point was calculated and as a contour of a supersonic part of the nozzle the form of a streamline $\psi = 0.9$ was used.

The calculation results for a case of dry air are listed in Table 3.

Table 3. Calculation results for a case of dry air

Fuel	k_v, m^{-1}	w_{max}, kJ/kg	w_{out}, kJ/kg
Benzole	0.41	23.6	9.6
Toluene	0.39	23.1	9.2
Kerosene	0.32	20.8	7.4

Thus, value of specific output power of laser radiation, which it is possible to expect at use as fuel benzole and toluene, and as an oxidizer and ballast gas - dry air, is ≈ 9 9.5 kJ/kg. At use of kerosene this value is approximately 2 kJ/ kg less.

By selection of fuel, except for power efficiency it is necessary to take into account production and toxic characteristics, and cost of components.

The application of benzole is inconvenient at low environment temperature. The freezing temperature is 5.5°C. For toluene this value is equal -80°C, and for kerosene -38°C.

The limit for room vapor concentration of researched fuel have following values: 5 mg/m^3 for benzole, 50 mg/m^3 for toluene, 300 mg/m^3 for kerosene.

From said it is possible to conclude, that the application of benzole is unwise, as it a little surpasses toluene according to power efficiency, but is significantly worse for toxic than toluene and is less convenient in operation.

As to selection between toluene and kerosene, the application of toluene provides specific power increase approximately 25 %. However use of kerosene has a number of advantages. At first, kerosene is approximately 3 times cheaper than toluene. Secondly, kerosene is fuel for airplane and engines. Therefore at the arrangement of the laser installation working on kerosene on a flight vehicle the additional system for fuel storage is not required.

As conclusion for this chapter it is possible to note the following.

The homogeneous laser working on combustion hydrocarbon fuel products is most reasonable for development of the mobile installation.

The specific requirements to the mobile laser installation, impose the limitations on use this or that fuel, oxidizer and ballast gas. The main limiting factor is the required long duration of continuous operation. The carried out estimations have shown, that summarized weight of fuel, oxidizer and ballast gas necessary for GDL operation, is very hight and is unacceptable for arrangement on a flight vehicle. At the same time, the main part of weight of fuel components used for GDL operation is the weight of an oxidizer and a ballast gas. In this connection, use of free air compressed by the aviation compressor, as an oxidizer and a ballast gas is the only way to organized operation of this installation.

The carried out comparative analysis of various fuel has shown, that most reasonable among them are toluene and kerosene. The application of toluene provides specific power of the installation approximately 25 % higher than that of kerosene. However, use of kerosene has a number of the advantages. At first, kerosene is approximately 3 times cheaper than toluene. Secondly, kerosene is the fuel for airplane and helicopter. Therefore at arrangement of the kerosene operating installation, on a flight vehicle the additional system for storage fuel is not required.

3.3. Selection of the powerinstallation for GDL

3.3.1. Selection of a schematic of the power installation

In the previous chapter devoted to the computational substantiation of the basic power characteristics, the capability of development of the mobile CO_2 - GDL with use of kerosene as fuel, and as an oxidizer and a ballast gas - atmospheric air supplied by the compressor is shown.

Besides, some GDL parameters, permitting to determine the required characteristics of the compressor are added.

In particular, the calculations have shown, that in the GDL the condition:

$$P_o / P_a = 20$$

should be realized, where P_o is the stagnation pressure of a working mix(mixture) in the GDL channel and P_a is the atmospheric pressure.

In view of channel losses from the compressor to the GDL inlet of the chosen compressor should have the pressure ratio not less than 22 - 23.

Temperature of inlet air in the GDL combustion chamber chosen at calculations is 720^0K, that corresponds the temperature range of compressed air for compressors with $\Pi c > 22 \div 23$.

For an estimation of necessary value of airflow rate through GDL it is rational to set specific energy extraction at known accepted power range of 100 kW. As was shown in the previous chapter, the GDL specific power with combustion of kerosene in air is 7,5 kJ/kg.

However this value can be reached at dry air only. At increase of humidity of atmospheric air the specific power begins to decrease and at humidity value 100 % (T = 25°C) it is 5,4 kJ/kg.

To achieve the higher specific power it is possible in principle to include air dryer in an air flow path; however, its large overall dimensions and power consumption can result in technical and operational problems and, besides the automatic control of dryer operation is necessary if air humidity and fuel have changed. It will be more conveniently to have certain losses of specific power at humid air, having compensated losses by increased air flow.

Therefore, having accepted average specific power of 6 kJ/kg, instead of 7,5 kJ/kg we shall receive an estimated value of an air flow - 16,25 kg / s.

Thus, for a operation of the laser installation in power range of 100 kW it is necessary to have the compressor with an air flow rate ≈ 16 kg / s and Пс >22÷23.

In essence task of air compression at flow rate of 16,25 kg /s and pressure in access of 20 atm does not contain unsolved technological problems, since compressor engineering is widely applied in industry. The task consists to find the optimal solution ensuring the reasonable cost of the installation, reliability of its activity and convenience in operation.

Let's consider three possible schematics of power installation.

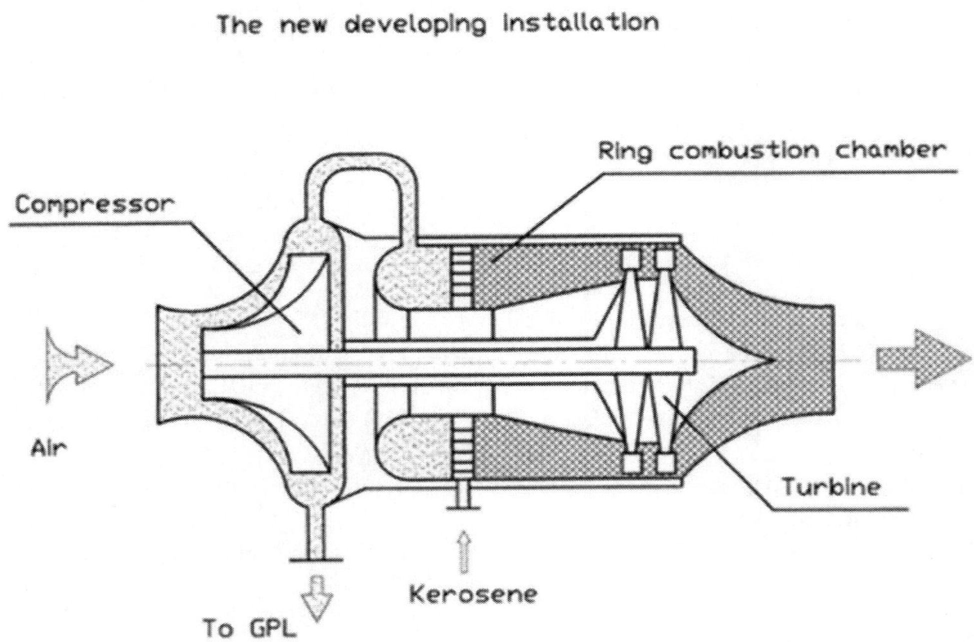

Figure 7.5. The new developing installation.

3.3.1.1. Development special power installation for the above mentioned parameters of the compressor

In this case the installation will consist (Figure 7.5) of the compressor itself, turbine for driving, gasgenerator for rotation the turbine, and also fuel supply system, control and regulation units.

As there is no installation for the mentioned parameters in industry, it is necessary to develop such installation from very beginning, to conduct preproduction activity and to find a producer. For all reason, the price of the produced installation will be high, and development - time consuming.

Besides, short order book for such installations (single production) will lead initially to unprofitable operation for any producer.

3.3.1.2. The installation of air compressor on aviation engine shaft

In general approach, the task can be decided rather simple. On the shaft of the aviation gas-turbine engine the compressor with the required characteristics is installed instead of the propeller.

The calculations show, that the shaft horsepower about 5,5-8 MW is necessary for compressor driving with the flow rate of 16 kg / s and Pк 23 kg / cm^2. It is power of the aviation gas-turbine engine of middle power class, a lot of which is in series production.

However, as engineering realization the task has a number of serious problems.

The nominal speed of the propeller shaft of existing aviation engines is in 5-8 thousand rpm range, and all high efficiency centrifugal compressors require operation speed of the order above.

For example, TB7-117 industry compressor being far from the best in efficiency already requires 30000 rpm.

The installation between the gas-turbine engine and compressor of the high-speed gear reducer (multiplicator) because of complexity in operation and large overall dimensions excludes such solution as acceptable.

Technically task is solved by the installation on the shaft of the engine of an axial multistage compressor (Figure 7.6).

But even the principal diagram general view demonstrates the technical irrationality of such solution, it practically two aero-engines, one of which a production one, and the second requires large design and manufacturing costs connected refurbishment for realization of rigidity of the shaft, organization of turbine side exhaust etc. Besides the overall dimensions and accordingly of weight of the installation are considerably increased.

3.3.1.3. Air bleed from aero-engine compressor final stage

As it is visible from the scheme Figure 7.7, if it is decided to take an air bleed for GDL after a final stage of the engine compressor, practically there is no necessity redesign the power installation with installing new aggregates. Moreover, the dismantling from the engine of the free running turbine operating for driving the propeller shaft for the reason of propeller nonuse, considerably simplifies a design of the installation as a whole.

This solution bribes by the simplicity, since the engines not only will be used with minimum modification, but also all control systems are kept, including the use of the control panel which is taken off from a flight vehicle.

Besides, from point of view of the power, dismantling of the free running turbine together with nozzle block allows to lower required power of the turbine of system of the compressor - free running turbine.

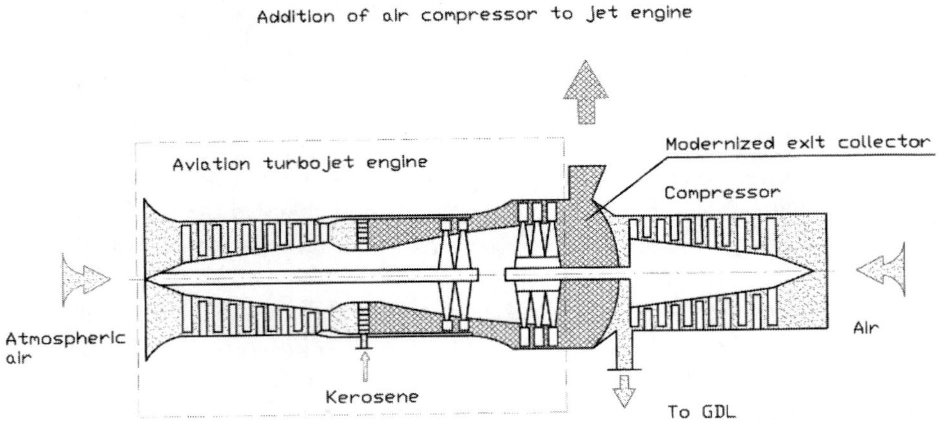

Figure 7.6. Addition of air compressor to jet engine.

The decrease of required power of the system offers the possibility to reduce quantity of required air and kerosene in the combustion chamber of the gas-turbine engine. With constant power rate of the turbine for the drive of the compressor and of the compressor flow rate, probably, it exist the possibility to take off a part of the air and to direct it to the combustion chamber of gas dynamic laser. At the same time, utilizing the characteristics of the combustion products down the main turbine (ρ, m, T), it is possible to organize an ejection of GDL gases down the supersonic diffuser for improvement of the gas dynamic laser overall performance, especially at start.

3.3.2. Selection of the aviation gas-turbine engine

With the purpose of optimal selection of an aviation-engine for this task it is necessary to formulate the initial requirements.

1. For obtaining air pressure in the GDL inlet not less than 20 atm and in view of losses in a flow channel the compressor of the engine Пc should be in range of 22-25, not less.
2. The GDL inlet air flow rate is 16,25 kg /s.

As for elimination of extreme engine power reduction the air bleeding from the compressor should not exceed 15-18 %, the total air flow rate through the engine should be not less than 100 kg /s.

3. The chosen aviation engine should widely be used by standard aviation operation, that enables to utilize engines which are decommissioned from flight vehicles, having passed prescribed operational life limit, but still are operable for ground installation. To buy them is much cheaper, than to buy new engines.
4. It is desirable to make a choice of domestic (Russian made) engines that enables in case of realization of the project to connect to activities, both engine developing company and manufacturer.

The Table 4 data of some Russian engines characteristics: Пс and G of airflow rate (kg / s) are given below.

Table 4. Пс and G of airflow rate (kg / s)

Engine	NK-12ST	NK-14ST	NK-14E	NK-16ST	NK-17	NK-36ST	PS-90A	PD-33
Пс	8,8	9,5	9,5	9,68	9,68	23,12	19,6	22,0
G (кg/s)	56,0	37,1	39,0	102,0	102,0	101,4	56,0	80,0

$$\lambda a = \frac{Wa}{a_*}$$

The analysis of the table results in a conclusion, that most reasonable for the laser installation could become the engine NK-36ST serially produced by Samara plant «Motorostroitel», which satisfies to all requirements presented above. It is obvious, this is not the single engine, which can be put to use for this task, but in advantage of it would be desirable to note the following.

The high performance engine NK-36ST of the Samara Technological complex "Motorostroitel," is developed in 1990 on the basis of an aero-engine NK-321 and is designed for the drive of the centrifugal supercharger in a structure gas pipeline pumping aggregate.

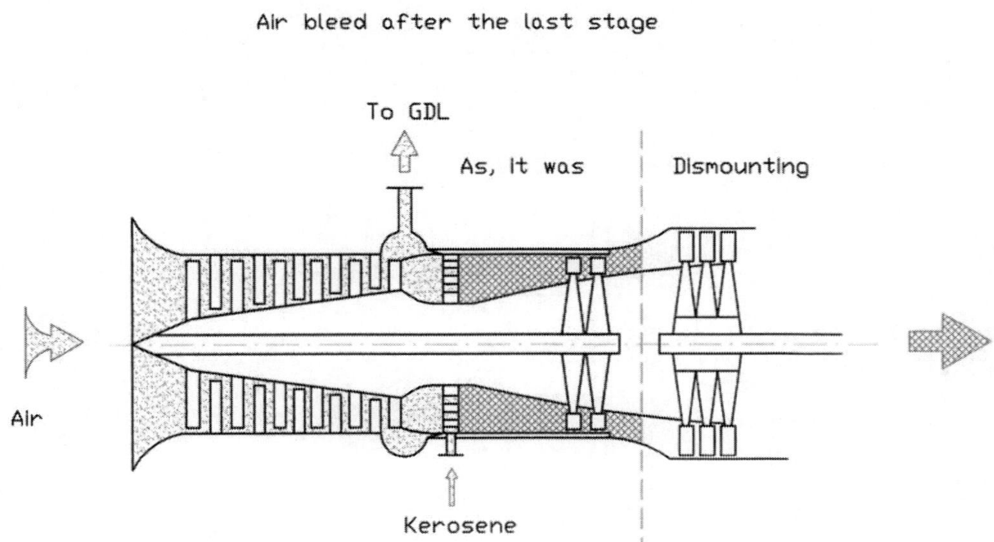

Figure 7.7. Air bleed after the last stage.

The modular design of the engine facilitates transport and assembly. The engine has the remote control panel and record of long life ground operation.

The general view of the engine is shown in Figure 7.8. In Figure 7.9 the design concept is shown.

Figure 7.8. General view of the engine.

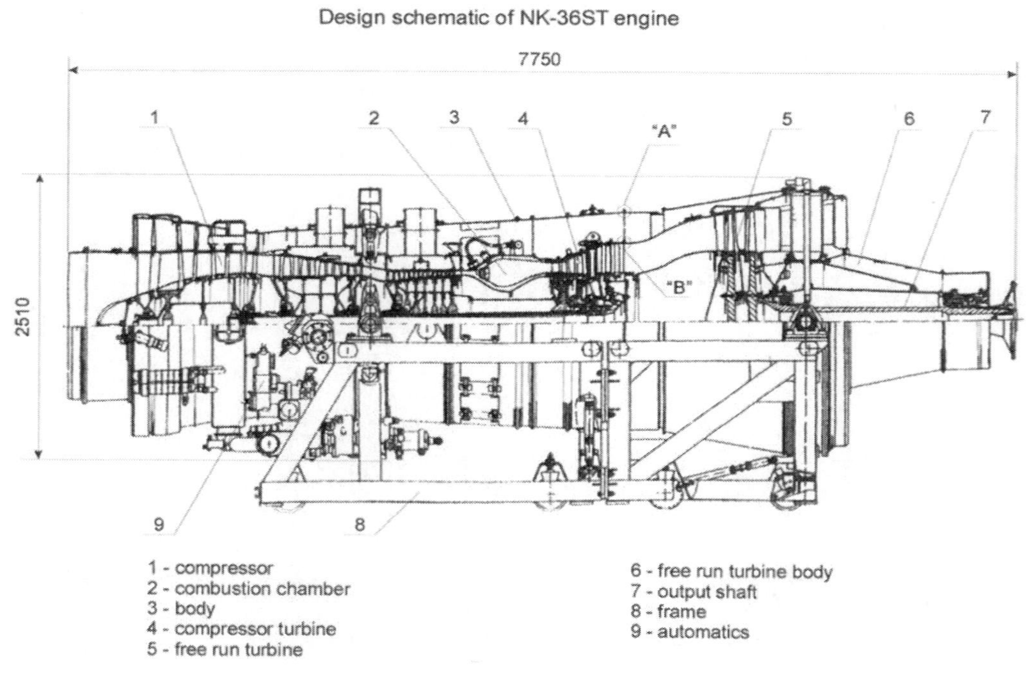

Figure 7.9. Design schematic of NK-36ST engine.

3.3. Gas dynamic assessment of the chosen aero-engine as the power unit for mobile GDL

In the previous chapter the task of GDL supply with oxidizer (air) by use of the NK-36ST engine compressor was considered. However one more function - use of engine exhaust gas as working medium for ejection of an GDL exhaust system is planed to the chosen engine.

The practice of development GDL with direct exhaust of combustion products in atmosphere has shown, that the aft diffuser of the laser produced the pressure recovery of gases up to atmospheric pressure requires for start significant the combustion chamber pressure increase (more than 30 atm), or decrease of exit pressure below atmospheric one. It is possible to ensure the last requirement, through installation of an ejector, for which the exhaust gases down the engine turbine are the working medium.

Gas dynamic estimation of the capability of the NK-36ST engine of GDL diffuser start support is shown below.

As it was already indicated, the basic adaptation of the NK-36ST gas-turbine engine is the elimination of design structure of the free running turbine, that allows to improve characteristics from the point of view of possible application in the laser installation, for which it is necessary to bleed air with the consumption $m_в$ = 16,25 kg/sec by pressure P ≈22 kg / cm^2 and temperature T = 720^0K.

Reasonable temperature for a GDL gas flow channel is T = 1600^0K at combustion-chamber pressure of 20 kg / cm^2. The given temperature is realized in the combustion chamber for fuel kerosene - air at mass ratio $K_к$ = 38,5, that is at kerosene flow rate m_n = 0,422 kg / s.

In the GDL working section it is necessary to ensure static temperature of combustion products Ta = 370 ^{0}K, that is realized velocity exhaust of combustion products through supersonic nozzles with velocity Wà = 1700 m/s at a Mach number Ma = 4,46, isoentropy index κ = 1,333, pressure Pa = 0,057 kg / cm^2.

For realization of a shock wave free flow of combustion products in atmosphere it is necessary to utilize the flat supersonic diffuser. One of major parameters of functioning of the diffuser in GDL operation conditions is the start pressure.

The start pressure (Pst) is determined by the intensity of a direct shock wave relation:

$$\frac{Pst.}{P\text{н}} = \frac{g(1/\lambda a)}{g(\lambda a)};$$

where Pн is the pressure behind the diffuser, and $\lambda a = \frac{Wa}{a_*}$ is the velocity factor equal to the ratio of exhaust velocity in working section GDL (Wa) to a critical velocity of a sound (a_*).

The value of a velocity factor λa is related to the Mach number by a ratio

$$\lambda a = \sqrt{\frac{X+1}{2} \cdot \frac{Ma}{\sqrt{1+\frac{X-1}{2}Ma^2}}}$$

and then to a Mach number Ma = 4,46 corresponds $\lambda a = 2{,}32$.

The function $g = \frac{Fcr}{Fa}$ determines the ratio of the geometrical area of critical cross-section of the diffuser to the area of gases exhaust and is according to expression

$$g(\lambda a) = (\frac{X+1}{2})\frac{1}{X-1} \cdot \lambda a(1 - \frac{X-1}{X+1}\lambda a)\frac{1}{X-1}$$

For direct shock wave the following ratio is valid

$$\lambda a \cdot \lambda_{vf} = 1;$$

where λ_{vf} is the dimensionless velocity factor after front of a direct shock wave

Thus, the ratio of pressure of start to pressure behind the diffuser is equal to

$$\frac{Pst}{P\text{н}} = 13{,}7.$$

The wind tunnel pressure of start is usually above their steady state operating pressure up to 10-30 %.

At the same time, in GDL laser conditions, when there is a involved gas volume in a working section (region of resonator mirrors) pressure of start of the diffuser can is still increasing to the value, at which the diffuser could not started.

Hence, it is necessary to lower exhaust pressure from the diffuser, that is possible to realized through an ejection of a gas jet, flowing out from the diffuser, by a gas jet flowing out after the turbine of the engine (Figure 7.10).

The designs of similar ejectors with maximum pressure recovery are known at high speeds of flow of gas in minimum cross-section. Such ejectors designs are characterized by smooth profile of an inlet part and with small cone angles (L) on an exit, which make value L = 6 – 10 degree.

The pressure recovery coefficient for such ejectors can reach value

$$\frac{\text{Pexit}}{\text{Pinlet}} = 0{,}955 \div 0{,}965 \text{ and higher.}$$

Thus the velocity factor value in minimum cross-section should is

$\lambda = 0{,}75 - 0{,}85.$

Thus such ejector inlet pressure of should be Pinlet. = 1,076 kg / cm^2 (Pexit. = 1,033 kg / cm^2), that can be provided with the gas-turbine engine without the free running turbine.

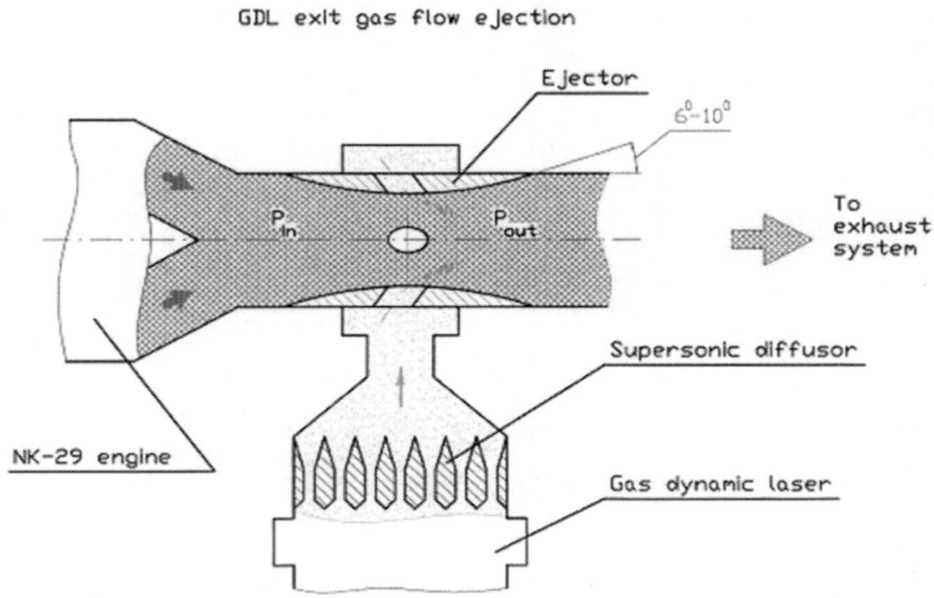

Figure 7.10. GDL exit gas flow ejection.

At a velocity factor $\lambda = 0{,}8$ in minimum cross-section and ejector inlet pressure Pinlet. = 1,076kg / cm^2 the static pressure in minimum cross-section becomes equal Pmin =0,73kg / cm^2, that allows to organize introduction of flowing out from the GDL diffuser gas in area of minimum cross-section of the exhaust device of the gas-turbine engine and by that to lower pressure behind the diffuser GDL at least down to Pexit = 0,8 instead of 1,033.

Such reduction of pressure will allow to lower pressure of start GDL down to a reasonable level of 18-20 atm.

3.3.4. Design solution for GDL power unit on the basis of the gas-turbine engine

In view of the technical plan generated by the authors of the present technical proposal, experts of the Samara complex «Motorostroitel», has developed on the basis of the NK-36ST gas-turbine engine the engine for a laser complex denoted as NK-29.

The NK-29 engine (Figure 7.11) represents the gas-turbine engine obtained as a result of adaptation of the production engine NK-36ST, which consists in following:

Dismantling of the free running turbine with the output shaft.

The dismantling is made through the joint "A" flange of a engine body (see Figure 7.9), through the joint "B" of the hot gas pipe down the turbine of the compressor and finally the free running turbine together with a turbine casing is removed from the engine. As the frame, on which the engine is mounted, consists of two not connected permanently sections, the removal of the second, turbine side section the does not represent complexity.

For organization of the smooth gas flow from the modified engine, and also for a fixing of the exhaust system to the engine, two conical adapters (casing) and internal fairing made of a stainless steel sheet are added.

The external adapter is terminated by a flange to be fixed with the engine ejector. The adapters and fairing were designed by «Motorostroitel», supplier in accordance to the aerodynamic characteristics of a gas stream and the strength requirements.

Ejector, which scheme is shown in Figure 7.10, is not the engine item. The ejector design is calculated and finally developed on the basis of engine, and laser complex exhaust system parameters and GDL exit characteristics (down the supersonic diffuser).

Ejector design represents a pipe from high-temperature steel having symmetrical narrowing in the middle part ahead of critical cross-section. The pipe can be manufactured from three parts with flange or welded connections.

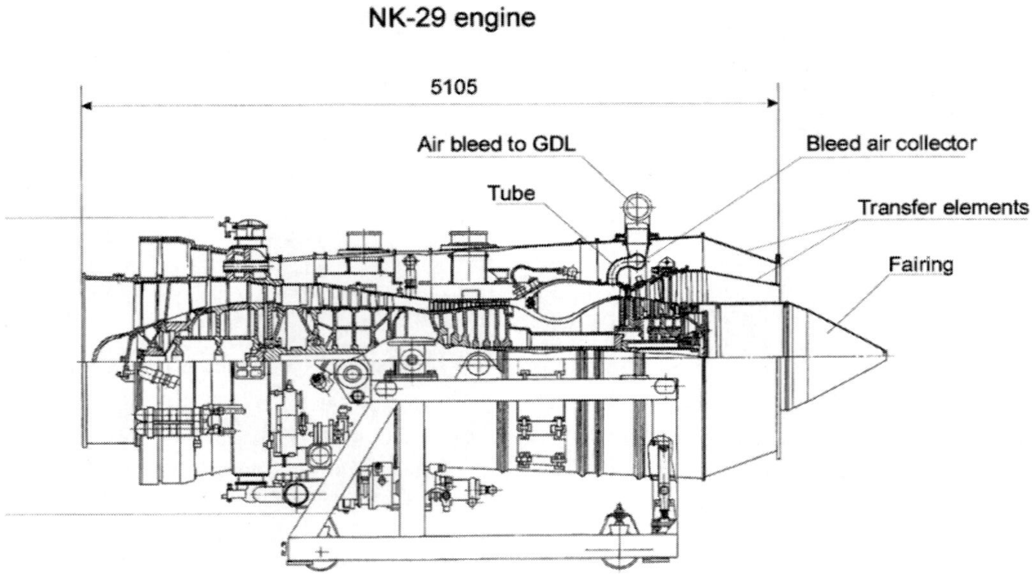

Figure 7.11. NK-29 engine.

The middle narrowing part can be turned from a thick-walled billet or from a rolled ring and has mating collars for different side collector for convenience of assembly for welding. The orifices connecting cavities of exhausts of the engine and GDL are located around the perimeter of critical cross-section. The collector is manufactured by stamping of a sheet, has the welded branch pipe and serves for uniform distribution of gas after GDL on all cross-section of an ejector. Overall dimensions of an ejector and final design will be submitted after fulfillment of general layout of laser complex in the final report.

The adaptation of the engine with removal of the free running turbine considerably reduces overall dimensions and weight of the installation. For a clarity of estimation of advantage in overall dimension reduction of the modified engine the presentation on Figs. 7.9 and 7.11 is made in the same scale.

Air bleeding to GDL

The next considered problem is the air bleeding after the compressor to the GDL

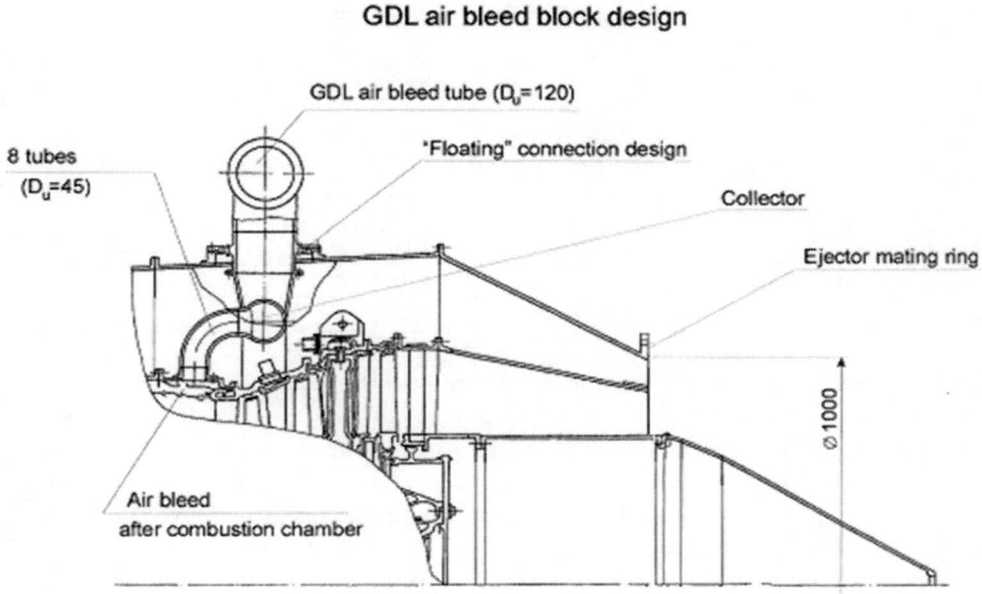

Figure 7.12. GDL air bleed block design.

During searching of the engineering solution for the engine adaptation the aviation specialists has offered to realize the air bleeding from the exit of the combustion chamber second contour, not from the compressor final stage exit, i.e. from a channel of a cooling flow of the combustion chamber. This solution is connected to the fact, that combustion chamber of the gas-turbine engine is one of the most thermal stressed items of the engine, and the cooling parameters are results of long development stage. Therefore the air bleed behind the compressor, reducing the air flow of the combustion chamber cooling flow, will change steady state thermal status, that can result in undesirable consequences for engine operation.

The collector ring for air bleeding with a branch pipe to the GDL supply was recommended to position in a zone of the exit of cooling flow of the combustion chamber, and for that the chamber and collector are connected by angular branch pipes (Figure 7.12).

The angular branch pipes, 8 pieces in quantity, can be manufactured of a pipe or stamped sheet from two halves and have leak tight flanges to cooperate with the envelope of chamber cooling flow channel. The angular branch pipes are fixed by welding to the collector ring, which has the terminal flanged tap. To this tap the branch pipe to supply air to the GDL is fixed. This branch pipe passes through an outer shell of a engine body and has "floating" attachment to it excluding deformation of designs at a thermal expansion of contacting elements.

It exists the experience of such bleed schema, however particular design should pass a number of development tests with the purpose of experimental selection of channel pressure reduction to ensure calculated bleed air flow rate.

All other elements of the engine remain without change, including all start and cutoff automatics and operation control.

Thus NK-29 engine, obtained as a result of adaptation of the NK-36ST engine will require minimum development, and manufacturing costs.

The basic performances of the engine NK-29 are below adduced.

Fuel	- Kerosene TC-1
Power, MW	- 25
Air flow, kg / s	- 107,2
Compressor pressure ratio, Пс	- 23,12
CC gas temperature, ^{0}K	- 1420
Gas exit temperature, ^{0}K	- 698
Propellant consumption, kg / s	- 1,434
Engine weight, with frames, kg	~- 5000

All parameters of the engine are indicated for a nominal mode.

The NK-29 engine is single mode, as by activity of the laser installation is required neither decrease, nor increase of the power, that provides reliability and life value growth.

The NK-29 engine should be manufactured by aviation plant irrespective of, whether it is refurbished through adaptation of the NK-36ST engine or again assembled with use of furnished engine parts from production engines.

Thus, as a result of the carried out research work not only the optimistic results for possibility of adaptation of the gas-turbine engine for GDL gas dynamic complex were obtained, but also design activities by definition of general realization of such engine are executed in practice. And, what is very valuable, this activity is executed by the experts of aero-engines developing - producing company.

3.4. The mobile laser installation

3.4. 1. The pneumahydraulic scheme (PHS) of the laser installation

PHS of the laser was developed from conditions:

1. Assurance of required calculated parameters in all PHS elements;
2. Automatic control of physical processes;
3. Use of minimum set of commands and control units;
4. Design simplicity of the scheme and safety in operation.

PHS is shown in a Figure 7.13. It includes directly gas channel - gasgenerator (GG), nozzle unit (NU), working section, supersonic diffuser and exhaust line.

Besides, the GDL air-supply system, consisting from supply line, air valve and disk pressure reducer; the fuel supply system including pipelines, fuel pump, valves and adjusting flow restrictors are included into PHS. To PHS the aerodynamic window also included.

Shown on the scheme the optical resonator and protective shutters are not PHS components, but their functions are interconnected and they participate in operation process.

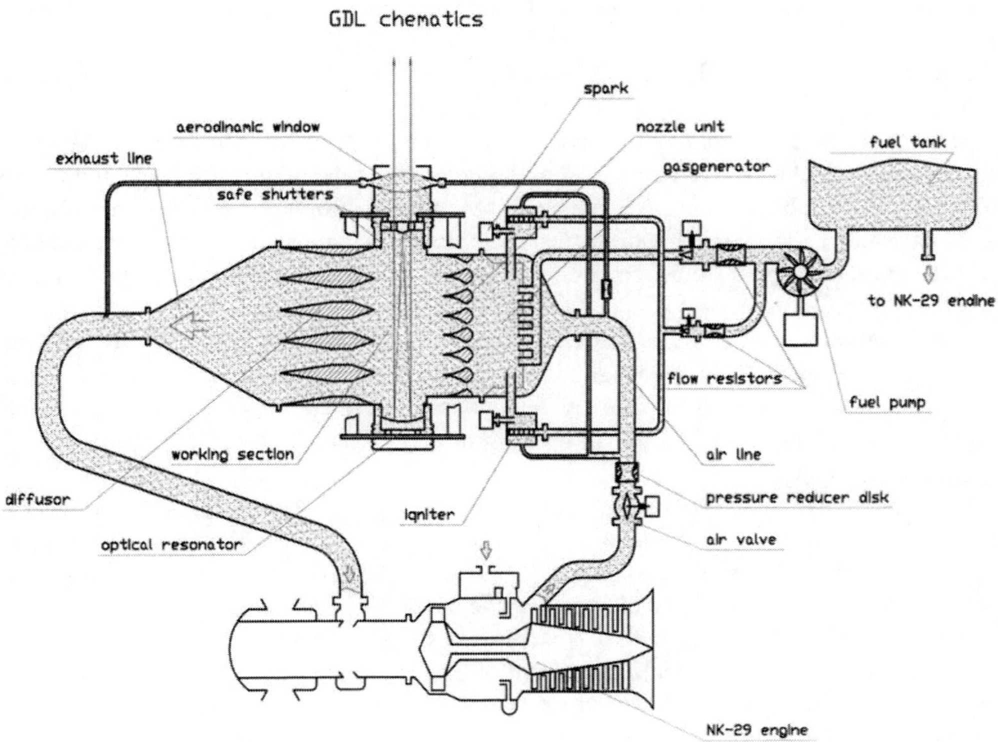

Figure 7.13. GDL schematics.

For automatic control of all cycle of activity GDL (start, steady state mode and shut down) and parameter monitoring, PHS is equipped with the measuring equipment producing signals through electronic units on functional control units of an automatics and on the operator console.

The measuring and control wiring schematics are not added in the report, as they do not require special development for the reason of ordinary professional level.

The required electrical voltage is 27 volts and power source - the electric generator of the aviation engine.

Start of the laser installation

The PHS element operation begins with opening of the air valve with pressing by the operator on the button "start" at aviation engine (further engine) being in operation status.

The valve opens for delivery half of air flow and hot air is supplied in the gasgenerator and igniters.

The command to open valve is a start command for GDL operation, all following commands to operation of elements of the scheme are made automatically, without interference of the operator.

As compressed air pressure is increased to 10 kg / cm^2 the pressure transducer gives a signal, according to which the electrical voltage is supplied to spark plugs of spark igniters and for the electric drive of a fuel pump GG. The valves of GG and spark igniter fuel supply are closed.

The lead of voltage application to spark igniter before fuel supply operation excludes a capability explosive combustion during start process.

At 20 kg / cm^2 fuel pump exit pressure, the command is given for opening spark igniter fuel supply valve. The high pressure fuel supply guarantees the high quality fuel atomization in injector, thus reliable ignition. Igniters begin functioning, forming "forced" plumes directed against each other. The igniter plumes heats up sensing elements of temperature detectors installed in bodies of prechambers and on reaching of 950°K temperature the signal is given on opening of the main fuel valve.

Fuel is supplied to the GG injector first row and being dispersed forms with air a fine mixture, which is ignited by spark igniter plumes. The complete burning of fuel has taken place on the injector second row. The GG temperature increases and at reaching value 1050-1100°K the commands are given to full opening of the air valve and valve of air delivery in the aerodynamic window.

The generator is forced to a steady state mode through idle transfer mode. The working gas is accelerated in the nozzle block up to a velocity M = 4, and after the working section (the resonator zone) is decelerated in the supersonic diffuser and downstream through an exhaust collector leaves the gas channel of the laser. The gas dynamic channel flow is stabilized, providing along all length of a working section static pressure equal 0,12 kg / cm^2. The vacuum sensors give a signal on opening of protective shutters and spark igniter cutoff (closing of the fuel valve and deenergizing spark igniter plugs). The air flow through switched off igniters protects the igniter plugs from an overheating during all operation cycle.

The opening of mirrors by shutters the optical resonator begins to form in an active medium of a working section an electromagnetic laser field with extraction of radiation. The GDL is functioning.

The GDL steady state mode provide a constant pressure of gas at the engine compressor exit and that at fuel pump exit, as well as constant hydraulic resistance of air pressure reducing disks and fuel flow restrictor.

The GDL is equipped with a system of emergency protection, which on an unacceptable falling or overshot GG temperature closes the fuel valve instantly, switches off a fuel pump and forms a command on the engine shutoff.

The GDL emergency shutoff can be executed by operator pressing the button "emergency shutoff", with similar command sequence on actuators.

The normal (nominal) cutoff of the laser installation happens in the following order:

By operator pressing the button "shutdown", protective shutters and the GG fuel supply valve are closed, and the fuel pump is switched off. The supply of the GG fuel ceases and during 7 ÷ 10 seconds gas channel purging by pure air has taken place, then the command on closing of the air valve is given and the engine is shutdown.

The GDL cycle sequence of start and shutdown is shown in Figure 7.14.

As it can be seen according the cyclogramme, each next command on functioning of the aggregate is given after fixing the results from the previous command. For this reason, at a presentation of these results on the display of the operator it is possible to control the start sequence manually, having disconnected an automatics of control.

The ranges on time between commands and their fulfilment are determined at designer operational development GDL.

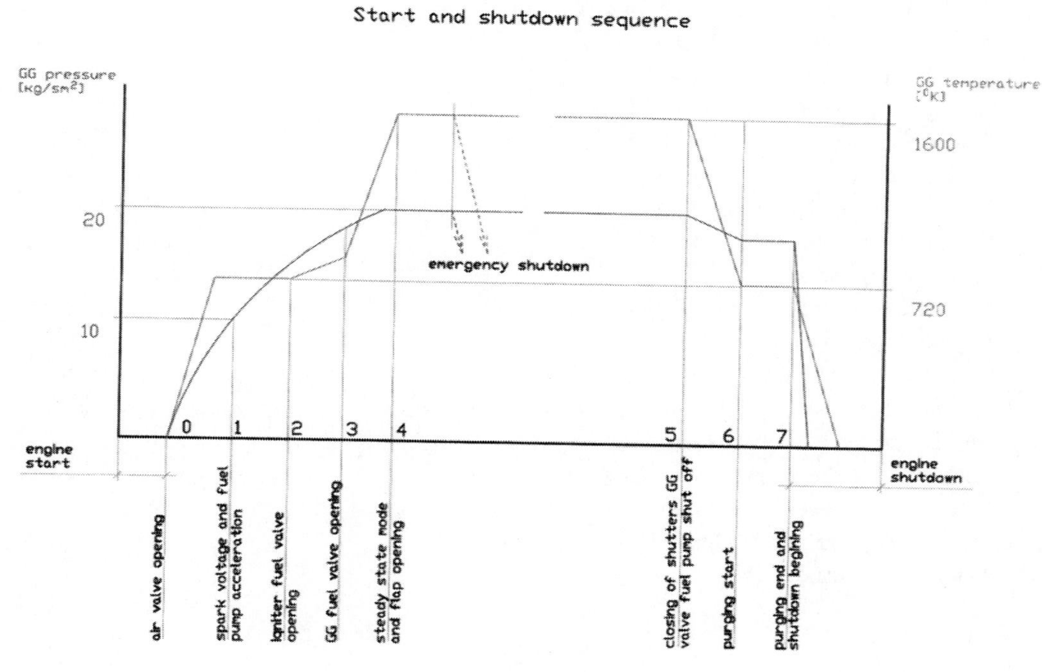

Figure 7.14. Start and shutdown sequence.

3.4.2. GDL –general design features

On the basis of computational examinations presented in the previous reports, the GDL general design features is generated according to recommended pneumohydraulic scheme.

At GDL designing authors tried to decide a task of the greatest possible simplification of a design with the purpose to cut cost (in general GDL is in category of expensive products). For this purpose, firstly the single mode scheme of the laser excluding necessity to refurnish of the laser with a complex control system, was chosen. Secondly, the completely uncooled design despite on a rather high operation temperature of gases is offered.

The last solution is stipulated by the following

1. Absence mobile and the rotating parts in GDL design – there is no necessity to use greasing materials;
2. Absence of a strict limitation on weight, and, hence, on thickness of walls and weight of details;
3. Capability to limit the duration of continuous operation;
4. Presence on the market of heat resisting structural materials, beginning from special alloys to structure ceramics.

The GDL general view is shown on Figure 7.15.

GDL units are the following:

1. Gasgenerator (GG) with spark igniters.
2. Nozzle unit (NU) with a working section (gas part of the resonator).

3. Supersonic diffuser.
4. Optical resonator with mirrors.
5. Chamber of the resonator.
6. Aerodynamic window.
7. Fuel pump.
8. Air valve.
9. Fuel valve (2 items).
10. Component supply units.
11. Attachment units.
12. Measurement system sensors.
13. Cable configuration.

For the base sizes determining all remaining overall critical dimensions of main assemblies, the sizes of cross section of the nozzle unit were chosen.

According at the summarized flow rate of working gas 16,647 kg/ s (air - 16,25 kg / s, kerosene - 0,422 kg / s), temperature - 1600 ^{o}K and pressure of 20 kg / cm^2 we obtain the area of critical cross-section F* = 8524 mm^2.

At adopted nozzle area ratio

F/F* = 25
F = 213100 mm^2.

Ratio between the height of a nozzle blade and width of the nozzle unit (across a flow) was chosen from experience of improvement of gas dynamic lasers and rational relationship between diameter of mirrors and distance between them.

In opinion of the authors, optimal values for the obtained area are:

height of a blade H = 130 mm;
width of the nozzle unit L = 1600 mm.

These sizes determine overall dimensions of the generator of gases and supersonic diffuser coupled with the nozzle unit, and, hence all GDL design coordination.

All mating places of aggregates are made as detachable flanged coupling with seals, that provides GDL high maintainability at operation.

In following chapter the design features of main assemblies will be described more in detail which determine GDL specificity and require design-technology improvement.

These are:
The generator of gases with igniter;
The nozzle unit with a working section;
The supersonic diffuser;
The aerodynamic window;
The optical resonator.

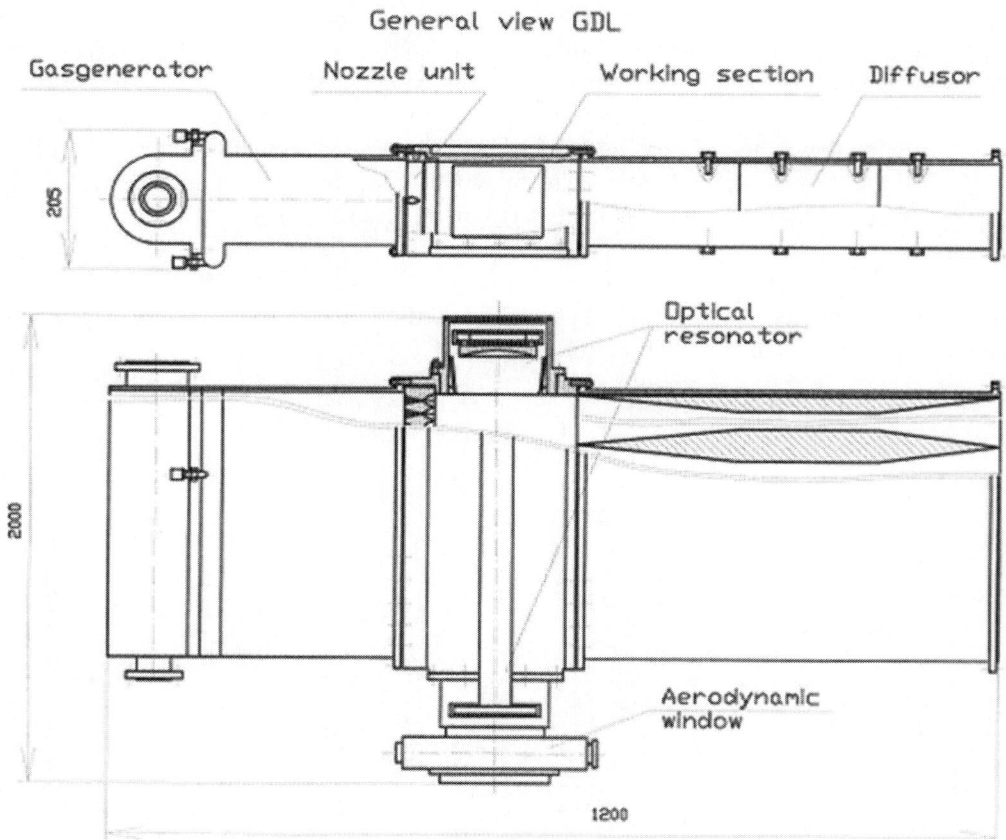

Figure 7.15. General view GDL.

All remaining elements of the scheme GDL do not represent engineering complexity and do not require special designer studies. In general, these elements will be utilized as serially produced by other industry branches (fuel pump, valves, air valve, the measuring equipment, the electric drives) or those are projected the elementary assemblies, which design is obvious and is determined by their assigning.

Those are:

1. Propellant component supply units - pipelines of a necessary configuration and diameter, branch pipes, flow rate controlling orifices and pressure reducers
2. Attachment units – general frame, mounting brackets of aggregates, plates for attachment of sensors etc.
3. Resonator chamber - usual casings shielding the mirrors and ensuring air-tightness of the resonator cavity. The chamber is fixed to flanges of the working section and is connected to racks of the resonator through elastic elements. The chamber has removable covers for an optical adjustment of the resonator, one of covers (an output mirror chamber) has a window docked with the gas dynamic window.
4. Protective shutters - (Figure 7.16) two-folding type rotated around of axis for opening and closing, located between windows of a working section and resonator

chambers. Basically shutters can be of any design and can be fasten or on a flange of a working section, or on a flange of the resonator chamber. In this case their attachment to flanges of resonator chambers is supposed.

5. Adapter - part of a gas line for flow transfer and modification from rectangular cross-section down the diffuser to round cross-section for mating with GDL exhaust pipe.
6. Multicore wire cable having wire for control and measurement transmission. The cable has thermal insulation and is fixed on the installation.

The GDL is assembled on the frame, which through rubber anti-vibration mountings is installed on a separate platform. The shape of the platform and fixing to a vehicle places are determined under design condition of this vehicle.

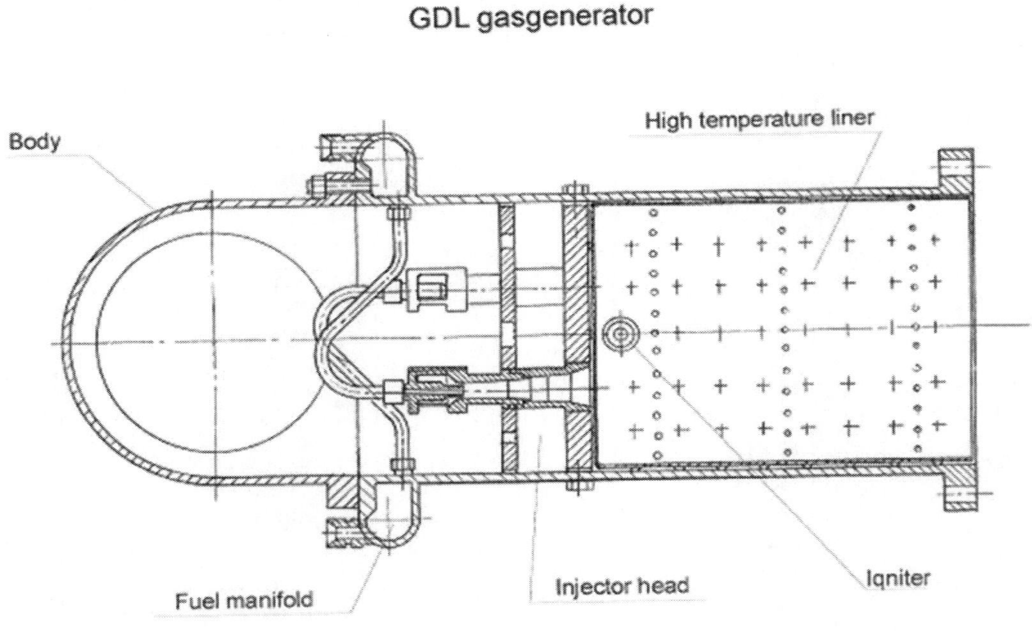

Figure 7.16. GDL gas generator.

3.4.3. Main GDL assemblies

3.4.3.1. Assigning and design

In the given chapter the conceptual approach to designing of GDL main assemblies is considered. The present operation on research of a capability of GDL development for the solution of a specific task did not assumed detail design-technology study of parts, assembles and installation as a whole, however authors made initial studies of main assemblies with the purpose to show the reality of their development, definition of structural and technological peculiarities, and estimation of their dimensional and mass characteristics.

3.4.3.2. Gasgenerator (GG)

The generator of gases is intended for complete (without soot forming) burning of kerosene in hot air for obtaining a working mixture of required of a chemistry, temperature and pressure. For uniform of all mentioned parameter flow at nozzle unit inlet GG is configured of rectangular cross-section identical to cross-section of the nozzle unit.

The philosophy of designing of combustion chambers of aviation gas-turbine engines having good technical tradition and wide experience of operation was included in GG design.

All aviation gas-turbine engines combustion chambers have a body (cold part) and heat jacket (hot part). The entry of air is organized through injectors and orifices in the heat jacket located on all of its surface.

The similar GG design is offered by the authors (see Figure 7.16)

The GG consists of body of the rectangular shape, heat jacket, burner plate, air manifold and two igniters.

The body has two flanges for mating with the nozzle unit and air manifold and inside has mating surface for the heat jacket with a burner plate. On lateral walls of the body there are orifices with flanges for the igniter installation. From above and from below to a body fuel collectors are welded with exit pipe connections directed inside of bodies to feed the injectors.

The burner plate has two walls with the installation sites for 28 injectors located in two rows. The first wall (flow down direction) has orifices for air entry in intermediate space. Air through orifices in injectors reaches the atomization zone.

The pipe connections of injectors are connected to pipe connections of a fuel manifold by tubes with nipple joints. The tubes are bent for compensation of thermal stresses, and their length is increased at the expense of connection of the lower row of injectors with the upper fuel manifold and on the contrary for same purpose.

New GG design feature is the application of supersonic injectors. In Figure 7.17 in1:1 scale the design of such injector is shown. The total number of injectors is 28 with 14 pieces in row. The injector has two mixing zones of fuel with air. The first zone (A) represents a Laval nozzle where air accelerates to the speed of sound with the introduction of fuel in the region of critical cross-section, the second zone (B) used for pressing of fuel drops of from walls of the nozzle and increases mixing efficiency. Air in the second zone enters from intermediate cavity of a burner plate. This injector design is patented and already have found broad application in various power plants. The air flow rate through injectors is 6,25 kg / s, what corresponds $\alpha \approx 1$.

The remaining air flow (10 kg / s) is directed through orifices in the heat jacket. Jacket is made of a thin-walled heat resisting sheet and is welded by spot welding to the burner plate. At calculated value of a residence time of mixture length of GG "hot" part is chosen 180 mm. In a heat jacket three rows of low pressure reduction orifices are made, which diameter will be updated during GG experimental improvement.

The GG gas collector is the adapter from a round flange of inner diameter = 150 mm to GG rectangular cross-section.

Igniters are introduced into a design for increase of reliability of mixture ignition as reliable ignition is problematic at the large flow rates and velocities of gas with direct ignition by spark igniters.

The igniter design is traditional, it consists of body having air inlet, one centrifugal fuel injector, electrical spark for ignition and prechamber with the nozzle.

The pressure difference along the nozzle provides necessary length of a plume in GG.

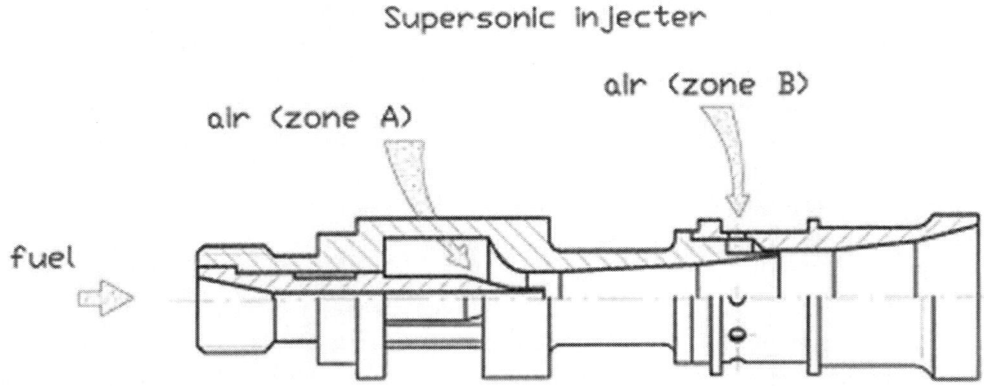

Figure 7.17. Supersonic injector.

On GG body there is the position place for pressure and temperature transducers, and on a prechamber -for temperature detector.

3.4.3.3. Nozzle Unit (NU)

NU provides acceleration of a working mixture up to supersonic velocities (M > 4) for gas temperature reduction down to value 300-350^{0}K, ensuring obtaining, population inversion of laser levels of a molecule CO_2.

In connection with that the working section structurally is the part of NU structure, the NU second function is the maintenance of the supersonic uniform expansion of working mixture in resonator zone.

Structurally NU (Figure 7.18) consists of body, complete set of nozzle blades and elements of attachment of nozzle blades in the body.

The body is made as one unit with a working section, has a mating place for the set of nozzle blades and transfers in working section of length 200 mm. With the purpose of gas dynamic losses reduction on walls, which can result in deceleration of a supersonic flow a top and bottom wall of a working section extend down streamwise at angle + 3^0.

Thus, the cross-section of a flow down blades 130 x 1600 transfers in the flow cross-section in front of the diffuser 142 x 1600. In NU body in a zone of a working section there are places of vacuum sensors installation.

The NU supersonic channels are formed by profiles of two neighbor nozzle blades (Figure 7.19 (a). The sizes of a channel, and, hence, geometry of blades were chosen by results of calculations presented in the previous reports and in view of structurally - technological peculiarities of blades.

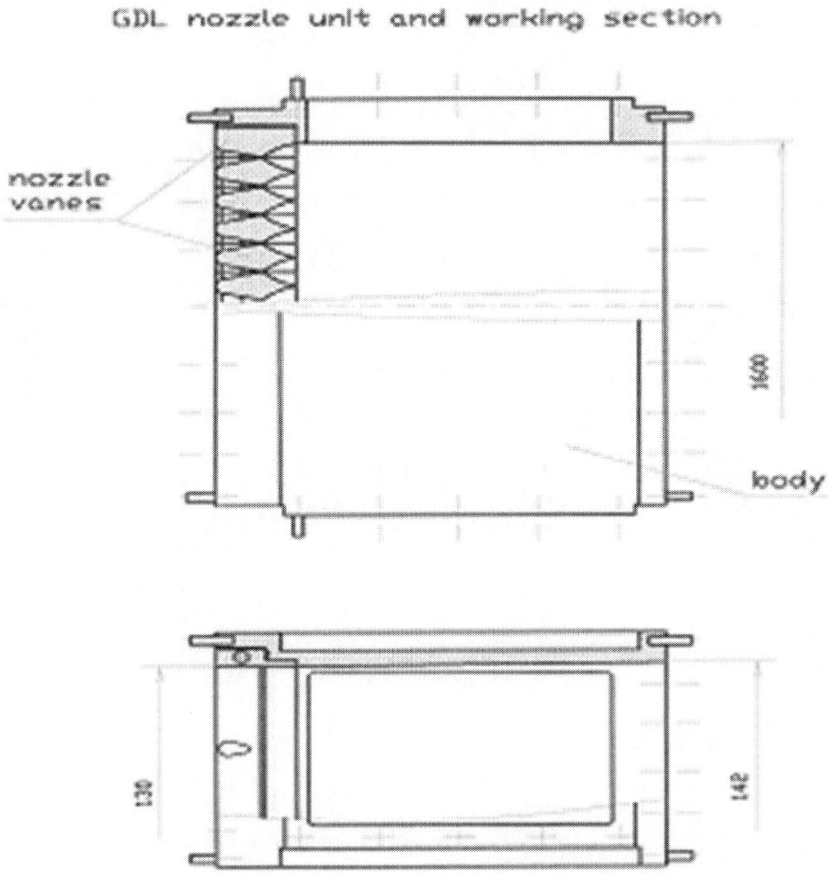

Figure 7.18. GDL nozzle unit and working section.

In particular size of critical cross-section is chosen (according to experience) h_{cr} = 0,8 mm. At the smaller sizes h_{cr} the laser characteristics are improved, however blades technology factors are poorer and the error of manufacturing of blades on the size h_{cr} in the greater measure has an effect on flow non-uniformity of various channels.

The expansion ratio is chosen 25, as is optimal from the earlier presented diagrams.

In this case size of the exit nozzle is 20 mm.

The coordinates of supersonic profile are determined on calculation method of nozzles for liquid rocket engines, the nozzle exit is performed at angle 4° between a center line of a blade and tangent to the profile of the nozzle. It is made for thin wall length reduction at nozzle exit. The obtained length of a supersonic part is 18 mm.

The subsonic part of a nozzle blade is made as of the low flow resistance envelope and has length of 25 mm (for increase of weight and rigidity of a blade). Critical cross-section has a straight line strip of 1 mm width.

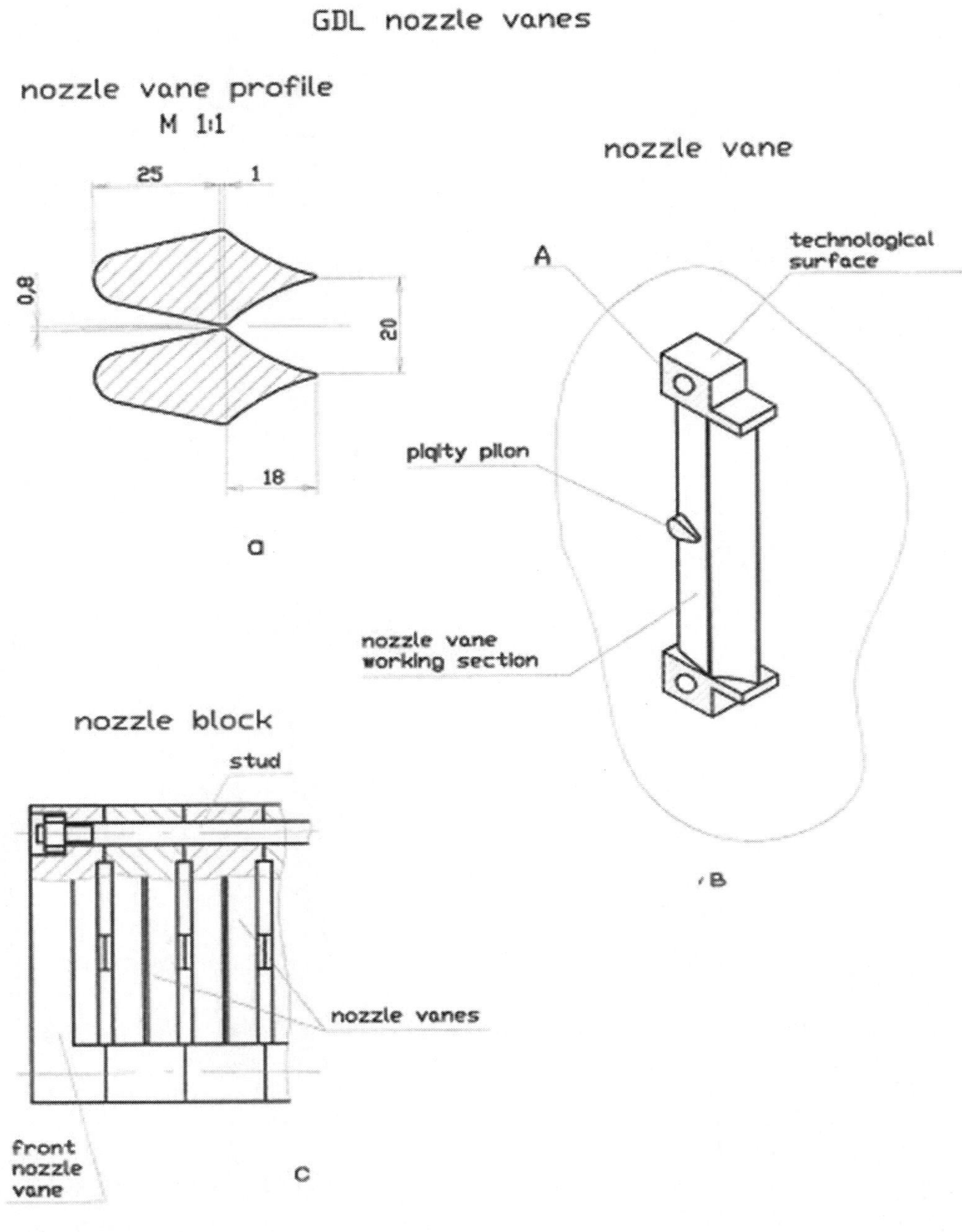

Figure 7.19. GDL nozzle vanes.

Thus, having set the thickness of exit edge of 0,5 mm, we determine all remaining sizes of a profile of cross section of a nozzle blade (see Figure 7.19 c).

Beside, the selection of the thickness of exit edge in many respects depends on the material of a blade and technological capabilities of producer. Basically this thickness should come nearer to zero for elimination of axial wake behind each blade, however acute edge is technologically difficult to maintain along blade height edge, is inclined to destruction at any mechanical loads or oscillatory processes by operation. Therefore thickness of an edge is set in view of mechanical characteristics of the chosen material of a blade.

Nozzle blade as design unit (Figure 7.19) consists of a profile part two technological regiments and pylons for rigidity.

For classical of supersonic expansion nozzle form the critical cross-section of the nozzle should pass in a supersonic part through an acute edge. Thus all acute edges of critical cross-sections of blades should with high accuracy coincide with each other at the of blade assembly process in formation of package (Figure 7.19 c). For this purpose for assembly base orifices the precession orifice type in regiments of the blade is adopted. At blade set assembly with precision studs coincidence is provided of all decisive points of a profile of the neighboring blades. The size of critical cross-section h = 0,8 mm (between a critical point of a profile and regiment plane) is provided through final machining of all surfaces lying in a plane "A" (Figure 7.19 c).

The rigidity pylons exclude stretch of blades, and, hence, change of the size of a profile (especially on critical cross-section) in GDL operating time.

For tight leak connection of a package of blades in NU, surface "Á" of assembled with studs package is machined for obtaining high quality surface.

Quantity of blades in the package - 81 pieces.

3.4.3.4. Supersonic diffuser

The purpose of use is overexpanded supersonic flow pressure recovery to close to atmospheric value through transforming in subsonic flow.

The supersonic diffuser should have an inlet confusor (supersonic flow), critical cross-section (sound flow) and exit part (subsonic flow). In gas dynamics the technique of calculation of axisimmetric diffusers is state of art. For this case diffuser has rectangular cross-section 142 x 1600 mm, therefore it is most rational for reduction maximum size to divide it in a number of channels with the help of internal contoured pylons.

As design (see Figure 7.20) the diffuser represents a body of the box-shaped form, inside which the pylons are installed at a inlet angle 6° and exit angle 8° by screws.

3.4.3.5. Optical resonator

The optical resonator as the structure is perimeter confined frame, mirrors and of the resonator adjustment gear. The design concept of the resonator is shown in Figure 7.21.

The frame consists of two identical lateral metal racks. The rack has a window for transiting a laser beam, a place for attachment of adjustment gear and elastic lip. The elastic lip seals an interface of place of transiting of the rack through the chamber of the resonator and exclude transfer of vibration from working GDL to the optical resonator. The racks are connected among themselves by pipes of 200 mm diameter made from carbon-polymer materiel. It provides at sufficient rigidity, small weight, excludes temperature resonator deformation having practically to zero thermal expansion factor.

For elimination of acoustic resonance inducing effect on pipes they are filled by rubber sponge.

The mirrors of the resonator are made from high purity copper with protection of a working surface by of gold layer. A reflection coefficient of mirrors is 98 %.

One of mirrors (solid piece) has a concave spherical surface. On other, flat, side of a mirror there are places of attachment of adjustment gears through which the mirror is fasten to the rack of the resonator. Along the axis of the mirror the through hole ∅ = 1 mm for resonator adjustment by the helium - neon laser.

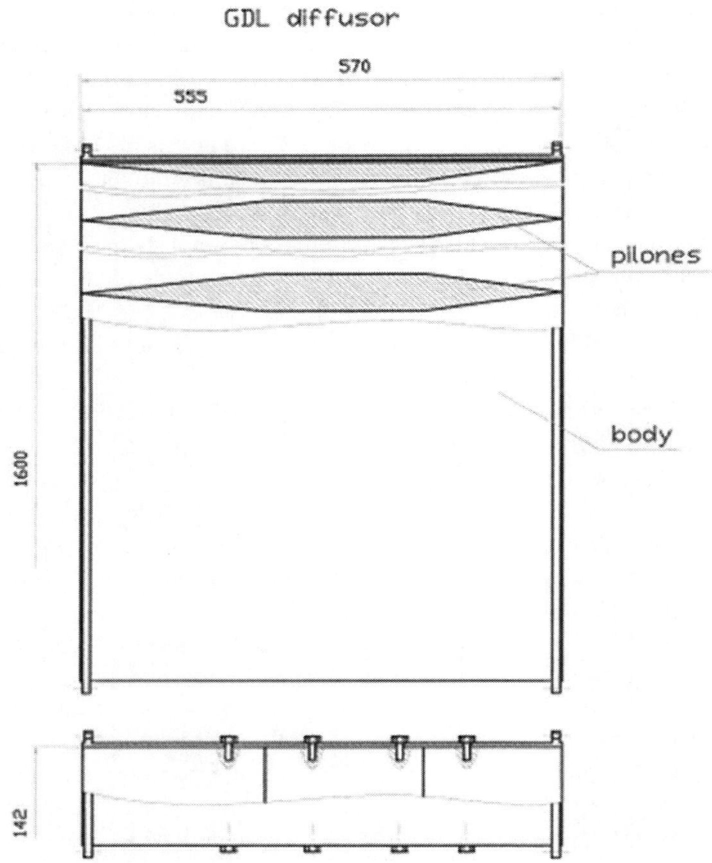

Figure 7.20. GDL diffusor.

The second mirror (translucent) has more complex form.

Directly mirror there is a convex spherical surface, around of it a ring slot is formed for transmission of laser radiation from the resonator. The elements of an adjustment and attachment of a mirror to the rack are placed on the external ring holding a convex mirror with the help of four flat thin pylons.

Convex mirror, the pylons and external ring are made as one piece.

As adjustment gears it is supposed at the first development phase to utilize the simple elastic holders with adjustable screws, which are used in laboratory lasers.

In future, if the laser installations will find broad application it is necessary to develop automatic adjustment gears. Some companies had conducted already activities in this direction.

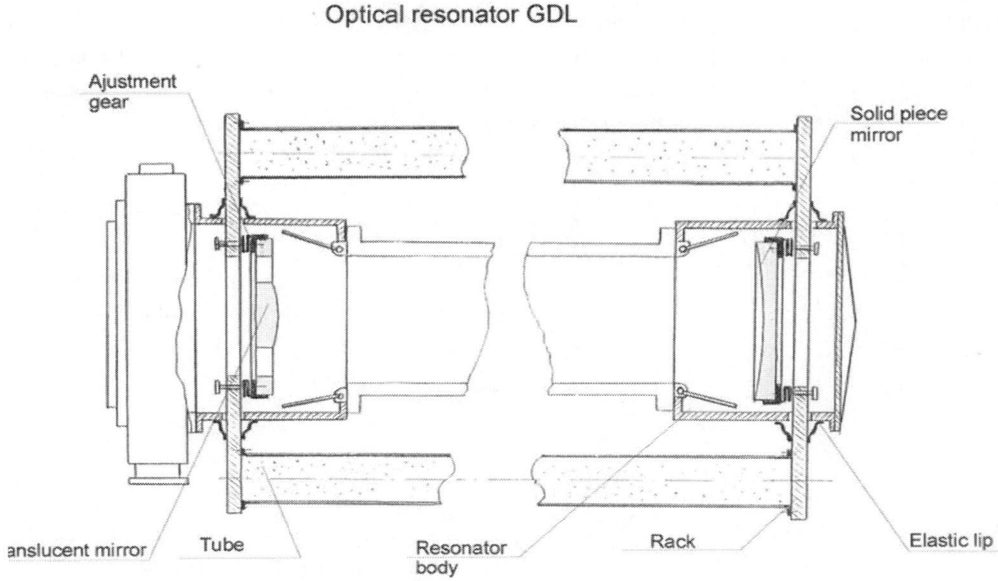

Figure 7.21. Optical resonator GDL.

3.4.3.6. Aerodynamic window

The aerodynamic window serves for extraction of a laser beam of ring cross-section from the resonator in an environment. By laser operation at rather small altitudes atmospheric pressure is much higher than pressure of gas in the resonator, that can result to leak of free air in the chamber of the resonator and reduce the characteristics of the laser. The window from a transparent material can not be utilized for the power laser, because of a partial absorption of radiation by the material of such window will cause it superheating and collapse. In the aerodynamic window the orifice for a transmission of radiation is overlapped by a supersonic jet. The pressure differential between ambient atmosphere and chamber of the resonator is organized through a system of shock waves along a bending of a supersonic jet.

The most economic schemes of aerodynamic windows, in which the working gas flow rate is minimum, are the windows realized through system of waves of compression and expansion with free swirl.

In presented case the cross section of a laser beam has the form of a ring with external diameter 125 mm. The inner diameter depends on a gain factor of an laser optical system. It was shown, that the optimal gain factor is in range M = 1,5 - 2. Taking into account, that with increase of the area of the aperture of the window the flow rate of working gas of the aerodynamic window is increased proportionally of this area, and, that the requirements to an angular divergence are not too high in case of application of the laser for disposal of a petroleum film, we shall accept M = 1,6. Then the inner diameter of beam cross section is equal 78 mm. To guarantee free transmission of a beam through aperture of the window it has a little bit more cross section than beam and represents a ring orifice with external diameter 130 mm and inner 73 mm.

The scheme of the aerodynamic window is shown in Figure 7.22. Basic elements of the window - ring nozzle and ring diffuser. The ring nozzle is inclined in the side of window exit in a direction of a gradient of pressure. The ring diffuser is formed by two conic surfaces inclined in the opposite directions. Thus, the jet of working gas which is coming out of the nozzles, is directed to the part of high-pressure zone. Under an operation of a pressure differential the jet of gas deploys and enters in the ring diffuser. In the diffuser the supersonic flow of working gas expands, transformes in subsonic flow and then will be rejected in an exhaust GDL line.

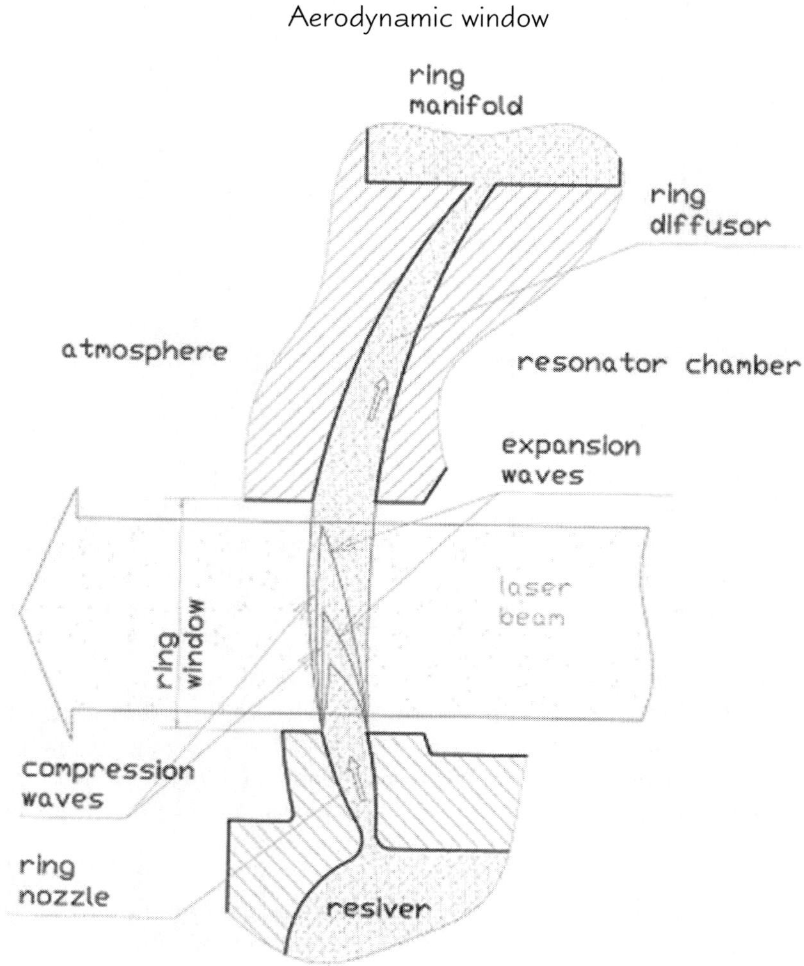

Figure 7.22. Aerodynamic window.

The pressure across a jet of working gas of the aerodynamic window changes from atmospheric ~ 1 kg / cm^2 dowm to chamber pressure of the resonator ~ 0,12 kg / cm^2. The change has taken place in waves of compression on boundary of a jet and ambient atmosphere, in gas jet in case of special contouring of the nozzle and in waves of expansion on boundary of a jet on the part of resonator chamber.

In this application ring nozzle of the aerodynamic window is countered so, that a Mach number in exit nozzle gas flow changes from ~ 3 at boundary of a jet with atmosphere up to

3,5 at boundary of resonator chamber. Therefore within jet the pressure changes approximately twofold. In waves of compression and expansion the pressure changes in ~ 5 times.

The scheme of flow of working gas through the aperture of the aerodynamic window are shown in Figure 7.22.

In Figure 7.22 it is possible to see the scheme of the aerodynamic window working gas supply. The gas in the receiver of the nozzle is suppled through flat pylons supporting the nozzle and a receiver. The pylons intersect a zone of propagation of laser beam. Their thickness and arrangement correspondent to thickness and arrangement of pylons supporting an output mirror of the resonator, that the shading of a beam was minimum. The collector ring established behind the diffuser of the aerodynamic window, has an exhaust branch pipe connected to a tube of gas removal down the diffuser.

The aerodynamic window prevents air leak in the resonator chamber at the atmospheric pressure to resonator pressure ratio ~ 10. Air will be utilized as working gas of the aerodynamic window with the flow rate ~ 2 kg/sec. The feed pressure of working gas is ~ 20 kg / cm^2.

3.4.3. Exhaust device of the laser installation

During gas-turbine engine operation the hot gas down the turbine exhausts with a high speed from the aft nozzle. At use of the engine on direct assigning (aircraft engine) the aft jet produces the additional thrust for the engine, is not limited to space and in a curtain degree is limited on acoustics.

During the use of the gas-turbine engine for the laser installation used on ground or on water vehicles of the requirement to a system of an exhaust of gas are much harder:

1. The acoustic oscillations of a jet should be minimized, as the installation can be applied on water reservoirs near to populating areas.
2. The exhaust should not create directional thrust vector (especially for the installations located on water vehcles), which can give all installation the non-authorized movement.
3. The exhaust jet of the engine should be mass of aft ejector working medium, ensuring operation of GDL supersonic diffuser.

Last problem is considered explicitly in the previous report, estimated performances and schematic of such ejector are added there.

To minimize acoustic oscillations (force of a sound reduction) during the exhaust from the engine, behind an ejector the exhaust silencer is to be installed (Figure 7.23).

There are three types of silencers:

Active, reactive and combined.

The silencers of an active type are based on an absorption and dispersion of oscillatory energy at the expense of moving against the frictional forces of gas and such devices work on a pressure differential. For analyzed case they are not applicable since all excess pressure will utilize for ejection.

The silencers of a reactive type work on a principle of the acoustic filter blocking transition of oscillations of certain frequency (dangerous to human) in a system behind silencers.

The combined silencers combine principles of active and reactive damping and also require some jet overpressure for obtaining a pressure difference.

For the given project the reactive silencer is chosen. It is necessary to note, that in a design of such silencer it is possible to utilize a cascade principle and to provide adjustment of its elements on some characteristic frequencies, that provides smoothing of oscillations in more broad band of frequencies. Structurally scheme of such silencers is constructed on a combination of tubular resonators acting as of acoustic capacities.

One of varieties of such silencers is the device developed by *"Pulseless control"* company (USA) for damping compressor acoustic oscillations. The device of such type has the greater selectivity and allows considerably lower noise level.

The such silencer development consists of stringent selection, both supply lines, and silencer elements, for example cutting of acoustic tubes after the experimentally obtained values of frequencies and amplitudes caused by the given gas jet.

It is well-known, that such acoustic filters can pass oscillations up to boundary frequencies

$$f = \frac{1}{\pi}\sqrt{\frac{1}{MaCa}}; Ma = \frac{\rho l}{F} Ca = \frac{V}{\rho c^2},$$

where:

$Ma = \frac{\rho l}{F}$ acoustic mass of a tube;

$Ca = \frac{V}{\rho c^2}$ - acoustic capacity of the chamber (volume V);

ρ - density of gas;

l - length of a tube;

F - the area of a tube;

c - rate of propagation of acoustic oscillations (velocity of a soun sound)

Thus, after obtaining the experimental data under the characteristics of acoustic oscillations at an exhaust of combustion products there is no difficulty to calculate and to design an acoustic silencer reducing noise level.

And the last requirement to the system of an exhaust of the laser installation - the exhaust gas thrust vector zeroing is determined by the customer.

Depending on place and method of arrangement of the laser installation the flow of exhaust gases can be divided, for example, on two opposite flow directions. It is possible to divide the exhaust around the circle to some weaker flows (this version is shown on applied figure) and at some stationary arrangement of the installation is permissible not to zero thrust vector, and to direct it in direction suited for the customer.

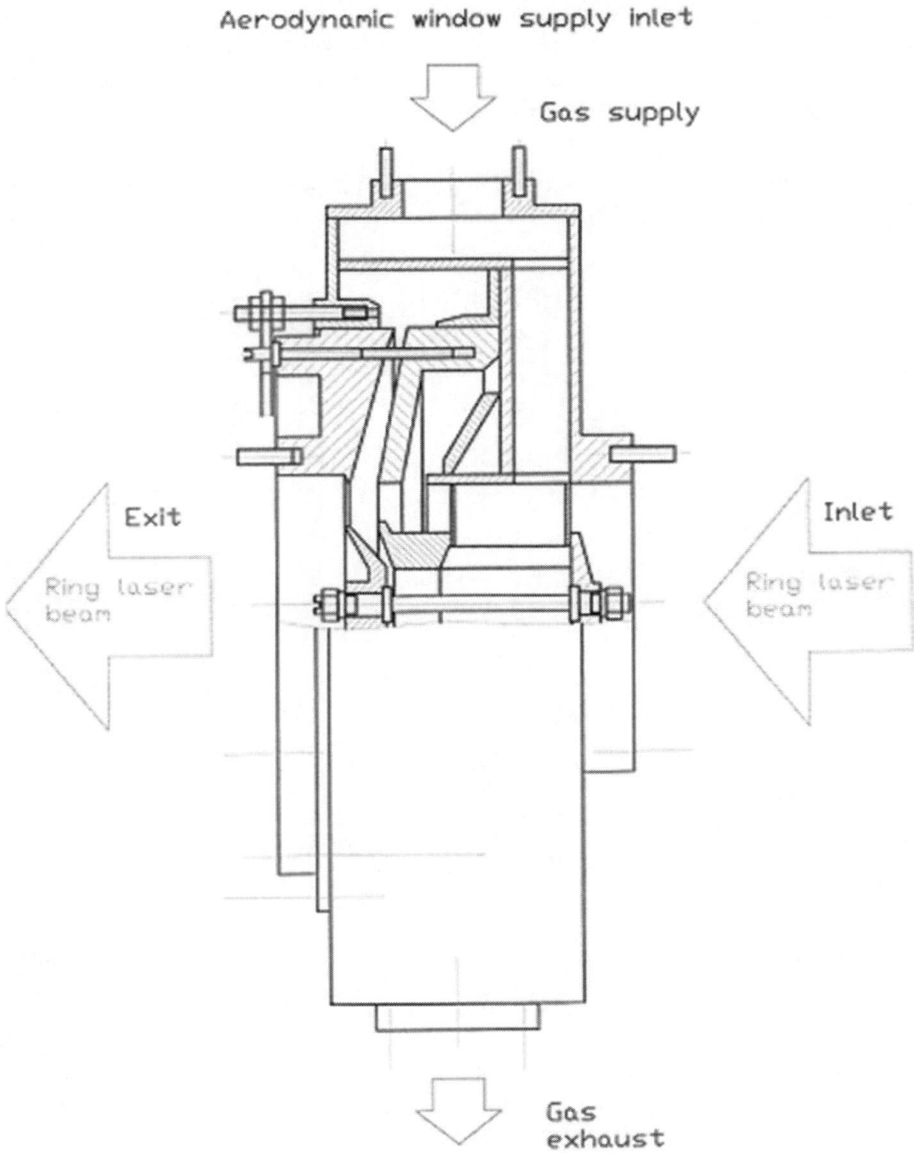

Figure 7.23. Aerodynamic window supply inlet.

As explained, the exhaust device of the laser installation can represent by design shown on figure.

All subsystem consists of three sequentially installed aggregates:

Ejector, silencer and exhaust collector.

- *The ejector* consists of bodies with a collector of gas supply from GDL and inner cone. The walls of a body and collector form cone – inverse cone - calculated profile, in minimum cross section region of which - is organized supply after GDL.

- *The silencer* represents the multisectional acoustic chamber, has outside and internal walls. The internal wall has a number of orifices connecting the chamber with a gas flow. The sections are divided by walls and in each section the acoustic tubes, selected on the sizes according to the acoustic characteristics of an exhaust system (acoustic adjustment of a silencer) are established.
- *The exhaust collector* represents spherical container with a number of opposite directed branch pipes, through which the gas expires.

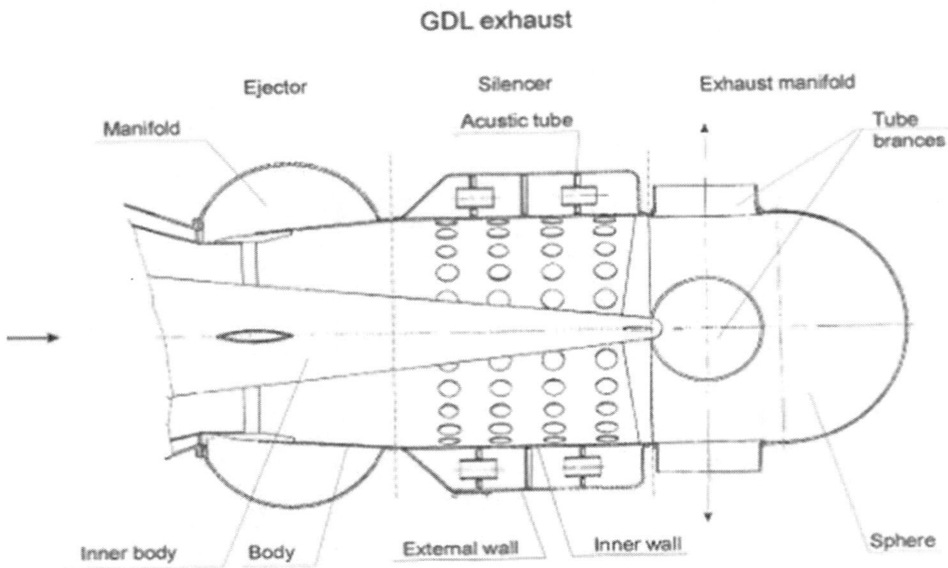

Figure 7.24. GDL exhaust.

As it was indicated above, this part of an exhaust system (Figure 7.24) is added as version, since its substantial design is depending on place and operation conditions of the laser installation.

3.4.4. Other systems and aggregates of the laser installation

The laser installation includes:

- Directly GDL - autonomous installation (is described above); with the systems
- Engine NK-29 - power unit ensuring GDL by air and the electric power. (was considered in the previous report).
- Exhaust device playing the role of a gas ejector, reducing installation noise level and zeroing (if necessary) an exhaust jet thrust of the installation. In the report its design and description are added.

Besides, in installation enter the systems and aggregates, which design in the report is not considered, as they do not represent technological problems and can be in most cases borrowed from serially made design assemblies and products of industry.

The following are included to them:

- Container for kerosene storage. It is a tank with a volume ensuring a required operating time of the installation.
- Fuel and power supply system and gas tubes. The pipes, adapters, flanges etc can be selected from used in aerospace industry. For compensation of temperature deformation and the eliminations of transmission of vibrations lines connecting GDL and aviation engine, are connected through bellows or flexible sleeves.
- External optical system. By development of power lasers, which radiation was transmitted to significant distances (hundred and thousand meters) usually the external beam forming optical system with the purpose of beam divergence reduction is used.

In presented installation and at use GDL for clearing of water surface, distances between GDL and surface of a water is no more then several meters. Therefore employment of complex and expensive telescopic external system is not profitable. It is enough to place on a course of a laser beam a flat rotary mirror, with which it is possible to scan a beam on a surface of water.

The control panel. The engine NK-29 is supplied with the control panel for engine start and shutoff. For GDL operation control it is necessary to finish the existing console by introduction the additional printer circuits on formation of commands on GDL control elements and monitoring display. Apparently, it will be by a workplace for the second operator.

3.4.5. Gas dynamic laser installation

In view of all computational investigations, design studies and engineering estimations within the present report works the provisional structure of the autonomous laser installation for clearing of surfaces of reservoirs of petroleum recordings is produced.

The general view of such installation is shown in Figure 7.25.

As it is visible from figure, the installation represents modular assembly.

The engine has an own fixing frame, autonomous subsystems and producer - supplier.

GDL also represents the complete aggregate, the developer and manufacturer of which is not connected to other suppliers.

The same concerns to fuel storage system (is not shown in figure).

Task of the installation developer as a whole is the coordination of these units both on arrangement on a vehicle, and organization of interaction during operation.

For example, relative arrangement of the engine and laser can be arbitrarily, depending on overall dimensions and device of a vehicle, however in any case the fulfillment of following conditions is necessary:

- The engine and GDL should not have of rigid connection on a frame, the GDL frame should have vibration absorption plates.
- The engine and GDL should be placed as close as possible, as the unreasonable elongation of gas lines connecting them, will result in to additional losses because of pressure reduction.

- On fire protection control requirement fuel storage container should ne placed on distance from the engine and GDL, and is separated from them by a fire-prevention wall.
- For protection of the attendants the remote control panel should be born for limits of the installation, and input part of the engine and line of a beam are protected by special fences. The control of a scanning mirror also is carried out for limits of the installation.
- The direction of exhaust gases should not intersect structural items of the installation.
- As the development of the laser installation does not assume development of testing site for development testing, prior to deliveries the development installation should be assembled, on which all designer and operational problem solutions will be find.

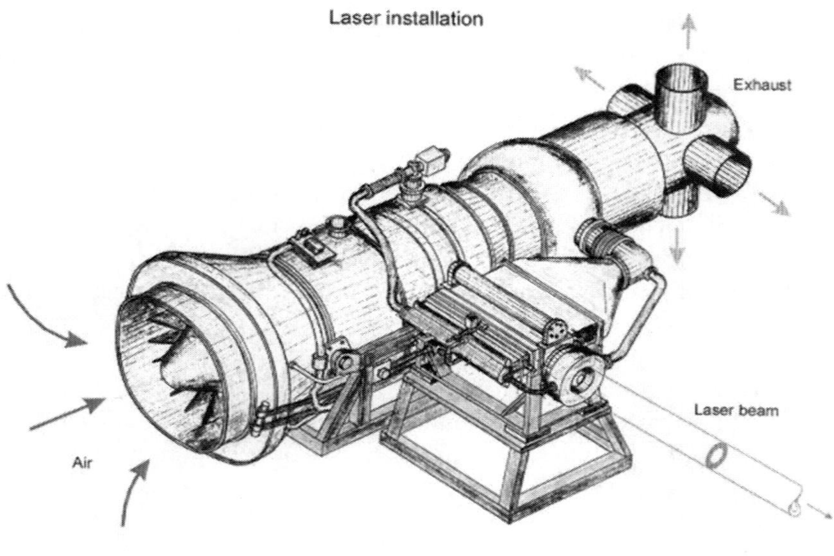

Figure 7.25. Laser installation.

3.5. Estimftion cost of designing and manufacturing of a prototype mobile gas dynamic CO2- laser installation

The present estimation of the of manufacturing of a prototype of a laser is preliminary and outgoes from conditions of necessity of mandatory realization of following milestones of activities:

- issue of the preliminary design on a laser installation;
- issue of design documentation;
- issue of master schedules and extension of the fabrication documentation;
- preproduction activity;
- manufacturing of gas dynamic laser;
- manufacturing of the unit of an exhaust and silencer;

- adaptation of an aero-engine NK-29 and assembly of it with the unit of an exhaust and silencer;
- development tests of an air generating installation;
- manufacturing of subassemblies and parts of general assembly of a laser installation;
- assembly of a laser installation;
- realization of development and final tests of a laser installation.

Except for the cost of the basic operations , the cost of manufacturing of the laser installation is:

- NK-29 aviation engine. The engine is produced serially and supplied by the Samara Technological complex - "«Motorostroitel". The cost of the engine 2,000,000.00 USD (price of 2005);
- An estimated cost of single laser unit is 14 000 000.00 USD.

At serial production of 10 laser installations per year using standard items and assemblies return costs efficiency can be one year, and the cost of production installation can be reduced down to12,000,000.00 USD. The sale profit of mobile laser installation can be as much as 15 - 25 %.

3.7.4. Conclusion

Besides, the bench gas dynamic laser was restituted by on their own costs and some demonstration operations with scanning the beam over water surface covered with petroleum film were carried out. The experimental confirmation of film effective burning or gathering was obtained.

The next project realization phase can be the development of the installation preliminary design, in case of the customer objective and specific technical tasks, formulated by him.

The preliminary design will include complete study of a design of all elements of the installation, technological works, will define all suppliers of the developing systems and subsuppliers from obtaining from them of the consent on terms and conditions.

In the preliminary design, schedules of all phases of installation development will be worked out: issue of the working documentation, technology modification, manufacturing of a prototype, prototype development and beginning of installations deliveries.

Chapter 17

3.8. Laser Therapy for Treating Tuberculosis

3.8.1. A Bit Scary Statistics

For a start, as a preamble, here are some excerpts from various sources on the problem of tuberculosis (TB), which are quite strange to read at the beginning of the third millennium, since many people confidently believed that there has been enough time to solve this problem during the past two millennia:

- Tuberculosis symptoms and signs: cough, loss of weight, chest pain, fever, night sweats. If untreated, 50 percent of patients die within five years;
- More than any other infectious disease, TB kills approximately 1 million women per year. Each year, TB kills 100 000 children. Tuberculosis is the most common cause of orphanage;
- Untreatable bacteria can destroy the progress of TB control achieved in the last 50 years. There are no drugs to combat some resistant TB bacteria (in developed countries 50 million people may be infected);
- The majority of people infected with TB never become sick because their immune system prevents the development of TB mycobacteria. Only 5 to 10 percent of those infected develop TB. Scientists today do not know exactly why some infected people develop TB and die, while others do not;
- At least one person is infected with TB every second, 1 percent of the world's population are infected each year. Untreated persons infect on average 10–15 neighbors during a year. For major cities, this figure is considerably higher. *The most susceptible to infection are prisons, the Army and the Navy, where the concentration of people living together for a long time is the greatest.*

According to the WHO over the past two centuries, TB killed about a billion people. The WHO warns that unless we take urgent action, in the next 10 years, TB will kill an estimated 30 million people and infect 90 million people. Further, by 2020 a billion people will have been already infected: 200 million people will be sick and 70 million people will die. So much for the White Plague (because of the extreme pallor seen among those infected)!

3.8.2. Laser Treatments for Tuberculosis

Currently, we know two approaches to fabricating laser systems for the TB treatment. They are based on excimer lasers and installations making use of important benefits of miniature solid-state diode-pumped lasers. The peak of the bacteriostatic activity of the generated laser radiation in various forms of TB lies at a wavelength of 265 to 266nm, and in this case, the effectiveness of its action is equal to unity. The wavelength of 248nm, which is emitted by an excimer laser, is closest to this range. At this wavelength, the interaction efficiency amounts to 0.8, which requires a proportional increase in the irradiation time. For a solid-state Nd:YAG laser (fourth harmonic) the radiation wavelength is equal to 266nm. The interaction efficiency of this wavelength is 1.0. The pulse energy with an average output power of 10mW (equal to the product of the energy in a single pulse by the pulse repetition rate). The power is determined experimentally in cultures of bacteria for the exposure time of 10 to 15min.

For an excimer laser the pulse repetition rate is no more than 100Hz; therefore, the energy of a single pulse is less than 0.1mJ, which can lead to tissue burn at a pulse duration of 5–10ns. To ensure a 'soft' effect on tissues, it is needed to reduce the pulse energy by one or two orders of magnitude, which is possible, but requires a proportional increase in the exposure time. At these energies (0.1mJ) an optical fiber is usually damaged. An optical fiber requires a high optical purity of working surfaces which is problematic when administering it in various cavities. The destruction of the output end of the fiber may cause penetration of small glass fragments into the patient. In a solid-state laser the pulse repetition rate is maintained at a level of 10000Hz; therefore, the energy of a single pulse is only 0.001mJ. That is why the soft tissue burn and destruction of optical fibers in the case of solid-state lasers is fundamentally impossible.

Now let us say a few words about the service life of some elements of lasers. In the case of an excimer laser, the main element is a gas tube, whose service life is about 1000–2000 hours at a cost of about 1000 USD. High-energy pulses can also lead to an early failure of optical elements. In the case of a solid-state laser, the essential element is a laser diode, whose service life amount to least 5000 hours at a cost of 700–800 USD.

The presence of hazards when working with lasers discussed is as follows. An excimer laser has in its design a significant amount of harmful gas. Besides, this laser requires a high voltage for its operation (about 10kV). The design of solid-state lasers is free of hazards, whereas the cost of the components of these lasers is approximately comparable with that of excimer lasers.

As for the additional conditions of production of the laser sources discussed, excimer lasers, apart from standard a optical-mechanical and an installation sites, also require the presence of a vacuum site, equipped to work with poisonous gases. The production of solid-state lasers requires only a standard optical-mechanical and an installation sites.

3.8.3. Amulet Semiconductor Laser Apparatus

An Amulet semiconductor laser apparatus with a fiber for introducing radiation in the affected area through an injection needle is intended for the treatment of destructive forms of pulmonary and bone TB that are resistant to conventional medical treatment, as well as to shorten the treatment of common forms of TB by topical exposure of the infected surface to ultraviolet (UV) radiation with a wavelength of 266 nm, which has very strong bactericidal and bacteriostatic effects. UV radiation in this case has low intensity and only affects the microflora without any damage of the living tissues of the human body. The typical time of UV irradiation of the affected area is 5 to 15min. In this case, the traumatic effect is absent.

The Amulet apparatus is designed to treat patients with tuberculosis affecting lungs, bronchi, bones and joints, to cure diseases associated with suppurative infections and other inflammatory processes, with the abnormal healing process, with the immune system variations and instability of the capillary circulation. In addition, the apparatus can be used in endosurgery, phthisiology, otolaryngology, traumatology, stomatology, treatment of burns, gynecology, therapy, surgery, urology, proctology, and dermatology.

The main technical characteristics of the Amulet apparatus:

Emission wavelength (nm)	266
Average output power (mW)	2.5, 5, 10
Pulse repetition rate (kHz)	10 – 20
Pulse duration (ns)	5 – 7
Beam diameter (mm)	<0,5
Radiation outcoupling	Fiber
Fiber diameter (μm)	400 – 800
Power consumption (220V/50Hz) (W)	<200
Service life (h)	>2000
Cooling	Air
Dimensions (mm)	450×165×290
Weight (kg)	<12

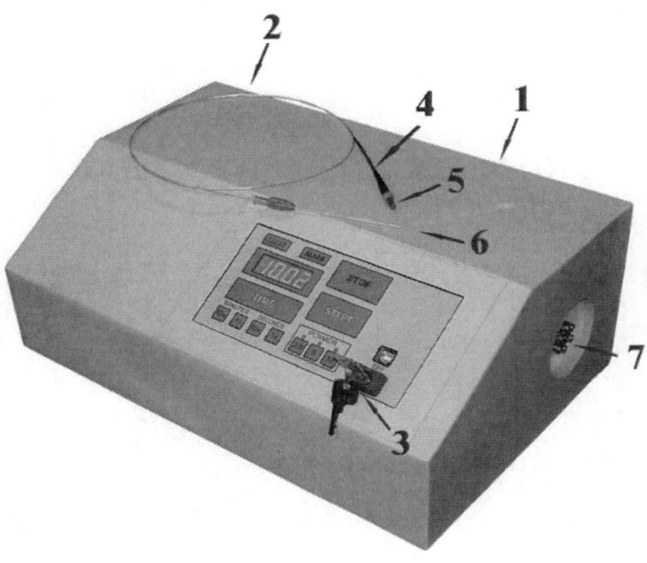

Figure 8.1. General appearance of the Amulet apparatus.

3.8.4. Clinical Trials of the Amulet Apparatus

In accordance with the decision of the Commission on Tools, Instruments, Apparatus and Materials used in General Surgery of the Committee on New Medical Technology of the Ministry of Health of the Russian Federation (Protocol No. 3 dated 22 March 2001) and relevant orders of the Main Military Medical Directorate of the Ministry of Defence (Department of endoscopy and surgical department for treatment of patients with complicated forms of TB of the lungs), the Director of the Research Institute of Phthisiopulmonology, I.M. Sechenov Moscow Medical Academy (Department for the treatment of patients with bone and joint tuberculosis) and the Director of the Central Research Institute of Tuberculosis of Russian Academy of Medical Sciences (Department of pulmonary tuberculosis), we performed clinical trials of a prototype of a therapeutic UV solid-state diode-pumped laser (Amulet), produced by LLC Energomashtehnika. The laser emits middle-level low-energy UV radiation with a wavelength of 266nm and intensity of up to 10mW/cm2 at the fiber output.

The use of low-energy laser radiation in medicine is very beneficial in the treatment of TB patients because of its ability to cause an analgesic, an anti-inflammatory and an anti-edema effects, as well as physiological and reparative regeneration of tissues.

The effectiveness of the therapeutic UV solid-state diode-pumped laser was tested in 7 patients with bronchial tuberculosis, 4 patients with laryngeal tuberculosis, 7 patients with chronic tuberculous empyema, 14 patients with osteoarticular tuberculosis complicated by surface wounds.

During the clinical trials, we improved the optical fibers which made it possible to increase the area of laser radiation scattering both in the case of endobronchial and transthoracic introduction of the fiber. Pathologically altered tissues were irradiated as a part of complex therapy of TB, including 4–5 anti-tuberculosis medications. In the process of clinical trials we monitored the presence of TB mycobacteria in abnormal tissues and performed histological studies of inflammatory changes. The TB-affected tissues of the patients were irradiated through a bronchoscope. The TB process in these patients had the following localization and character:

- Bronchial TB in the middle lobe, fistula;
- Bronchial TB in the middle lobe, ulcer;
- Bronchial TB at segmental (B6) bronchi, infiltration;
- Laryngeal TB with the destruction of epiglottis;
- Pulmonary TB, cavern and thoracic fistulas;
- Mycobacterium TB was found in all the patients.

Exposure conditions:

- The intensity of the radiation is constant.
- The distance of the distal end of the optical fiber from the irradiated surface is 1-2cm.
- The exposure time is 30s.
- The number of laser treatments is 8-12.

The use of therapy solid-state diode-pumped laser UV radiation together with specific and intensive health-improving restorative treatment at the end of clinical trials showed that the oronasal fistula was free of cheesy masses. In the course of the microscopy of bronchoalveolar lavage fluids, TB mycobacteria were net found. Besides, a decrease in inflammation of the bronchial mucosa was visually ascertained. The histological examination of biopsy material revealed rapid healing, formation of fibrosis in the bronchial mucosa with signs of hyalinization, and narrowing of the bronchial lumen. In patients with laryngeal TB, a decrease in swelling and redness of the mucous larynx above the vocal cords as well as the healing of erosive and ulcerative lesions were visually observed. During laser irradiation of TB empyema (10–12 laser treatments for each patient) the volume of purulent discharge and increased bleeding were observed, which we attribute to the reduction of inflammatory changes and improvement of local blood circulation.

In the case of bone TB, application of UV radiation leads to copious and purulent discharge of necrotic masses, which clearly demonstrates the bactericidal and bacteriostatic effect of UV radiation on flora. In all cases, after 6–12 irradiation sessions held daily, the wound surface was cleaned, and a pink granulation tissue covered the wound bed and its edges. Examination of swabs and inoculations from the wounds demonstrated a drastic decrease in the presence or growth of the colonies. This allows the surgical treatment of wounds and placement of secondary sutures, which greatly accelerates the healing of wounds.

3.8.5. Conclusion

Thus, the clinical trials performed allow us to make a conclusion that a therapy UV solid-state diode-pumped Amulet laser is effective in the treatment of TB-affected tissues and bones due to the bactericidal and bacteriostatic effect and stimulation of reparative processes.

INDEX

C

D

E

F

G

H

I

J

K

L

M

N

O

P